THE
COCAINE
CRISIS

THE
COCAINE
CRISIS

EDITED BY
DAVID ALLEN
National Drug Council
Nassau, Bahamas

PLENUM PRESS • NEW YORK AND LONDON

Library of Congress Cataloging in Publication Data

International Cocaine Symposium (1st: 1985: Bahamas)
    The Cocaine crisis.

    "Based on the proceedings of the First International Cocaine Symposium, held November 21–22, 1985, in the Bahamas"—T.p. verso.
    "Sponsored by the American Embassy in the Bahamas and the Ministry of Health of the Commonwealth of the Bahamas"—Introd.
    Bibliography: p.
    Includes index.
    1. Cocaine habit—Congresses. 2. Cocaine habit—United States—Congresses. 3. Cocaine habit—Bahamas—Congresses. I. Allen, David Franklyn. II. United States. Embassy. Bahamas. III. Bahamas. Ministry of Health. IV. Title. [DNLM: 1. Cocaine—congresses. 2. Substance Abuse—congresses. W3 IN13045 1st 1985c / WM 280 I61 1985c]
HV5810.I57   1987   613.8′3   86-30518
ISBN 0-306-42482-7

Based on the proceedings of the First International Cocaine Symposium,
held November 21–22, 1985, in the Bahamas, cosponsored by the American Embassy and the
Bahamas Government.

Grateful acknowledgment is made to the following publishers for their kind permission to use
material from these publications:

"The Implications of Crack," by Dr. Sidney Cohen, was originally printed in the *Drug Abuse and Alcoholism Newsletter,* July, 1986, and is reprinted with the permission of the Vista Hill Foundation.

"Cocaine: Drug Epidemic of the '80s," by Dr. Arnold Washton; portions of this manuscript are adapted from previous articles by the author published in *Advances in Alcohol and Substance Abuse,* Vol. 6, 1985, and reprinted with the permission of Haworth Press, New York, New York.

"The Self-Medication Hypothesis of Addictive Disorders: Focus on Heroin and Cocaine Dependence," by Dr. E. J. Khantzian was originally published in the *American Journal of Psychiatry,* November, 1985, and is reprinted with the permission of the American Psychiatric Association.

"Epidemic Freebase Cocaine Abuse: A Case Study from the Bahamas," by Prof. James F. Jekel, Dr. David F. Allen, et al.; portions of this manuscript were originally published in *The Lancet,* March 1, 1986, and are reprinted with the permission of the publishers.

© 1987 Plenum Press, New York
A Division of Plenum Publishing Corporation
233 Spring Street, New York, N.Y. 10013

Printed in the United States of America

This volume is dedicated to all those human service professionals including psychiatrists, psychologists, researchers, counselors, nurses, social workers, and others who struggle unceasingly as wounded healers to bring hope and meaning to those whose lives have been devastated by cocaine.

Two years ago I would have said that the cocaine outbreak would go away. After all, stimulant epidemics have relatively short lives. The first cocaine rampage just one century ago subsided after it became clear that the new safe panacea was neither safe nor a panacea. A lesser flare-up in the first decades of the twentieth century also receded. The "speed" (methamphetamine) craze began about 1970 and had essentially disappeared by 1975.

However, predicting from past experience is hazardous when a problem contains entirely novel elements. This is true today of cocaine. Attempting to peer into its future is difficult, perhaps because too much depends upon decisions and actions not yet made or even contemplated...

SIDNEY COHEN
Cocaine: The Bottom Line
The American Council for
Drug Education, p. 12, 1985

FOREWORD

Carlton E. Turner

Director, Drug Abuse Policy, and
Deputy Assistant to the President
The White House
Washington, D. C.

The First International Drug Symposium sponsored jointly by the American Embassy and the Bahamian Government provided the first hands-on forum
for sharing ideas about the treatment of cocaine addicts who freebase. This
is sometimes known as smoking cocaine in the "rock" or "crack" form.

This forum was essential, because too many psychiatrists, psychologists,
and other medical professionals have refused to acknowledge the known and,
I must add, historical danger of cocaine. In fact, some perpetuate the myth
that cocaine is OK. They reason that cocaine is so expensive that only the
wealthy will be able to afford it, and they will know how to use it recrea-
tionally and responsibly. This cavalier attitude in the midst of the world's
second cocaine epidemic has made it nearly impossible to convince people
that cocaine kills. I hope this book will convince some doubters. Unfortunately, some will only be convinced by a grave marker.

The world's first cocaine epidemic was in the 1880's. At that time,
cocaine was legally distributed at a low cost and was of high purity. Then,
just as today, people, and even Sigmund Freud, initially believed that cocaine was innocuous: a great drug without any side-effects. Cocaine was in
wine, fountain drinks, and other tonics. And, as today, cocaine cost the
lives of many people, including health professionals. Ultimately, society
rejected cocaine.

Ignorance of the horrible legacy of the first cocaine epidemic led to
the problems we face today; cocaine knows no geographic boundaries. It
knows no socio-economic boundaries. It knows no boundaries for corruption,
destruction, or seduction.

This book is about understanding cocaine, and if used appropriately,
it will help the world understand that cocaine addiction cannot be conquered
in isolation, and cannot be dealt a fatal blow with rhetoric. Communities
of the world must recognize that all countries - producing countries, trans-
shipment countries, and user countries - are vulnerable to drug abuse.

There is much talk today about educating the young not to use drugs,
yet the largest number of cocaine and marijuana users is found in people
between 20 to 40 years of age. These people have made a choice and educating them, in my opinion, will be impossible without concurrent strong

enforcement and a concomitant demand by society that cocaine use will no longer be tolerated.

Simply put, we must highlight the medical consequences of cocaine use. This book does! Cocaine and other illegal drug use has become a major threat to the security of nations, a major health problem, a major threat to the tranquility of the law-abiding, civilized world. Cocaine destroys families by deteriorating interpersonal relationships and financial stability. It corrupts normally incorruptible people.

We in this hemisphere do not need to generate more scientific data to understand that cocaine kills. It does. We know it. We must communicate it, and we are, through this book.

We must focus on the user. We must stop cocaine from killing more of us. Without the money supplied by the user, there would be no need for massive law enforcement outlays. Without the money supplied by the user, there would be no production in South American and other places. Without the money supplied by the user, there would be no corruption of governmental entities and private organizations. Without the money from the user, the cocaine traffickers would not be able to put contracts on ambassadors and kill enforcement agents. Without the money from the user, corporations would not lose billions of dollars through lost productivity and increased accidents from drug use in the workplace. Without the money from the user, the banks would not be facing fines for laundering money.

We can talk about stopping the cocaine trade, but the supply will not stop as long as users are willing to pay. Society must understand the horror of cocaine and must accept the fact that the price for one small white line or one small rock can also have another price, the cost of a life.

Before any drug is made available to the public, the industrialized world requires that pharmaceutical manufacturers spend millions of dollars to prove that a drug is safe. On the other hand, illicit drugs are accepted as safe, and society becomes a guinea pig until such time as the negative health consequences are recognized by all. Little concern is given to premature deaths, aborted fetuses, wasted lives, or destroyed families. And yet there seems to be no outrage against the drug dealer or user. Yet there is total outrage against any pharmaceutical firm whose product, when misused causes one or two people to die.

Additionally, when one talks about drug abuse, it is the only health issue on which the media insists on covering both sides. Indeed, we must be wary of some press coverage. Certain journalists have complemented their coverage of the devastating effects of crack with lyrical descriptions of the crack "high." Intellectually, this is wrong. Health-wise, this is wrong. Communication-wise, this is wrong.

Democracies of the world must understand that there is need for eradication of the drugs as close to the origin as possible. There is a need for interdiction of the drugs. There is a need for strong law enforcement. There is a need for communication of the research data, as well as detoxification and treatment. Prevention and education, yes, are needed, but the only solution to the problem we face today is to hold the user accountable for his or her action. Until that is done, the drug epidemic is a steamroller, rolling along inexorably toward destruction with narco-dollars providing the financial push. Perhaps the United Nations conference in 1987 will be the world forum for all to unite against drugs. But rhetoric alone will not work, because when you buy drugs, you buy terror.

This book will help you decide on which side of the aisle you sit, on the side with the philosophical view that it is everyone's right to use drugs, or the side with firm convictions that drug use must stop and stop now!

FOREWORD

The Hon. Dr. Norman R. Gay

Minister of Health
Commonwealth of the Bahamas

Although much publicity has been given to the use of freebase (crack) cocaine in the past few months in the United States, the Bahamas has suffered from this form of addiction since early 1983. While referred to as crack in the United States, it is called rocks or base in the Bahamas. Straining both the socio-economic and human resources, this epidemic of ready-to-smoke freebase cocaine has had a massive impact upon our country. Nevertheless, we are resolute in our determination to conquer it. The ancient proverb, "Where there is tragedy, there is opportunity" is at work among us. Reacting to their fear of becoming victims, the lifestyle of many people is changing. The national awareness of the dangers of this heinous addiction is awakening all sectors of the community, challenging them to fight back as well as to say "NO" to the drug.

Animal experiments show freebase cocaine is the most dangerous form of addiction so far encountered by mankind. Rats in experiments, when given the alternative choice of unlimited food, drink, sex, or freebase, will choose to take the freebase until they die. The probability of a person becoming addicted to freebase after the first try or "hit" is approximately 80% in the Bahamian experience. This most unusual property compared to the known first-time use compulsion of all other drugs, including heroin, which is less than 20%, is coupled with the rate of deterioration (both mental and physical) where the norm is 3 to 5 years, yet freebase cocaine may lead to deterioration occurring in as little as 3 to 6 months. Herein lies the startling tragic scenario.

Recognizing the uniqueness of the Bahamian experience and the devastation of a freebase epidemic, the Ministry of Health of the Bahamas and the American Embassy in the Bahamas organized the First International Drug Symposium on cocaine in Nassau, in November, 1985. At this meeting, experts from different countries shared their experience with the treatment of cocaine addiction and were also exposed to the expertise gained by Bahamian professionals during the cocaine freebase epidemic in the Bahamas. This Symposium, in bringing together experts from the producer countries of Peru, Bolivia, and Colombia, the trans-shipment countries of the Caribbean, and the consumer countries of the U.S. and the U.K. provided a meaningful, mutually educational experience. Indeed, this book is the result of the on-going process of international dialogue and cooperation initiated at the Symposium.

In a continuing effort to further define the type and scope of the Bahamian drug problem, Professor James Jekel of Yale University, working closely with a team of Bahamian researchers, validated the seriousness of the cocaine freebase epidemic in a research paper published in The Lancet of March, 1986. This paper was tabled at the Health Ministers' Conference at Lancaster House in London held that same month. Many of the Health Ministers expressed concern at the extent of cocaine abuse on this side of the Atlantic. For some of the Ministers, the very existence of freebase cocaine was a new phenomenon.

In countries currently unaffected by this problem, at this time there is as yet, even at governmental level, little or no appreciation of the dynamics of ready-to-smoke freebase, or crack, cocaine; hence, the sense of urgency or serious attention to the problem is not perceived as necessary. Based on the present rate of spread, if freebase (crack) cocaine becomes available in any country, a freebase epidemic is likely to follow. The voluminous supply is not going to go away. The fantastic economics involved make the user countries inescapable targets.

In spite of the emphasis on enforcement, much more must be done by local and international law enforcement agencies to stamp out drug trafficking. This requires a total commitment by the community in general and the government in particular, for the effectiveness of law enforcement depends on the society it serves.

The seriousness of cocaine, freebase or crack, addiction is such that governments should now consider mandatory treatment of confirmed addicts. No longer can we permit the wholesale destruction of young people in the prime of life through this addiction; no longer can we permit the untold misery suffered by our families; no longer can we permit the incalculable losses to our societies of valuable human and financial resources. The onus must be on all of us in society to provide meaningful treatment programmes and to require addicts to use them to the fullest and with the best of intentions to recover and return to their society, their family, their life itself.

The solution to the cocaine problem is not simple; it requires hard work and long-term commitment. Simultaneously, however, we must work toward prevention by developing good healthy living habits. Our total life-style needs to be re-examined. We need to improve nutrition, to restrict abuse of legal drugs such as alcohol and tobacco, and to develop effective exercise patterns. Creating healthy bodies would increase self-esteem and generate greater interest in caring for ourselves.

Finally, it is my wish that this book on this most important subject will encourage countries, families, and individuals to join the war against cocaine. It is my strong belief that this Bahamas, this small nation, having gained unique and invaluable expertise through its paralyzing experience with freebase cocaine, will be able to make a major contribution to the war against drugs and the development of a better, more caring, drug-free world.

ACKNOWLEDGEMENTS

First of all, I would like to extend my sincere appreciation to all the contributors who worked so hard to make this publication a reality. To be able to gather the fruits of the expertise of these highly talented and knowledgeable individuals has been an exciting and rewarding experience for me.

Secondly, I would like to acknowledge most sincerely the excellent work of Mrs. Barbara McKinley-Coleman. She worked extremely hard in editing, typing, and proof-reading the manuscript. Without her tremendous efforts, the publication would not have been possible.

Special thanks are due to Mr. Charles Dawson, Public Affairs Officer of the United States Information Service in Nassau, who has been extremely supportive and worked very hard behind the scenes to make the First International Drug Symposium a reality and this publication possible.

I would like to extend my sincere appreciation to Dr. Brian G. Humblestone who did a superb job at the First International Drug Symposium as Chairman. Through his leadership the participants were afforded an open atmosphere of freedom and congeniality to discuss their research, their work, and to share their hopes for the future with their colleagues in the field of drug addiction.

Also I would like to thank the Hon. Dr. Norman R. Gay, Minister of Health, Commonwealth of the Bahamas, who has provided strong support and encouragement in the development of the First International Drug Symposium and numerous activities in the Bahamas in the War Against Drugs. It has been most rewarding to work with a fellow practitioner as Minister of Health who understood in depth the problems of cocaine addiction. He has contributed greatly to the War Against Drugs in the Bahamas.

I would finally like to thank my wife, Victoria, who gave me encouragement, and valuable advice in developing the manuscript.

Dr. David F. Allen,
Chairman, National Drug Council
Nassau, Bahamas
September, 1986

CONTENTS

INTRODUCTORY CHAPTERS

Introduction.. 3
  D. F. Allen

History of Cocaine.. 7
  D. F. Allen

Modes of Use, Precursors, and Indicators of Cocaine Abuse............. 15
  D. F. Allen

The Implications of Crack... 27
  S. Cohen

THE CONSUMER SOCIETY: THE AMERICAN EXPERIENCE

Cocaine: The American Experience..................................... 33
  E. M. Johnson

Cocaine: Drug Epidemic of the '80s................................... 45
  A. M. Washton

The Self-Medication Hypothesis of Addictive Disorders:
    Focus on Heroin and Cocaine Dependence.......................... 65
  E. J. Khantzian

Psychotherapy for Cocaine Abusers.................................... 75
  K. M. Carroll, D. S. Keller, L. R. Fenton, and F. Gawin

Public Health Approaches to the Cocaine Problem:
    Lessons From The Bahamas.. 107
  J. F. Jekel

THE TRANS-SHIPMENT SOCIETIES: THE CARIBBEAN EXPERIENCE

The Bahamas and Drug Abuse... 119
  B. G. Humblestone and D. F. Allen

Epidemic Freebase Cocaine Abuse: A Case Study From The Bahamas........ 125
  J. F. Jekel, D. F. Allen, H. Podlewski, S. Dean-Patterson,
    N. A. Clarke, and P. Cartwright

Drug Abuse 1975-1985: Clinical Perspectives of the Bahamian
    Experience of Illegal Substances................................ 133
  N. A. Clarke

Cocaine Psychosis.. 139
  T. Manschrek

Cocaine and the Bahamian Woman: Treatment Issues..................... 145
  S. Dean-Patterson

Treatment Approaches to Cocaine Abuse and Dependency
    in The Bahamas... 161
  T. McCartney and M. Neville

Cocaine Update... 167
  M. H. Beaubrun

THE PRODUCER SOCIETIES: THE SOUTH AMERICAN EXPERIENCE

Somatic Disorders Associated with the Abuse of Coca Paste and
    Cocaine Hydrochloride.. 177
  F. R. Jeri

Coca Paste Effects in Bolivia... 191
  N. D. Noya

Cocaine and Basuco: An Overview of Colombia, 1985.................... 199
  A. Perez Gomez

WORKING TOWARDS THE FUTURE: A CHALLENGE FOR CHANGE

Cocaine Addiction: A Socio-Ethical Perspective....................... 207
  D. F. Allen

Epilogue: A Vision of Hope.. 221
  D. F. Allen

About the Editor.. 225

Contributors.. 227

Bibliography.. 229

Index... 245

INTRODUCTORY CHAPTERS

INTRODUCTION

David F. Allen

National Drug Council
Nassau, Bahamas

When I began my psychiatric residency at Boston City Hospital in 1970, the first patient I interviewed described an idyllic scene of little multi-coloured rabbits which he saw jumping up and down on the office floor. Upon further investigation, I learned that he had been snorting large quantities of cocaine daily for a number of years. Little did I dream that this incidence of cocaine psychosis was the forerunner of what I would be seeing with increasing frequency when I returned home to the Bahamas ten years later.

In 1983, mental health professionals in the Bahamas were becoming alarmed at the increasing numbers of patients seeking treatment for cocaine addiction. Unlike users who snorted powdered cocaine, these patients described an entirely new process which they called freebasing. Heating the cocaine hydrochloride powder in water with baking soda produced crystallized cocaine pellets called freebase or rocks in the Bahamas, and crack in the U. S. The cocaine rocks were then smoked in a water pipe or a home-made "camoke." The ensuing high quickly became an obsession, demanding the person's time, money, and energy. In essence, then, the cocaine freebase epidemic in the Bahamas preceded the cocaine crack epidemic in the U.S. by two years or more.

With soaring hospital admissions for cocaine treatment, the Bahamian government appointed a National Task Force to investigate the situation and to recommend strategies for the containment and rectification of the impending drug epidemic. As Chairman of the Task Force, I saw the devastating effects of the addiction in society manifested in the destruction of lives, broken homes and families, decreased productivity in the workplace, and increased crime leading to fear and a changed lifestyle for the populace.

In 1985, Dr. Carlton E. Turner, Deputy Assistant to President Reagan and Director of Drug Abuse Policy at the White House, visited the Bahamas and was struck by the ravages of the rapidly spreading freebase cocaine epidemic. Recognizing in the Bahamian situation important future trends in drug abuse, prevention, and treatment, he spearheaded the impetus to organize an international conference on cocaine designed to bring together clinical experts from the producer countries of Colombia, Peru, and Bolivia, the trans-shipment countries of the Caribbean such as the Bahamas and Jamaica, and the large consumer countries such as the United States. The goal of the conference was to share experiences and knowledge in order to improve the management and treatment of the cocaine addicted person. Sponsored by the

American Embassy in the Bahamas and the Ministry of Health of the Commonwealth of the Bahamas, the First International Drug Symposium on Cocaine
became a reality in November, 1985. Held under the expert guidance of the
distinguished Chairman, Dr. Brian G. Humblestone, the Conference was a tremendous success in promoting fruitful dialogue between drug treatment experts
from South America, the Caribbean, and the United States. Experts from each
region presented aspects of cocaine addiction found in their respective
countries. Those from the U. S. focused on the problems associated with
snorting cocaine hydrochloride, and learned from the Bahamas about freebase
and freebasing, phenomena not prevalent at that time in the U. S. The South
Americans discussed the ravages of the coca paste (basuco) addiction. Recognizing the enormity and complexity of the problem, each expert expressed the
need for a thorough scientific understanding of the problem in order to
adequately cope with it.

Shortly after the Conference, the freebase or crack epidemic hit the
United States, with the sudden appearance of the ready to smoke cocaine rocks
or pellets. Wreaking havoc first in New York City and California, the new
form of the drug has now spread to other parts of the U. S., causing a public
outcry and eliciting pledges from the White House to intensify the "War on
Drugs." With the growing numbers of addicts coming forward for treatment,
there has been a heightened need for more and better treatment facilities
skilled in treating the person addicted to cocaine.

In the wake of the cocaine crisis, there is still little understanding
among professionals about cocaine addiction in general and crack or freebase
addiction and its treatment in particular. To continue and expand the dialogue begun at the First International Drug Symposium, this collection of
papers by experts in the field of cocaine addiction is presented to provide
a deeper understanding of the international problem from a clinical perspective.

The first section of the book is a general introduction to the cocaine
crisis. Papers on the history of cocaine, the various modes of use, and
the precursors and the indications of cocaine abuse provide an overview of
the major issues. Implications of crack cocaine are explored by Dr. Sidney
Cohen.

The second section deals with the experience of the United States, the
major consumer country. Elaine Johnson of the National Institute on Drug
Abuse gives a comprehensive overview. Dr. Arnold Washton, co-founder of
the 800-COCAINE Hotline, vividly charts the mushrooming cocaine crisis in
the United States. The self-medication hypothesis of drug addiction is ex-
pounded by Dr. E. J. Khantzian of Harvard. Psychotherapy approaches to the
treatment of the cocaine addict are presented by Dr. Frank Gawin and his
associates from the Yale addiction centre. In the final chapter of this
section, Prof. James Jekel of Yale considers the public health approach to
the problem of widespread cocaine addictions.

The third section dealing with the Caribbean trans-shipment countries,
draws on research done in the Bahamas, a major trans-shipment centre for
cocaine enroute from South to North America, and where the first epidemic
of freebase or crack appeared. The Bahamian experience is introduced by
Drs. Humblestone and Allen and is followed by an epidemiological analysis
of the existing epidemic by Prof. Jekel of Yale. Dr. Nelson Clarke provides
a clinical perspective and Dr. Theo Manschrek of Harvard discusses cocaine
psychosis using his researches as a Fulbright Fellow in the Bahamas. Recognizing the problem among women, Dr. Sandra Dean-Patterson examines the
special needs and issues of women in treatment. Drs. Neville and McCartney
discuss new treatment approaches which involve acupuncture and cognitive
behavioural learning. Prof. Michael Beaubrun, pioneer in Caribbean psychia-

try, ends this section with a discussion of cocaine addiction in Trinidad.

The South American experience with the coca paste, the first step in the production of cocaine hydrochloride, comprises the fourth section. Prof. F. Raul Jeri of Peru examines a number of case histories of coca paste addicts. Prof. Nils Noya of Bolivia describes the coca paste, or pitillo, problem in that country, and Dr. Augusto Perez-Gomez discusses the coca paste (basuco) problem in Colombia.

The book ends with a socio-ethical analysis of the cocaine crisis.

It is often easier to state a problem than to offer insight into its solution. Dr. Sidney Cohen, a contributor to this book and a well-known expert in the field of drug addiction, claims the outlook concerning the cocaine crisis is dismal. He states:

> The public health aspect of the cocaine story may no longer
> be the major adverse consequence, even though morbidity and
> mortality are mounting. It is the disorganizing impact of
> the many billions of "coca-dollars" on the producing and the
> consuming nations that produces a level of corruption, violence,
> and demoralization that damages everyone.*

Emphasizing that the control of supplies may no longer be a realistic goal, Cohen believes that much attention should be given to demand reduction and prevention. This includes special education programmes and people taking responsibility for their actions. In spite of burgeoning social problems and the shortcomings of the social system, responsibility for our own conduct remains the only barrier or defence against the onslaught of the cocaine evil. It is perhaps on this issue of whether we as adults are willing to take responsibility for our actions and teach our children to do the same that the war against cocaine will be won or lost. *

---

* Cohen, S., The Story of Cocaine is Scary, The Journal, November, 1985,p.2.

HISTORY OF COCAINE

David F. Allen

National Drug Council
Nassau, Bahamas

It has been said that history teaches two major principles: first, that things change slowly; and second, that things change rapidly. This situation is characteristic of the history of cocaine where the issues have drifted along at a slow pace, followed by crisis situations, then slackened off only to accelerate again. (1)

Cocaine is derived from the coca plant grown in the South American continent. It is found in two coca plant species. Native to the Peruvian Andes and Bolivia, Erythroxylum coca has a concentration of alkaloidal cocaine as high as 1.8%. Erythroxylum novogranatense is cultivated in the drier mountainous regions of Colombia and along the Caribbean coast of South America. The Truxillense variety of this species is grown on the north coast of Peru and in the dry valley of the Maranon River, a tributary of the Amazon in northeast Peru. Harvested for legal export, the leaves of the novogranatense are used for pharmaceutical purposes. (2)

Use by Earlier Civilizations

The stimulating properties of the coca leaf have been known for centuries. Recent archaeological finds in Ecuador indicate that human experience with coca leaves goes back at least 5,000 years, long before the Inca empire was established. These data suggest that coca leaf chewing was widespread and uncontrolled at that period, being used for religious rituals and combatting fatigue in persons working at high altitudes. (3,4) The best documented accounts of the early use of the coca plant comes from observers of the Inca civilization. To the Incas, the coca plant was of divine origin with the special ability to enhance communion between man and gods. They believed that they were the direct descendants of the god Inti, who had created the coca leaf to alleviate hunger, thirst, and to ease the burden of their life.

The Inca state was all-powerful, controlling every aspect of daily life, including cultivation of the coca plant, and the use of its leaves. Generally, only those with high social or political rank, i.e., the ruling classes were permitted to chew the leaves. On certain celebrated occasions, however, coca leaves were given to soldiers defending the nation, workers on imperial projects, or those being rewarded for special services to the empire. In the Inca civilization, the use of the coca plant was well regulated and the casual chewing of the leaves was considered a sacrilege.

The ritualistic use of the coca leaves involved chewing the leaves into a quid, or ball, and then adding tocra, a black alkaline substance the size of a pebble. This combination facilitated the release of cocaine from the organic material of the leaf, and aided the bodily digestion of it. Chewed coca leaves provided a subtle but long-lasting low-grade high with mild stimulation, reduced appetite, and increased physical stamina. Because the use of the leaf was strictly controlled, it is probable that the adverse consequences on the Indians and their society were minimized.

Sometime around AD 1000, there was a diminution in coca leaf use. It can be surmised that as this was the period of the beginning of the Inca subjugation of this area and the subsequent control of the peoples of those lands, that it was due to their influence that chewing the coca leaves diminished among the general populations of the region. Strict regulation by the Incas continued until the early 1500's when the ruling Incas became embroiled in a bloody fight for succession, paving the way for the de-control of the use of the coca leaves as well as the conquest by the Spanish conquistador, Pizarro. Coca leaf chewing was again widespread by the arrival of the Spaniards. The letters of Amerigo Vespucci in 1507 discuss the Indians chewing the leaves. And as with the Incas, they added ashes containing alkali to the cud to release the cocaine. (4)

In 1618, a manuscript by Don Felipe Guaman Poma de Ayala describes coca leaf chewing as an unauthorized social activity engaged in by the Indians when they were expected to be working. Other explorers reported that chewing coca leaves increased the endurance of the Indians, enabling them to work harder and eat less food at higher altitudes. (5) It appears that by 1569, Spanish entrepreneurs in South America recognized that coca leaves could be used to recruit Indian workers for intensive labour. Enforcing this concept by official sanction, King Philip II of Spain issued an imperial edict that coca leaves were necessary for the well-being of the Indian people.

Cocaine and Europe

With the widespread use of coca leaves in South America, it is surprizing that the pleasures and possible uses of the leaf were not transferred to Europe at this time. Coca leaves were in fact brought to Europe by Spanish explorers in small quantities throughout the 17th, 18th, and 19th Centuries. However, as a result of the deterioration of the coca leaves during shipping, the practice of chewing coca leaves was not adopted in Europe.   It is believed that if coca leaves, like tobacco leaves and coffee beans, had retained their potency during shipping, the impact of coca on Europe at that time would have been vastly different.

Although the German chemist Friedrich Gaedecke isolated alkaloidal cocaine from the coca leaf in 1855, it was not until 1859 that Albert Niemann of the University of Gottingen isolated the chief alkaloid of coca and named it "cocaine." Niemann reported that it had a "bitter taste" and a numbing effect on the tongue. (6) It was also in 1859 that Pablo Mantegazzo published his essay praising the ability of cocaine to reduce fatigue, increase strength, elevate spirits, and increase sexual desire.

Between 1863 and 1865, the Corsican Angelo Mariani developed Vin Mariani, a mixture of coca and wine which quickly became popular, being enjoyed by many celebrities, including Robert Louis Stevenson, Henrik Ibsen, and Thomas Edison. Indeed, Mariani was given a gold medal and cited as a benefactor of humanity by Pope Leo XIII, a frequent imbiber.

Further understanding of the local anaesthetic properties of cocaine developed in 1880 when Vassili von Anrep, a Russian nobleman and physician at the University of Wurzburg, noted that he could not feel a pinprick after the subcutaneous administration of cocaine.

Although Professor Anrep was one of Sigmund Freud's instructors, Freud took an interest in cocaine not as a local anaesthetic, but rather as a stimulant of the central nervous system and as an aid to overcoming morphine addiction. Freud, in 1884 was a house officer in neurology at a hospital in Vienna and tried cocaine himself, becoming enthusiastic about its therapeutic qualities. He claimed that a small dose lifted him to the heights in a wonderful fashion. As a result, he published "On Coca" describing and praising the effect of cocaine for the treatment of fatigue, impotence, and depression. As a result, cocaine use exploded in Europe, being praised by physicians as a "wonder drug" for a whole variety of problems and was particularly used for the treatment of morphine, opium, and alcohol addictions.

Obviously, this sent ripples through the medical community where the reaction was mixed. Increasing numbers of patients began to describe addictive aspects of cocaine, and in 1885, Albrecht Erlenmeyer accused Freud of unleashing the Third Scourge of Humanity, after alcohol and the opiates. Freud was badly affected. Although Freud had dropped the concept of using cocaine to treat morphine addiction, he still enjoyed his personal use of cocaine. Eventually, as more of his patients described the increasing addictive potential of cocaine and under mounting pressure from colleagues, Freud abandoned his praise and his use of cocaine, the "wonder drug." Unfortunately, the description of the beneficial effects of cocaine given by Freud and his associates was not to be superceded for nearly 90 years. (7,8)

Cocaine in North America

The popularity of the concept of cocaine as a local anaesthetic as well as a possible treatment for narcotic addiction spread quickly among professionals in the United States. No doubt the work and testimony of Sigmund Freud and his associates had some impact on the international popularity of cocaine for these uses. A number of scientists, including Dr. William Halstead, experimented with cocaine to clarify the overwhelming claims made for the drug from across the Atlantic. Halstead, a noted surgeon from Johns Hopkins University, became addicted to the drug, making a ruin of his life and practice. Wishing to help him, his friends sent him on a trip around the world, believing that he would forget or lose his appetite for cocaine. However, after returning from his trip, Halstead returned to seeking the pleasures of the drug, remaining addicted to it to the end of his life.

In 1885, John Styth Pemberton of Georgia introduced French wine Cola to compete with Vin Mariani and sold his Cola as the ideal nerve and tonic stimulant. In 1886, he developed Coca-Cola, a popular soft drink made from a syrup consisting of coca and caffeine. Cocaine was further popularized when Dr. W. A. Hammond, a leader of the American Medical profession, endorsed the virtues of cocaine use. As a result, increasingly the "lower classes" particularly in the dry states during Prohibition, started to use cocaine instead of hard liquor. To compound the situation, bars began to add a pinch of cocaine to a shot of whiskey and cocaine was also sold door to door.

Meanwhile, medical circles became more aware of the problems of cocaine use and of its failure to cure opiate and alcohol addictions. In fact, in 1891, 200 cases of cocaine intoxication and deaths were reported.

As time went on, the association of cocaine use with alcohol became accepted. More threatening to the status quo was the belief that black persons acquired extraordinary powers through using cocaine, giving them the de facto ability to defy laws and commit increasingly violent crimes. In fact, in 1903, cocaine was removed from Coca-Cola because of the fears that the

Southern politicians had about cocaine's effects on blacks. Dr. Harrison Wright, the U. S. representative to the Shanghai Conference in 1910, reported that the use of cocaine by southern blacks "is one of the most elusive and troublesome questions confronting law enforcement and is often a direct incentive to the crime of rape"(9,10).

However, importation of coca increased and it is estimated that Americans consumed as much cocaine in 1906 as they did in 1976, with only half the population. But in 1906, the increasing fear of crime and the effects of addiction led to the development and subsequent passage of the Food and Drug Act. This Act required the elimination of cocaine from patent medicines and soft drinks and heralded the beginning of the end of the free and easy distribution of cocaine.

With unemployment rising, cocaine use peaked in 1907, and 1.5 million pounds entered the country. The situation, especially acute in New York, led to the passing of the New York Anti-Cocaine Bill, which clamped down hard on the widespread use of cocaine. In 1908 importation of coca leaved dropped to half the 1907 level and public pressure increased against cocaine users. Indeed, users were considered to be "cocaine fiends" and the best thing for them was to let them die. Although cocaine use did decline in the general population, its popularity among intellectuals and writers increased, seemingly due to its reputation for stimulating creativity.

By 1914, forty-six states had passed laws regulating the use and distribution of cocaine. This set the pace for the passage of the Harrison Act which provided for a complete ban on cocaine, although it was incorrectly classified as a narcotic in the Act. With the passage of the Harrison Act, cocaine became less available and its price subsequently increased. The Act was amended in 1919, placing even stricter controls on cocaine and opium, whereupon cocaine was scarce and its price rose to $30 per ounce, three times higher than a decade earlier. Prohibitively expensive, cocaine use became restricted to the bohemian-jazz culture and the ghetto, where it was a symbol of affluence.

In 1922, an amendment to the Narcotic Drug Import and Export Act made the importation of cocaine totally illegal and established stricter penalties for violations of the law. The Uniform State Narcotic Act, passed in 1932, regulated the sale of cocaine in each state. The concentrated and cumulative effects of these laws, coupled with an increasing public awareness of the dangers of the drug, led to a further decline in cocaine use, a trend which continued until the 1960's.

According to Stone, et al. (11), there was a general breakdown of repression and loosening of restraints during the 1960's in contrast to the constricting socio-political atmosphere of the 1940's and the rigid conformation of the 1950's. With its widespread experimentation with marijuana and hallucinogens, the atmosphere of the 1960's was fertile ground for the re-emergence of cocaine.

In the 1970's cocaine hydrochloride, the white powder form of cocaine, once again became a symbol of affluence and glamour. Increasingly a main attraction at swinging parties, it was mentioned in films and used by intellectuals and professionals. In keeping with the ambition of the culture to be cool, confident, and successful, cocaine euphoria seductively and temporarily provided these feelings. By the 1980's the demand for cocaine had become extensive.

Cocaine in South America

The vast land mass and climate of the South American Continent, particularly in Colombia, Peru, and Bolivia, is ideally suited for the cultivation of the coca plant and has spurred a multi-billion dollar business. Organized, wealthy, and international, the cocaine cartel has become a powerful alter-bureaucracy which influences individuals, private organizations, businesses, and even Governments.

Processed as cocaine hydrochloride, a white powder, cocaine is exported from South America through trans-shipment zones such as the Bahamas, and smuggled by air or sea into North America. Although South American govern-ments expressed concern regarding the trafficking of cocaine, it was not until the ravages of the local coca paste epidemic became evident that the problem became urgent. (12) Through local efforts and international cooper-ation with the United States and other governments, much has been done to reduce drug trafficking through the destruction of crops, interdiction of shipments, education of the consumers, and economic assistance to the peasant no longer producing the plants. However, in spite of these efforts, coca fields in South America flourish, and the flow of cocaine to the outside world continues unabated.

The Bahamas

The Bahamas is an archipelago of 700 islands and cays stretching between Cuba and Florida. Cocaine from South America enters the Bahamas via boats, planes, etc. It is dropped near isolated cays and unused airstrips through-out the islands. Small planes and fast boats then smuggle it into the U.S. Allegations of major drug rings existing in the Bahamas led to the appoint-ment of the Royal Commission of Inquiry. Their report confirmed the Bahamas as a major cocaine trans-shipment area. In addition, it drew attention to the increasing local consumption of freebase cocaine and the concomitant problems of the addiction on individuals and society alike, bringing untold misery and destruction in its wake. (13)

Responding to the problem, the Government appointed a National Task Force which published an in-depth description of the problem and called for an on-going National Drug Council to organize a comprehensive onslaught against drugs. (14) Through education, rehabilitation, law enforcement, and the creation of alternatives, much has been done to stem the tide of the epi-demic. Although reduced by the combined efforts of Bahamian and American law enforcement agencies, large scale trafficking still occurs.

The United States

The increasing demand for cocaine has caused much concern. It is reli-ably estimated that 100 tons of cocaine are smuggled into the U.S. each year. According to John Lawn, administrator for the U.S. Drug Enforcement Adminis-tration, there has been a 350% increase in cocaine use in America. (15) It is claimed that 5,000 persons try cocaine for the first time each day. Twenty-eight percent of Americans between 18 and 25 years of age have tried cocaine. Seventeen percent of high school seniors have experimented with cocaine.

More frighteningly, the same type of ready-to-smoke freebase cocaine which has had such a devastating effect on the Bahamas is now wreaking havoc in certain parts of the U.S., especially in New York, Detroit, Florida, and California. According to Arnold Washton, 33% of calls to the 800-COCAINE hotline deal with crack (ready-to-smoke freebase) cocaine. The tragic cocaine-linked deaths of star athletes Len Bias and Donald Rogers have stunned the U.S. as well as other parts of the world.

Responding to the challenge of the drug menace, President Reagan pro-claimed a war against drugs, and in 1982 appointed the South Florida Task

Force under Vice President Bush. In many ways this Task Force has been successful, although the drugs interdicted represent only a small percentage of
the total cocaine being smuggled into the States. In 1983, Government agents
in South Florida seized 6 tons of cocaine and in 1985, the figure rose to 25
tons. The drastic drop in the price of cocaine indicates the massive growth
in supply, in spite of the efforts of law enforcement agencies. While in
1982, dealers in Florida were paying $60,000 per kilogramme of cocaine,
today they are paying 40% less.(16) The general consensus in America is that
although interdiction is crucial, major emphasis must be placed on understanding the reasons for the enormous growth in demand for drugs and developing
and implementing meaningful educational programmes to reduce that demand.

CONCLUSION

It is evident from this brief history that the status of the coca plant
and its derivatives has changed immeasurably over the last several thousand
years. For nearly 4,700 of those years, use of the coca plant was mainly
restricted by geography, by method of use, by purpose of use, and by population of use, and all of these were, for the most part, strictly regulated
by the ruling and religious classes.

However, with the advent of new technologies a little over 100 years ago
it became possible to extract the alkaloid content from the leaf and from the
alkaloid, cocaine hydrochloride salt, and with further refinement, the freebase. As the forms of cocaine were perfected, the methods of using cocaine
changed to keep pace with the pursuit of ever greater effects from the drug.
From chewing leaves to adding coca paste to tobacco or marijuana and smoking
it as a cigarette, the search for a greater high led to snorting or injecting
the hydrochloride powder and then onward again to inhaling the freebase vapour in special pipes.

The evidence of the dangers of cocaine use were realized soon after the
introduction of the powder. Laws banning the importation and use of cocaine
other than for legitimate medical purposes were passed in recognition of
those dangers. However, the inevitable time lag between recognition and
legislation meant ruined lives, destroyed families, and deaths. The costs
to society were incalculable. Medical services were overstretched and
precious resources, both human and financial, were drained striving to provide medical help for those disabled by the drug.

The stage is now set for cocaine epidemics on an unprecedented scale
worldwide. Science teamed with modern transportation technologies, and
entrepreneurial skills matched with effective marketing techniques mean that
the ultimate high is now freely available, attractively packaged, easy to
use, low in cost, and deadly.

It is said that those who do not learn from history will be doomed to
repeat it. We must improve our knowledge and our technologies to deal with
the untold misery this new/old situation brings in its wake. We must improve
our laws and enforcement capabilities to curb the new international cartel
spawned to deal in cocaine and death. We must change attitudes and outlooks
so that drug taking becomes unattractive, even if the high is appealing.

Not only must we look to the past and its lessons, we must rapidly
develop responses to the frightening speed with which cocaine addiction
occurs and spreads. If we cannot learn, develop, and respond, in depth and
urgently, we may be doomed, not only to repeating history, but to losing our
brightest and best hopes for our futures as individuals, as families, as
nations.

REFERENCES

1. R. Ashley, "Cocaine: Its History and Effects," St. Martin's Press, New York (1975).
2. E. Van Dyke, R. Byck, Cocaine, Scientific American, 246:128-146 (1982).
3. Ibid.
4. N. Stone, M. Fromme, D. Kagan, "Cocaine, Seduction and Solution," Clarkson N. Potter, Inc., New York (1984).
5. C. Van Dyke, R. Byck, op.cit.
6. G. Austin, Cocaine USA, in "Perspectives on the History of Psychoactive Substance Use, NIDA Research Monograph No. 24," National Institute on Drug Abuse, Rockville, Md. (1978).
7. C. Van Dyke, R. Byck, op.cit.
8. G. Austin, op.cit.
9. D. Musto, "The American Disease: Origins of Narcotic Control," Yale University Press, New Haven (1973).
10. G. Austin, op.cit.
11. N. Stone, M. Fromme, D. Kagan, op.cit.
12. Please refer to the articles in this book by F. R. Jeri, N. Noya, and A. P. Gomez.
13. J. Jekel, D. Allen, H. Podlewski, S. Dean-Patterson, N. Clarke, P. Cartwright, C. Finlayson, Epidemic Cocaine Abuse: A case study from the Bahamas, Lancet, I:459-462 (1986).
14. D. Allen, et al.,"Report of the National Drug Task Force," Bahamas Government Publication, Nassau (1984).
15. J. Lieber, Coping with Cocaine, The Atlantic Monthly, January, 1986.
16. J. Lieber, Ibid.

MODES OF USE, PRECURSORS, AND INDICATORS OF COCAINE ABUSE

David F. Allen

National Drug Council
Nassau, Bahamas

INTRODUCTION

   The complexity of the cocaine problem is compounded by the different
methods of using cocaine, the widespread vulnerability of the general population, the subtlety of its outward symptomology and therefore the difficulty
of detecting the presence of the addiction. This chapter provides discussion of, first, the modes of use; second, the precursors of abuse; and
third, the indicators of abuse.

MODES OF USE

Coca Leaf Chewing

   As previously discussed in the chapter on the history of cocaine, chewing
the coca leaf was the earliest form of cocaine abuse. In this process, the
leaf is chewed in conjunction with a lime-based mixture which facilitates the
release of cocaine from the organic material of the leaf and aids the body
in digesting it. Providing a subtle but long lasting effect, chewing coca
leaves leads to mood elevation, mild stimulation, increased physical stamina
and reduced appetite. Used for centuries by the Indians of South America,
chewing coca leaves was employed in rituals, celebrations, and physical
exercises. Although now overshadowed by other methods of use, chewing coca
leaves is still practiced by many rural peoples of South America, in particular, Peru. (1,2)

Coca Paste

   Coca paste smoking is the predominant form of cocaine abuse in South
America. Coca paste is an intermediary product in the manufacture of cocaine hydrochloride and contains 40% to 90% cocaine sulphate, as well as
other coca alkaloids and contaminants from the refining process such as kerosene and sulphuric acid. The paste is put into marijuana or tobacco cigarettes and then smoked. The cigarettes are called "basuco" in Colombia and
Peru, and "pitillo" in Bolivia.

   Initially, there is a period of intense euphoria which leads to compulsive use followed by persistent dysphoria, paranoid delusions, and multifold
hallucinations. Associated with overriding anxiety and depression, the addiction leads eventually to cardiovascular problems, malnutrition, muscle

spasms, tremulousness, and convulsions. Death may occur from accidental overdose, cardiac arrhythmias, and violent behaviour, including suicide.

For further discussion of this phenomenon, please refer to the papers in this volume by Jeri, Gomez, and Noya.

Snorting Cocaine Hydrochloride

In this method, the cocaine hydrochloride powder is laid out in lines on a smooth surface such as a mirror. Using a soda straw or a rolled-up dollar bill, the powder is inhaled through the nose, producing a feeling of elation which lasts from ten to thirty minutes. The high rises and subsides gradually. The efficiency of absorption of cocaine by this method can vary from 30% to 60% depending on the quality, quantity, and duration of use. This is not a benign or harmless form of cocaine abuse. If practiced regularly, snorting cocaine can have destructive consequences, such as poor concentration, loss of motivation, poor work attitudes, depression, and paranoid ideation.

Snorting causes irritation of the nasal membranes leading to ulceration, bleeding, nasal sores and chronic sinus congestion. In serious cases, perforation of the nasal septum may occur.

Cocaine hydrochloride is very soluble and can be absorbed through any mucous membrane including the eyelid, mouth, anus, or vagina. Cocaine suppositories used per vagina or anus may prove fatal due to the vast absorption of cocaine in the blood in a short time. A fatal dose of cocaine results from the quantity entering the blood per unit time, not necessarily from the amount in the blood. (3)

Intravenous Cocaine Hydrochloride

Intravenous cocaine has a very high dependence potential due to direct entry into the blood and the subsequent rapid rate of delivery to the brain. With the threat of AIDS from contaminated needles, however, a number of cocaine addicts have changed from IV use to the freebase method.

The complications of this mode of abuse include phlebitis, hepatitis, and possibly AIDS along with the other sequelae of cocaine abuse discussed in the section on freebase.

Freebasing Crack or Rock Cocaine

Ready to smoke freebase is the fastest growing form of cocaine use in the United States (where it is called "crack") and the Caribbean (where it is referred to as "rocks"). It is the most efficient way of delivering a potent form of cocaine to the brain in a very short time. The high addiction potential of this method creates an instant demand for the drug. The change to freebase cocaine may be seen as a marketing strategy by cocaine traffickers to increase their sales through increased demand for the drug.

The freebase form involves dissolving the white cocaine hydrochloride powder in warm water and adding an alkali, for example, sodium bicarbonate (baking soda). This leads to the precipitation of cocaine crystals called base, rocks, or crack.

The cocaine base is sold as rocks in the Bahamas while in the U. S. several base rocks may be put into a vial and sold as "crack." Cost comparisons are difficult because of weight, impurities, etc., but as of July, 1986, persons who have used crack in the U. S. and smoked rocks in Nassau claim that a $20 rock in Nassau costs about $75 in the U. S.

The crack or rock is then ignited and smoked in a water pipe. In the Bahamas, a home-made pipe called a "camoke" is used. A camoke is made by stretching foil paper over a glass, making small holes in one side of the foil and a large hole at the other side for the mouth. The crack or rock is placed upon cigarette ash over the smaller holes. The cocaine is ignited while the user inhales (wafts) through the larger hole. Using this method, 80% pure cocaine vapour reaches the brain in 8 seconds, causing a rush or high with intense euphoria, pounding heart, and a warm flush. The rush, sometimes described as "orgasmic", "Christmas" or "Joy," lasts for about two minutes followed by a glow that continues for ten to twenty minutes.

According to Cohen, the resultant euphoria is extreme, much like that produced by direct electrical stimulation of the reward centres of the brain. It is believed that cocaine stimulates the release of dopamine at the synaptic junctions and then blocks its re-uptake, making it more available and hyperstimulating the receptor cells. The reinforcing memory of the first high causes the user to seek it again. However, due to the depletion of dopamine from the first intense high, the subsequent highs are less intense, of shorter duration and associated with increasing dysphoria on the crash. On the subsequent highs, addicts describe being able to see the first high, but being unable to reach it.

The resultant dysphoria from frequent use leaves the person unhappy, unsatisfied, and with terrible mood swings. This state in turn becomes a negative reinforcement to use cocaine to escape the "blues." As use continues, a well-defined cocaine withdrawal depression develops. The addict is sad, angry, anxious, and on edge. As a result, he is continually forced back to cocaine for relief of the depression.

In the Bahamas, we have seen persons who have freebased up to five days straight with little food, sleep, or hydration. After a cocaine binge of many hours or days, the addict experiences a state of anhedonia. The dopamine stored in the pleasure centres of the brain is depleted and the pleasure threshold is raised for the simple joys of living, such as reading, walking, etc. The depletion of dopamine causes the receptors on nerve cells to become supersensitive, producing a biologically based craving for cocaine similar to hunger or thirst. This intense craving drives the addict to seek relief in further cocaine abuse. It must be stressed, however, that the longer the user is off cocaine, the greater the chance for the dopamine supplies to be replenished and the pleasure threshold to be reduced.

In summary, freebase or crack cocaine is extremely addictive and produces compulsive use due to its (a) Quick absorption rate in the brain; (b) The positive reinforcement of the memory of the first, intense high; and (c) The negative reinforcements of the deepening dysphoria after the subsequent highs, the withdrawal depression, and finally the prevailing sense of anhedonia.

Usually freebase cocaine is used in secluded quarters protected by armed guards with an elaborate warning system against intrusion by law enforcement officers or other unwanted persons such as rival dealers. In New York, these places are called "crack houses," in California, "rock houses," and in the Bahamas, "basehouses" or "cocaine camps." The freebase or crack is made, sold, and used in these houses where other activities such as prostitution, suggestive dancing, and sexual orgies may take place.

Symptoms of Freebase Abuse: Physical Symptoms

1. Respiratory System. Inhalation of pure cocaine vapour into the lungs leads to increased incidences of bronchitis with black blood exudates, pneumonia, tuberculosis, and chest pain. Even after recovering from freebase

addiction, the addict may require respiratory therapy for up to a year before the lungs function effectively.

2. **Gastrointestinal System.** Cocaine vapour stimulates the appetite centre in the brain leading to anorexia and loss of weight. This of course is associated with vitamin deficiencies of thiamine, pyridoxine, and ascorbic acid, and increased incidences of infections.

3. **Cardiovascular.** Cocaine has a stimulating effect on the heart which may lead to arrythmias (e.g., ventricular fibrillation), coronary heart disease and severe palpitations. One addict described a sensation of his heart "jumping out of his skin" after freebasing.

4. **Central Nervous System.** Cocaine, upon crossing the blood-brain barrier, may irritate the brain substance and cause severe convulsions. These are very threatening experiences, but do not deter the addict from further use. In one particular case, the disintegration of a plastic sheath being used to transport cocaine in the stomach of a courier, unleashed a lethal dose of cocaine into his body. The man became extremely violent, with severe convulsions, and eventually died. The vasoconstrictor effect of cocaine may also cause hypertension and lead to cerebral haemorrhages.

5. **Skin.** Freebase addicts experience severe itching over the body, especially in the limbs. Repeated scratching may cause a line of sores along the limbs. Colloquially this severe itching is called the "cocaine bug."

6. **Pregnancy.** Work by Chasnoff et al. has shown that cocaine-using women have a higher rate of spontaneous abortion than women in other groups. Also there seems to be an increased rate of abruptio placenta. Infants exposed to cocaine had significant depression of interactive behaviour and poor response to environmental stimuli. (4)

7. **Glandular System.** Initially cocaine causes hypersexuality which may be followed by impotence. In some cases, the hypersexuality is manifested in a polymorphous sexuality where "anything and anybody becomes a turn-on." Some female crack addicts describe a galloping sexuality which leads them to associate cocaine and sex. In some cases the person may end up in prostitution or other base sexual habits in order to support their habit.

8. **Death.** Cohen (5) has outlined a number of ways a cocaine addict might die. These include:

a. Overdose: In binge-freebasing, an addict may increase the size of the rock with higher concentrations of cocaine entering the blood and brain. This may lead to depression of the respiratory centre or ventricular fibrillation of the heart.
b. Hypersensitivity to cocaine: Rapidly broken down by liver and blood esterases (enzymes), small amounts of cocaine may cause death in persons with congenital absence of esterase.
c. Major convulsions: may lead to death through inhalation of vomitus.
d. Coronary heart disease: The vasoconstrictor effect of cocaine leads to decreased coronary blood supply and coronary pathology.
e. Depression of the respiratory centre by the combination of cocaine and heroin (speedball). Though cocaine is a stimulant, in large doses it depresses the respiratory centre like heroin.
f. Homicide: The paranoid ideation and behaviour of freebase or crack addicts may lead to murder. Our experience in the Bahamas leads us to believe that freebase releases aggression with severe violent consequences.
g. Accidental death: Cocaine gives drivers a false sense of competence which allows them to take foolish risks which may in turn lead to traffic fatalities.

h. Immune deficiencies: Persistent malnutrition and emaciation of
    some people leads to severe viral and bacterial infections such as
    Hepatitis B and tuberculosis.
i. Suicide: After a binge, the addict may become so severely
    depressed that he or she kills him or herself.

Symptoms of Freebase Abuse: Psychological Effects

Initially freebase cocaine provides the user with a sense of well-being,
confidence, and drive. In some way, it produces the prototype of the desired
image of western life, i.e., slim, cool, confident, and a feeling of success.
As one young freebase addict stated, "When I have a hit, I feel like I own
the world, and as I walk down the street, it all belongs to me."

In snorting, this sense of well-being could continue for months or years.
But with freebasing, the user starts showing signs of psychopathology within
three to six months. A number of patients have shared with us that they
snorted for years without ill-effects, but that they developed anxiety and
paranoid thinking after three or four months of using crack or rock cocaine.

The first signs of emotional decompensation are dysphoria and anhedonia.
Losing interest in the basic pleasures of life, the addict seeks refuge in
more cocaine use. Eventually the dysphoria leads to a cocaine withdrawal
depression manifested by insomnia, weakness, sadness, social withdrawal, and
loss of interest in work. In this state, the addict is on edge and spends
his time contemplating how to get more crack. Addicts have said that the
drug becomes synonymous with gold jewelry, money and electronic equipment
and that as craving intensifies, stealing these objects occurs automatically.
They will steal from friends, close family members, or anyone else, even
while feeling apologetic for doing so! It's as if their brain is on automatic pilot for freebase and they must have it, regardless of the consequences. This may validate the biologically based craving mechanism of cocaine.

Compulsive crack use is associated with ethical fragmentation, in which
the user defies superego and any other moral influences and enters an amoral
wasteland. At this point, anything goes: prostitution, promiscuity, sado-
masochistic behaviours, stealing, lying, bestiality, etc. This phenomenon
occurs even in the most well-put-together people with excellent family backgrounds, good education, and professional training. A breakdown in physical
health related to reduced attention to instinctual needs often accompanies
ethical fragmentation. In this condition, referred to as instinctual fragmentation, many addicts forego normal instinctual pleasures such as sex,
ingestion of food, and even water because of preoccupation with cocaine intoxication. In certain cases, patients lose considerable weight due to lack
of self care and their overall condition deteriorates to danger level.

Further psychological decompensation leads to paranoid ideation and behaviour. This may vary from being slight, e.g., "peeping" the feeling of
being watched by a policeman while hitting, to being more severe, as the
addict, who fearing being killed by a basing colleague, murders him instead.

In the Bahamas, chronic abusers describe various types of cocaine psychoses as "high base craziness" and "low base craziness." Having had the
opportunity to study their behaviour in base camps, my understanding of
these phenomena is as follows: "High base craziness" involves a sudden experience of widespread auditory hallucinations when everything in the environment of the addict speaks to him. For example, a young man described
how, after freebasing, he went to his closet to get his clothes, but his
suit asked him, "What do you want?" Afraid, he walked toward the door,
which told him, "Get back!" Retreating, he then heard the sofa say, "If
you sit on me, I'll kick your ..." With a sense of impending doom, intense

anxiety, and mounting panic, the young man ran to the hospital, where he received help.

This is a very frightening experience, and recovered addicts continue to dread the experience even after treatment.

In "low base craziness," the addict searches the ground on his hands and knees seeking crack cocaine in a ritualistic fashion. In a state of delusion the addict picks up any object even remotely appearing to be crack and tries to smoke it in his pipe or camoke. In this process, the addict picks up pieces of stone, lint, wood, garbage, debris, and other unsanitary materials. Visiting a base camp in the woods, I observed crack addicts living in makeshift homes made out of wooden boxes. Outside one of these structures, there was an unkempt, malnurished and emaciated woman attempting to smoke crack. After a while she was on her knees clearing away brush and picking at debris on the ground. Unable to communicate clearly, she continued in her ritual and indicated that I was a police officer who had come to take her away.

Whether these types of psychoses are distinct entities remain to be seen. Dr. Theo Manschrek of Harvard and a team of local researchers are presently investigating these phenomena.

Low-Low Depression

Besides the more common cocaine withdrawal depression, a more severe depression is found in crack addicts who have used increasing quantities of cocaine over a period of 8 to 10 months. In this condition, which I describe as a low-low depression, the user relates becoming very depressed with persistent thoughts of suicide. According to them, the depression is like a thick dark cloud which descends on them and death appears to be the only way out. The phenomenon occurs shortly after a freebasing binge yet it seems that even close friends find it difficult to assess the severity of the situation.

A young man who made a number of attempts at treatment, persisted in returning to crack use. From being well-dressed and groomed, he became less interested in his personal appearance and concomitantly moved from prestigious jobs to low status jobs. One night during dinner with his family, he asked to be excused. An hour later he was found hanging dead in the closet of his room. He left a note: "I love you very much, but I can't beat cocaine."

It is not known whether this low-low depression is a separate entity. Perhaps due to the extreme depletion of dopamine in neuro-reward centres, psychological collapse occurs, in which a person becomes suicidal, homicidal, or manifests some other bizarre behaviour to cope with or defend against the sense of impending doom. Obviously much research is needed in this area.

Symptoms of Freebase Abuse: Social Effects

Besides the physical and psychological consequences of cocaine addiction, the social impact on the addict's life is devastating. Freebase cocaine is a no-barrier drug and it creates havoc in persons from all segments of the socio-economic spectrum. Once the user is hooked, the mind is on automatic pilot for cocaine. Hence every sleeping dream, every waking thought, every motivation and ambition is how to obtain more cocaine. The addict will sell all his belongings, steal, prostitute him or herself, or even kill for more cocaine. A young executive secretary rushed into a cocaine treatment clinic begging for help because she had just sold her $10,000 car for $80 and the cocaine had only lasted for ten minutes.

Even after the addict has been off cocaine for a while, upon encountering persons he used to base with, he/she has a physiological buzz which simulates a cocaine high and drives them to seek more cocaine.

Another interesting phenomenon is that pushers of freebase or crack cocaine inevitably end up using the drug. Of course many addicts deal the drug to keep their own supply going. Hence young men and women   enter a lifestyle of crime: stealing, prostitution, and violent behaviour. Even if they want to go straight, it is hard to break away from their group without being threatened with blackmail, or worse.

Obviously, using cocaine does temporarily appear to give the addict some benefits socially, through a sense of community, protection, and in some cases, an association of glamour. However, the price the addict pays for these fleeting and dubious benefits far outweighs any hazily perceived benefits.

The effects on the family are catastrophic. In addition to family breakdown, freebasing leads to increased family violence, child abuse and neglect, and constant turbulence. For example, a young lady came to a drug treatment clinic to seek help for her husband. She claimed he sold all her china, furniture, and jewelry. One day she found the house had been shot up with bullets. After confrontation, he admitted to her that he had shot up the house because he wanted her to think that someone else was stealing their belongings. In another case, in a neighbourhood near a cocaine camp, a mother said that the children had had no breakfast that morning because the father took the eggs, bread, and milk and had sold them to buy cocaine.

Finally, crack cocaine is always associated with increases in crime of all types from stealing to murder. By making neighbourhoods unsafe and fearful, it impoverishes the social life of the community, decreases the property values, and makes children and adults in the community more vulnerable to addicts and addiction.

PRECURSORS OF COCAINE ABUSE

Recognizing the multifaceted aspect of cocaine addiction, it is more likely that a multiplicity of factors, rather than a single factor will be found as a precursor to drug abuse in a given individual. My experience of working with many freebase addicts is that the freebase form of cocaine is a no-barrier drug. Anyone, regardless of social, familial, or constitutional factors, is vulnerable to addiction. Approximately 70% of our patients admit they were addicted after experiencing their first high.

Nevertheless, numerous experts such as Cohen (6), Bry (7) and others, have discussed factors which act singly or together in contributing to drug abuse. It is important to stress in discussing precursors to drug abuse that association does not necessarily prove causation. Using a list of factors adapted from Bry, the following may act as precursors to cocaine abuse:

1. Poor Parent-Child Relationships (8)

Conflict in the family contributes to insecurity in the child and to communication problems between parent and child, increasing their vulnerability to drug abuse. In my experience, unfulfilled parental expectations in children is a heavy burden for the child, creating a huge gap between the child's expected and actual self. The child feels like a failure, becomes depressed, and is open to cocaine, which makes him feel successful and glamourous, with a sense of belonging, if only for a short time.

A young man, aged 19 years, reported that his father had great plans for him concerning education. However, he failed in school and was not able to obtain a high status job. Because his father was extremely angry with him, the patient lived with a deep sense of failure. Introduced to freebase cocaine, he said the highs made him feel cool, confident, and successful.

2. Low Self-Esteem (9)

Individuals with poor self-esteem have decreased ego strengths and a lower threshold for ambiguity, frustration, and dissonance. They experience higher levels of anxiety, passivity, and hopelessness, and find it difficult to resist drugs.

A small, inadequate-looking young man who admits to a lack of confidence claims that when he has his freebase, he feels on top of the world. To use his words, "When I have my hit, I walk down the main street as if I own everything. For that moment, I am great."

3. Psychological Disturbances (e.g., Depression) (10)

Persons suffering from severe psychiatric illness, i.e., DSM-III, Axis I diagnoses, may seek to relieve their pathos through mind-altering drugs. Khantzian suggests that cocaine addicts use their habit as a form of self-medication for underlying pathology. In our clinic, we are seeing increasing numbers of patients who, after two or three months off cocaine, manifest severe depression, manic episodes, borderline pathology and even overt schizophrenic decompensation. This underscores the need for the cocaine therapist to be sensitive to changing mental status examination and to be prepared to use appropriate psychiatric treatment when necessary.

A young man, aged 22 years, with a severe freebase addiction was admitted to the outpatient treatment programme. Initially, his mental status showed no clear psychopathology. After two months of being off cocaine, he became depressed and very passive. Eventually he became psychotic, manifested by bizarre posturing, extreme agitation, and animalistic grunting. Admitted to inpatient psychiatric treatment, he responded fairly well to psychotropic medication.

This was a painful experience for the staff, for it appeared that he was more compensated while on freebase cocaine. We in the Bahamas, are very concerned about this phenomenon. As the educational initiatives become more effective, and more persons are aware of the dangers of freebase cocaine, it appears that increasing numbers of chronic mental patients are using it. Whether this is because they are a special target of the pushers or their special vulnerability, it is too early to tell. The combination of mental illness and freebase cocaine presents with serious psychopathology, which may include violence (murder or suicide), extreme agitation, bizarre delusions and hallucinations.

4. Low Academic Motivation (12)

Studies indicate a correlation between low academic motivation and drug abuse. Obviously environmental factors (e.g., the lack of a good home situation) and psychological problems (e.g., attention deficit syndrome) often play a part.

5. High Experience Seeking Tendency (13)

Adolescents tend to be programmed for experiencing exciting events.

When such events are not available, they experience a "low." In this sense, perhaps a residual from the 1960s, feeling "high" is a positive value while feeling "low" is a negative one. Drugs are sought after to provide and maintain the desired high. As one young freebase addict said, "At first, being on 'base was one continuous party."

6. Low Religiosity (14)

Persons with an internalized religious faith which provides meaning for life appear to have less involvement with drug abuse. Often the faith provides a sense of community with other believers, providing increased support to avoid drugs. Our experience indicates that after or even during rehabilitation, a number of addicts drift towards some type of religious community. Whether this is to replace the addiction, to obtain a sense of community, or the search for existential meaning is not clear and requires further research.

7. High Family Substance Abuse (15)

The abuse of addictive substances by parents has a powerful and reinforcing influence on children. High family substance abuse is internalized by the family and acted out by the child: a nine year old freebase cocaine addict said that his drug abuse habit started when he was given a puff of freebase by his parents while they were getting high.

8. High Peer Substance Abuse (16)

Adolescents are influenced more by their peers than by their parents. Increasingly it becomes difficult to say "No" to drugs if peers are heavily involved in drug use. Numerous young freebase addicts admitted they started freebasing at a party when their friends were trying it. Many of them said that on the first high they knew they were hooked and went seeking the experience the following day.

9. High Community Availability of Drugs (17)

When drugs, particularly cocaine, are freely available, abuse is more likely to occur. The Bahamian experience has validated Cohen's (6) views that high availability, low cost, high quality freebase cocaine combine to produce a severe epidemic. High community availability also makes treatment difficult, because recovering addicts are often offered free cocaine upon returning to their neighbourhoods.

10. The Use of Other Drugs

About 80% of the freebase addicts in our clinic admitted starting their drug abuse with marijuana while 10 to 12 years old. Whether this is an association or a causation is not clear. However, in our perspective, the early use of marijuana may be indicative of future freebase addiction. (18)

According to work done by Bry et al. (7) the potential risk of drug abuse is directly proportional to the number of risk factors involved. Conversely, it must be emphasized that in certain drug infested areas, some adolescents with multiple risk factors do not use drugs. More research is necessary to elucidate the reason for this phenomenon.

INDICATORS OF COCAINE ABUSE

The detection of cocaine abuse is difficult because the addiction is usually far advanced by the time the known signs are evident. It is not uncommon for young persons to have been on cocaine for up to a year before it was noticed by their parents. In my experience a person may have snorted cocaine for up to two years or more before deterioration occurred. However the deterioration occurs more rapidly (i.e., in a 3 to 6 month period) with ready to smoke freebase or crack cocaine. Some of the signs which might indicate cocaine abuse in the family are:

1. Change in Personality

Any sudden change of personality may relate to cocaine abuse. For example, a quiet reserved person becomes loud and outgoing, or vice versa. Other changes include irrationality, argumentativeness, and increased combativeness.

2. Change in Behaviour

In the initial stages, the person on cocaine may be confident, productive and have improved performance at home, school, or at work. For example, a mother said that initially her son changed from being lazy to being extremely helpful around the house. Washing dishes, scrubbing floors, and mowing the lawn, he was indeed very useful. It was not until her television set was missing that she suspected drug abuse. Other indicative behaviours include staying out all night, missing school or work, extreme aggressiveness and even violence, suspicious or paranoid behaviour, or total isolation.

3. Ethical Fragmentation

Cocaine, particularly freebase, addicts appear to lose or detach themselves from superego or other ethical controls. As a result, they have no conflict in stealing, lying, widespread promiscuity or prostitution. Concepts of loyalty or team responsibility become of no importance.

4. Change in Associations

The addiction to cocaine, especially freebase, is communal which sooner or later becomes evident. The parent or spouse may notice an entirely new group of friends which may be accompanied by an open rejection of formerly close friends. Similarly, telephone calls may become suspicious or threatening. In certain instances, members of the user's family may be approached or accosted by pushers for monies owed to them for the purchase of cocaine.

5. Change in Work Habits

Cocaine addicts eventually become less productive with little interest in the quality of work or home work. In adolescents, the grades may slip and a once industrious child loses interest in school. It is important to stress that drug problems on the job or in school generally appear as low productivity, poor attitudes, chronic absenteeism, or tardiness and stealing rather than overt drug situations.

6. Physical Signs

There are a number of key physical signs which point to a possible cocaine addiction:

a. Extreme loss of body weight so that a normally well-built person becomes very thin. As is said in the Bahamas, "She/he was once a brickhouse, now she/he is a stickhouse!"
b. Intense itching in the limbs, called the "cocaine bug." One theory is that the cocaine vapour recrystallizes and irritates the skin.
c. Nasal septum irritation or sores are sometimes evident in persons who snort cocaine.
d. Convulsions. The freebase vapour may recrystallize and irritate the

brain substance, causing convulsion. This may be the first overt sign of freebase/crack addition.

e. Persistent hacking cough with dark sputum. Crack addicts have increased incidences of coughs, colds, pneumonia, and even tuberculosis. Characteristically, they have a pungent, dark coloured, blood-tainted sputum.

f. Scaly skin. The lack of nutrition and over-exposure of the body may result in avitaminosis associated with scaly skin, sores, and intense weakness.

g. Tired or blurry eyes. Freebase addicts may participate in cocaine binges that occur around the clock. As a result, they appear spaced out, with tired or blurry eyes.

7. Drug Paraphernalia Around the Home

As a cocaine addict becomes entrenched in their addiction, he or she becomes cavalier in leaving drug paraphernalia around. A parent or boss may find water pipes, crack containers or vials with multicoloured caps, pieces of rock or crack cocaine, baking soda (used to precipitate the freebase from the cocaine hydrochloride) and glasses with aluminium foil (i.e., home-made smoking apparatus for smoking crack). Parents should keep informed about the different types of paraphernalia because it is constantly being changed to avoid scrutiny by law enforcement officers and others. For example, in the Bahamas, freebase addicts have been adapting soda cans or pieces of pvc piping to smoke freebase.

SUMMARY

Obviously none of these signs is conclusive. If the parent or boss is suspicious, rather than deny the situation, they should have the person consult a known expert in the field. A good professional will perform a thorough clinical assessment followed by urinalysis. He/she will then form an opinion based on the composite analysis of the clinical assessment and the urine test. This is stressed because a urine test for drugs is a medical investigation and should only be used in tandem with a complete clinical assessment.

Once the diagnosis is made, the professional will discuss the various options for treatment. Parents or others should explore all the options and choose the one most suited to their particular situation.

The treatment of crack or freebase addiction is arduous and often frustrating for patient and family. Therefore once a programme of treatment has been chosen, the family should be involved in the therapeutic process for education, support, and insight. This in turn will encourage the patient and increases the probability of the successful outcome of treatments.

REFERENCES

1. C. Van Dyke, R. Byck (1982) Cocaine, Scientific American March, 1982, pp. 128-141.
2. S. Cohen (1985) "Cocaine, The Bottom Line," The American Council for Drug Education, p.11.
3. N. Stone, M. Fromme, D. Kagan (1984) "Cocaine: Seduction and Solution," Clarkson N. Potter, New York.
4. M. D. Chasnoff, W. J. Burns, et al, (1985) Cocaine Use in Pregnancy, New England J of Medicine, Sept. 12, 1985, Vol. 313, No. 11.
5. S. Cohen, op. cit.
6. S. Cohen, Drug abuse: predisposition and vulnerability, in "Psychiatry, the State of the Art, Vol. 6," P. Pichot, P. Berner, R. Wolf, K. Than, eds., Plenum Publishing Press, New York (1985).
7. R. Bry, Empirical foundations of family-based approaches to adolescent

substance abuse, _in_ "Preventing Adolescent Drug Abuse: Intervention Strategies, NIDA Research Monograph No. 47," National Institute on Drug Abuse, Rockville, Md.

8. R. H. Blum, et al., (1970) "Students and Drugs," Jossey-Bar, San Francisco.

9. H. B. Kaplan (1977) Antecedents of deviant responses: predicting from a general theory of deviant behaviour, _J. Youth Adolescence_ 6:89-101.

10. S. Paton, R. Kessler, D. Kandel (1977) Depressive mood and adolescent illicit drug abuse: a longitudinal analysis, _J. Genet. Psychol._ 131:267-289.

11. E. J. Khantzian (1985) The Self-Medication Hypothesis of Addictive Disorders: a focus on heroin and cocaine dependence, _Am. J. Psychiatry_, 142, p. 1239, 11 November, 1985.

12. G. M. Smith, C. P. Fogg (1979) Psychological Antecedents of Teenage Drug Abuse, _in_ "Research in Community and Mental Health, Vol. 1," R. G. Simmons, ed., JAI Press, Greenwich, Conn.

13. M. Zuckerman, R. S. Neary, B. A. Brustman (1970) Sensation-seeking scale correlates in experiences (smoking, drugs, hallucinations, and sex) and preference for complexity _in_ "Proceedings of the 78th Annual Convention of the American Psychological Association, Vol. 5," American Psychological Association, Washington, D. C.

14. F. A. Tennant, Jr., R. Detels, V. Clarke (1975) Some childhood antecedents of drug and alcohol abuse, _Am. J. Epidemiology_, 102:377-385.

15. D. B. Kandel (1978) Convergencies in prospective longitudinal surveys of drug use in normal populations, _in_ "Longitudinal Research on Drug Use: Empirical Findings and Methodological Issues," Hemisphere, New York.

16. D. B. Kandel, D. Treiman, R. Faust, E. Single (1976) Adolescent involvement in legal and illegal drug use: a multiple classification analysis, _Social Forces_, 55:438-458.

17. D. F. Allen, et al. (1984) "National Task Force Report on Drugs," Bahamas Government Publication, Nassau, Bahamas.

18. D. B. Kandel, D. Murphy, D. Karus (1985) Cocaine use in young adulthood: patterns of use and psychosocial correlates, _in_ "Cocaine Use in America: Epidemiological and Clinical Perspectives, NIDA Research Monograph No. 61," N. Kozel and E. Adams, eds., U. S. Government Printing Office, Washington, D. C.

THE IMPLICATIONS OF CRACK

Sidney Cohen

Neuropsychiatric Institute
UCLA
Los Angeles, California

At times the repackaging of a product can have marked effects on its sales, distribution system and consumption patterns. When a drug like cocaine is involved, improvements in availability, pricing and convenience of use can produce major shifts in its acceptability.

Crack is freebase, alkaloidal cocaine and it has been smoked for a dozen years or so. Its preparation from cocaine hydrochloride has customarily been a consumer activity although a few dealers prepared freebase for those regular customers who requested it.

PREPARATION OF COCAINE PRODUCTS

It would be well to review the various coca preparations because confusion about the nature of some of them exists. Coca leaves are ground up and treated with sulfuric acid and a solvent like kerosene or gasoline to form coca paste (basa). This contains 40-80 percent cocaine sulfate and many impurities. Coca paste is inexpensive and widely smoked in tobacco or marijuana cigarettes in the countries where the coca bush is grown. This is an inefficient way to deliver cocaine to the brain because the high temperature of the burning cigarette destroys much of the cocaine. However, because rather large amounts of coca paste are put into the cigarette the reaction is intense.

Coca paste is treated with hydrochloric acid and a solvent like ether or acetone to form cocaine hydrochloride. Purities of 95 percent can be achieved but dilutents are invariably added before sale to the consumer. Being soluble in water, it can be sniffed or injected. Attempting to smoke cocaine hydrochloride leads to a loss of some of the active material, therefore freebase was created for those who preferred smoking. Basic cocaine melts and vapourizes at a much lower temperature than the salt. Freebase is produced by treating the hydrochloride with an alkali like ammonia water, baking soda or sodium hydroxide. Sometimes the cocaine is extracted with a volatile solvent, at other times the mixture is simply heated and smoked. This procedure, which had been carried out largely by the user, gave rise to a paraphernalia industry which included the sale of chemicals, glassware and pipes.

ALONG COMES CRACK

What the ready to wear industry did to tailoring and fast foods did to cooking, crack (or rock) did to freebasing. The consumer is supplied with the ready to smoke material and the packaging permits the purchase of small amounts - as little as $10 to $20 worth. This opens up new consumer markets like juveniles and the indigent. The procedure is simple. Baking soda and water are added to cocaine hydrochloride, heated and dried. It is broken into small yellow-white clumps and the crack pieces are put in vials and are ready for sale and use. The adulterants that had been in the cocaine hydrochloride remain; in addition, the baking soda, now table salt, is also present. As with other methods of preparing freebase, the material is placed in a plastic pipe, heated until the cocaine vapourizes and inhaled. Although everyone speaks of "smoking" freebase, it is an incorrect description of the procedure. Nothing is combusted. The vapour of freebase is inhaled after it it has been heated, converting it from a solid to a liquid to a gas.

Freebase can produce a very compelling type of cocaine dependence because the very rapid onset and brief duration of an extreme euphoria keeps the consumer coming back for more. Crack may be inhaled during the waking hours every few minutes and, since it causes a profound insomnia, this behaviour may continue for days. The cocainist may be hunched over his pipe so persistently, complaints of back problems have been heard. The binge ends when the person is exhausted or convulsing or the cocaine is gone. Hundreds of thousands of dollars a year might be spent on crack by one person. Since it would be difficult to come by such large sums of money legitimately, the compulsive user is forced into stealing, dealing or some other criminal enterprise. A new wave of theft and burlaries in many cities is being attributed to the growing use of crack.

For the dealer, crack is a profitable way to deal cocaine. He might sell a gramme for $75. The same amount would yield 15 vials of crack at $20 a vial.

THE SIGNIFICANCE OF CRACK

Crack is not simply a new convenience form of freebase. It is competing with sniffing as the most popular form of using the drug. Its explosive growth during the past two years indicates not only that supplies are plentiful, but that potential customers are seemingly endless.

The marketing strategy is one of low cost, shifting, high volume operations. The manufacture of crack is also a small scale procedure in which only a kilo or so at a time is converted from the hydrochloride to the base and then broken down into small chunks suitable for the ultimate consumer. Although the operation is in small units, the swarms of adolescent sellers are only outnumbered by the buyers. The amount of money in the business is large but it moves into the cocaine network in relatively small amounts like thousands of dollars a day. Each crack domain may encompass a few blocks, making its disruption hardly worth the attention of the criminal justice system. No countrywide or even citywide head that can be chopped off is identifiable. Thus fortified rock houses may be phasing out. These fortified dwellings were a readily identifiable selling outlet for cocaine base and could be attacked with modified tanks or tractors.

Techniques or drugs that produce more intense and immediate effects tend to displace those that provide slower and more moderate effects. This partially explains the crack explosion that is under way. Crack is felt now and it rates a ten on the euphoria scale. But a second dictum is needed to further explain the compulsion to continue using. It might be stated as "Drugs that have a brief duration of action produce more drug hunger than those whose effects are prolonged." With crack, the high is measured in seconds or a few minutes and the desire to get back up there drives the individual with sufficient supplies to use all of it right then. The demand is built into the brain's reinforcing centres. Even the eventual anguish and depression that follow a crack run become just another reason to smoke more crack in the hope that these noxious feelings will vanish. Crack is very close to being the perfect drug to compel people to continue using it. It competes with intravenous cocaine or intravenous heroin and cocaine. These are dangerous drugs, used dangerously. It is probably impossible to remain a "social" crack user if crack is available. While people in psychological distress are more vulnerable to incessant crack use, all those who try the drug are liable to become crack dependent. In addition to its enormous dependence potential, crack is also capable of producing overdose complications like cardiac arrest or pulmonary failure.

Crack smoking is like the behaviour of migrating lemmings, who follow each other over a cliff to their destruction, driven by some unknown impulse. Crack smokers are similarly caught up in a compulsion and tend to have a similar fate. That so many humans cannot see beyond the immediate high at the not too distant consequences is a measure of the immaturity of our species.

We have, then, a drug whose supplies are difficult to control and whose psychologic characteristics make it entrapping to those who use it. Not only are there a number of immediate risks, but protracted use will also lead to paranoid, disorganized thinking - a considerable risk to the person and those in his or her vicinity. Observers of the crack scene believe that this drug could waste too many people, induce too many into criminal activities, corrupt others, and distort the economy of nations.

What is to be done? Of course, anything that can be done to reduce supplies and to treat those who have come to acknowledge their personal cocaine disaster must be continued and increased. However, supply reduction and treatment alone are incapable of resolving the overall problem. We are left with what is called prevention: changing behaviour so that those who have never used cocaine will not start using it and those who are using the drug will stop. Both efforts are exceedingly difficult to achieve. Educating a person, usually a young person, about the nature of cocaine in general and crack in particular, will not persuade many to refrain from trying the drug. What is needed is an entirely new attitude toward the use of drugs for pleasure. This will require not only efforts launched by schools, but a variety of programmes sponsored by numerous information resources. The message should be twofold: the use of drugs never solves problems and it makes them worse by adding a drug problem. The second message is that the pleasure drugs give is contentless. Therefore, the high is not integrated into life experience as natural pleasures are. The enjoyment of rewarding encounters with other people, books, activities or successes can be recalled and re-enjoyed. Chemical pleasure might be remembered but without re-experiencing it. It has no staying power. In addition the nature of the nervous system ought to be explained: that repeated highs will lead to repeated lows. Meanwhile, training in enjoying life, coping, improving independent decision making and enhancing self-esteem will produce a person for whom drug taking is not an option.

It is the so-called "social" users of cocaine that are an important part of the problem. They seem, at the moment, apparently not in distress because of their usage of cocaine, although outside observers may have a different evaluation. The fact that the 70 percent of "social" users contribute to the 30 percent who are clearly in trouble from their drug practices has not yet reached public awareness. Therefore, they will tell those in the vicinity that cocaine is harmless and that it's great. It is the "social" users who induce others to try the drug and perpetuate this outbreak of cocaine use. How can they be persuaded to desist?

It must be made obvious to everyone, including the occasional user, that cocaine is not a glamourous or a safe drug. The reality, as described above, is that it is a stupid drug in which to become involved. Many occasional users of cocaine know of the tragedies others have experienced with the drug. A peculiar sense of omnipotence permits them to assume that "it can't happen to me." No matter which approach is used - education about the course of the cocaine process or seeing friends in trouble with the drug - it is unlikely to alter the "social" user's drug taking. Although we have laws penalizing the possession or use of cocaine, they are rarely enforced. Would it be helpful to enforce these laws more vigorously? It may be worth trying on a local level to see whether any significant change in cocaine using behaviour occurs.

An educational appeal to the crack dependent person is worthless, as usually he or she would very much like to stop using, but is unable to do so.

SUMMARY

It is difficult not to be alarmed at the trends in cocaine use.

This article was originally printed in the Drug Abuse and Alcoholism Newsletter, July, 1986, and is reprinted with the permission of the Vista Hill Foundation.

THE CONSUMER SOCIETY: THE AMERICAN EXPERIENCE

COCAINE: THE AMERICAN EXPERIENCE

Elaine M. Johnson

National Institute on Drug Abuse
Rockville, MD

HISTORICAL PERSPECTIVE

Cocaine use is not a new phenomenon. Its early use by Andean Indians in South America and its later appearance in Europe during the nineteenth century where pharmacological researchers established its medicinal value have been exhaustively chronicled in the literature (Ellinwood and Kilbey, 1977; Byck, 1975; Mortimer, 1901; Musto, 1973). At least 50 years ago its anaesthetic properties, central nervous system actions, and its cardiovascular effects were known. It was used as an anaesthetic, as treatment for seasonal allergies, and later as a pharmacological adjunct in the treatment of opium, morphine, and alcohol addictions. It had widespread use in the United States and in the late nineteenth and early twentieth centuries as an ingredient in many patent medicines, tonics, and soft drinks.

Between the 1930's and the late 1960's, cocaine use all but disappeared from the American scene. The downturn in cocaine use was probably due to a number of factors, including the Depression and restrictions on the importation, manufacture, and distribution of cocaine (Musto, 1973). However, its value to the medical practitioner as an effective local anaesthetic and vasoconstrictor continues to be recognized.

As late as the 1970's, cocaine use was not recognized as a significant drug abuse problem. A case in point was the second report from the National Commission on Marijuana and Drug Abuse, published in 1973, which stated that on the basis of the available data, they could verify little social cost related to cocaine use in this country. Concurrently, the report of the Federal Strategy Council on Drug Abuse indicated that morbidity associated with current patterns of cocaine use did not appear to be significant. Indeed, this document stated that, at that time there were virtually no confirmed cocaine overdose deaths and that a negligible number of individuals were seeking medical help or entering drug treatment programmes for cocaine abuse problems.

Nevertheless, the adverse health consequences of cocaine abuse did not go unrecognized. For example, the 1975 White Paper on Drug Abuse issued by the White House offered this warning:

The effects of cocaine if used intensively - particularly if injected-are not well known...recent laboratory studies with primates, as well as reports of the effects of chronic cocaine injection during the early 1900's suggest that violent and erratic behaviour may result. For this reason, the apparently low current social cost must be viewed with caution; the social cost could be considerably higher if chronic use began to develop.

Scientists have known for a long time that cocaine had the potential for producing extremely strong psychological dependence, and that cocaine use can be fatal. In 1977, in the forward to a monograph on cocaine research produced by the National Institute on Drug Abuse (NIDA), "Cocaine: 1977" it was stated, "Despite obvious knowledge limitations, we do know a few things that are important: We know, for example, that cocaine can kill...Death sometimes occurs even when the drug is snorted rather than injected. We also know that cocaine is among the most powerfully reinforcing of all abused drugs." At that time, it was thought that limited availability of the drug and its extremely high cost produced a natural barrier to widespread use, especially among the young and vulnerable. With the increased availability of the drug and its lower cost, this barrier has unfortunately eroded and the drug is being used by all population groups throughout the United States as well as in other parts of the world.

There seems to be general agreement that a major upswing in cocaine use has occurred or is occurring in the United States. We still do not know the full scope of the problem, whether use has peaked, still on the upswing, or when it might level off. However, we do have evidence from national epidemiological surveys supported by NIDA which shed some light on the extent of the problem. Also, the adverse consequences recently reported in the popular media and the medical literature make the destructive properties of this drug abundantly clear.

INCIDENCE AND PREVALENCE

The National Survey on Drug Abuse, periodically conducted by NIDA, found that by 1985, approximately 22 million Americans had tried cocaine at least once in their lives. That is an increase of an estimated 16 million users from the number reported in 1974. Americans currently using cocaine (have used at least once in the past month) increased from 1.6 million in 1977 to 4.3 million in 1979. The number has remained stable at about 5.6 million through 1985.

In our society, more men than women use illicit drugs. However, the number of young (ages 12-17) people of both sexes who have ever used cocaine is almost equal. This may reflect a narrowing gap in illicit drug use (or at least cocaine use) for males and females.

Further examination of data from the National Survey on Drug Abuse demonstrates that the young adult group (18-25) is clearly the predominant cocaine using group. However, between 1979 and 1985, the trends for use in the past year (annual prevalence) among both youth (12-17) and young adults stabilized or decreased slightly while the trend among older adults (ages 26 and older) increased. This suggests a cohort effect, that is, those who began to use cocaine when they were younger than 25 have continued to use the drug. In contrast, any increases seen in marijuana use in this age group may be entirely the result of the continued use of the drug which started in adolescence. The risk for cocaine use continues much longer and thus far, starts later than marijuana use.

While young adults between the ages of 18 and 35 may constitute the largest age group of cocaine users in the United States, there is concern about new data from NIDA's national high school senior survey, which show increases in lifetime, annual and current use of cocaine by this very young population. Cocaine has been tried at least once by 17% of the class, the highest lifetime rate so far in this study, which has been conducted by NIDA since 1975. Thirteen percent have used the drug over the past year, up from 11.6% the previous year, and 6.7% of the class currently (at least once in the last 30 days) use the drug, up from 5.8% the year before.

NIDA-funded researchers of this survey, Johnston, et al., (1986), fear that many of this graduating class may use cocaine in the next few years, based on the records of past classes. For example, in the class of 1976, which the researchers have followed, only 10% had tried cocaine by their senior year, but 40% had tried it by age 27. College students showed an increase in use throughout their four years of enrollment.

According to this most recent survey, cocaine use is up in 1985 among males and females, college-bound and non-college-bound, rural and urban areas and all regions of the country except the south. Furthermore, nearly 80% of the seniors acknowledged the harmful effects of using cocaine regularly, but only about one-third saw much risk in experimentation. Moreover, more than half of the students stated that cocaine would be fairly or very easy to obtain and about half have some friends who use it.

Another major concern are the other findings of the survey which reveal that the five-year decline in drug use among America's high school seniors appears to have stalled in 1985. Marijuana use is no longer declining, as it had been since 1979, nor is the use of tranquilizers, barbiturates, alcohol and cigarettes. In addition to cocaine, phencyclidine and opiates other than heroin are also on the rise.

Dr. Denise Kandel (1985) has found that most people who use cocaine are heavy users of other drugs, particularly marijuana. The earlier teenagers start to use marijuana, the more likely it is that they will use cocaine.

Dr. Kandel first interviewed young people from New York state secondary schools at age 15-16 in 1971-1972, and interviewed them again in 1980 and in 1984. She found that cocaine is the most prevalently used illicit drug after marijuana. Cocaine was used by 37% of men and 23% of women by age 24-25, and by 43% of men and 28% of women by age 29. Kandel found that cocaine users are most likely to be unmarried; to have been involved in car accidents while drunk or "stoned;" to have been arrested; and to have suffered psychiatric problems. Kandel also has shown that many cocaine users began drug use with the "gateway drugs," marijuana or alcohol, and then continued to use them in conjunction with cocaine.

There has been considerable question concerning whether marijuana use contributes to the subsequent use of other drugs such as cocaine. NIDA-supported researchers (Clayton, 1982; Clayton and Voss, 1983; Johnson, 1973; Kandel, 1985), using secondary analyses of previous epidemiological surveys, have produced a central hypothesis which supports this theory. In the literature, this hypothesis is referred to on different occasions as the stepping-stone, drug progression, or developmental stages hypothesis. The nature of this research has focused on statistically assessing the proposition that the more extensive one's involvement with a drug at a lower stage of development the greater the likelihood that one will experiment with drugs at the next or subsequent stage of development.

In an examination of lifetime prevalence rates of data reported in the National Drug Abuse Survey in 1982, the following statistical relationships were found: Among those who reported using marijuana only once or two times only 2.1% have tried cocaine while 20.2% of those using marijuana 3-10 times have tried cocaine. For those who reported having used marijuana 100 or more times, 73.4% have tried cocaine. The direction of this progression, from marijuana to cocaine and not the reverse, is further supported by the findings that only three-tenths of one percent of the survey's respondents who had never used marijuana reported having tried cocaine.

Many of the people who become heavy users of cocaine do not fit the stereotype of a drug addict. They are often successful, well-educated, upwardly mobile professionals in their twenties and thirties. A majority of them are men, but a growing number are women. Until the etiology of cocaine dependence is completely understood, it will be impossible to determine in advance who is most vulnerable to cocaine addiction. Everyone who uses the drug must be considered at risk.

Crack

Due to the widespread reporting of a new form of cocaine now appearing in the United States, special mention should be given to this form of cocaine use. In November, 1985, an article appeared in the New York Times describing a new form of cocaine being sold on the streets of New York known as "crack." This new substance appears to be cocaine in the freebase form which is sold after it has been processed from the hydrochloride powder using bicarbonate of soda and water as opposed to the more volatile method using ether. In the spring of 1985, during a NIDA field investigation primarily focused on cocapaste smoking in South Florida and New York, the existence of crack was noted. However, at that time, the substance had limited recognition and availability and in fact was being confused with coca paste; since that time there has been major reporting in the popular media regarding this new form of cocaine. The principal trafficking areas thus far identified are New York City, Miami, and Los Angeles.

Crack is extremely addictive and becoming increasingly accessible to adolescents. If the practice of smoking this freebase form continues, a rise in the number of young people addicted to cocaine can be anticipated as well as an increase in the number of cocaine-related deaths among young otherwise healthy individuals. Because it is smoked rather than snorted, crack reaches the brain within seconds, resulting in a sudden, intense high. However, this reaction dissapates within minutes, leaving the user with an enormous craving to use more. This up-and-down cycle leads to addiction much more quickly than if the drug is snorted.

HEALTH CONSEQUENCES

There had been a long-held belief in the general population that cocaine did not create dependence because dependence was associated with physical addiction: physical addiction was taken to mean that a drug had to produce physical withdrawal symptoms and increased tolerance, for example, like heroin. Those ideas on dependence are obsolete. Science has recognized that the essence of dependence is the drive that leads to compulsive use: the degree to which need or desire for a drug controls behaviour. While dramatic withdrawal symptoms occur with some drugs that lead to compulsive use, others like the amphetamines, cocaine, and tobacco, lead to less obvious changes when the drug is stopped. But it is not the drama or the intensity or the ease of measurement of the physical signs of withdrawal that determine whether a drug creates dependence. Johanson (1984) provides clarity on this issue:

> Two components are essential in the experimental assessment of the
> dependence potential of any drug. The first is the demonstration
> that the drug will be voluntarily self-administered by the experi-
> mental subject or, in the terminology of behaviour analysis, that
> the drug has positive reinforcing properties...A second component
> ...is the demonstration that at doses which are voluntarily admin-

istered there are toxic consequences to the organism. (Dr. Johanson concluded)...there is strong evidence demonstrating that cocaine is a drug of extremely high dependence potential.

The health consequences of a particular drug can only be understood by studying its pharmacology. In 1975 the National Institute on Drug Abuse initiated a series of studies on the pharmacology of cocaine in humans. These studies have yielded important findings on the relationship between dose and route of administration, and between blood levels of the drug and physiologic or psychologic effects (Fischman et al., 1976; Resnick et al., 1977; Van Dyke et al., 1978; Fischman and Schuster 1980, 1981, 1982; Van Dyke et al., 1982; Fischman et al., 1983). This research is summarized in the Institute's first triennial report to Congress (NIDA, 1984).

A statement from this document provides a very clear message about the highly reinforcing nature and resulting health consequences of this drug:

It is clear from studies and observations over the past few years that cocaine is a drug that is addictive. Tolerance develops rapidly to cocaine, so that repeating the same dose causes a progressively diminished response (Fischman and Schuster, 1981). Although the phenomenon has not been documented under controlled conditions, clinical observations suggest that physical dependence develops to the point that repeated doses are required to prevent the onset of a withdrawal syndrome (Siegel, 1982). ...Although controlled studies in humans that would clearly document these attributes have not been carried out, the weight of the data from the short-term controlled laboratory studies with humans and animals, when considered together with clinical observation, indicates cocaine is an addictive drug.

Recently, with the highly publicized deaths of two professional athletes due to cocaine abuse, considerable question has been posed in the popular press concerning toxic levels of cocaine which produce death. There is considerable scientific evidence from both animal and clinical studies that repeated high doses of cocaine produce a predictable severe behavioural and physiologic toxic state (Post et al., 1976; Epstein and Altshuler, 1978; Foltin et al., 1981; Stripling and Hendricks, 1981; Branch and Dearing, 1982; Siegel, 1982). From these studies and others it is known that cocaine stimulates the heart and causes vasoconstriction which results in increased heart rate, force of contractions, and increased blood pressure and, in general, an increased workload on the heart. Because the heart is sensitized to circulating neurotransmitters, an abnormal rhythm may develop (arrythmia or fibrillation) which can compromise the oxygen supply not only to the body but to the heart itself. An additional problem can arise if the coronary vessels also become constricted to even further diminish the heart's supply of oxygen. In the case of abnormal rhythms, ventricular fibrillation may develop which is fatal because the heart ceases to pump blood. Some individuals appear to be very sensitive to the effects of cocaine on heart rhythms and exhibit changes of rhythms at relatively low doses of cocaine. However, toxic states for all persons who use the drug cannot be predicted. The variability is not clearly attributable to prior experience with cocaine or other drugs, genetic differences, health status, or other easily identified factors (NIDA, 1984). Such factors as the dose of the drug, the purity of the drug, impurities in the drug, and if the drug is used in combination with another drug or alcohol, also influence its toxicity.

Over the last decade, a number of cases have been reported in the scientific literature describing heart attacks associated with cocaine use in relatively young people (under the age of 40) and often involving intranasal use of the drug. In many instances, there was no previous history of heart disease.

For example, Wetli and Fishbain (1985) described clinical and/or autopsy findings in seven patients in whom documented non-intravenous cocaine abuse appeared to be related to pathologic cardiovascular events. The patients, six men and one woman ranged in age from 20 to 37 years old. The men admitted to chronic cocaine use; the woman claimed that her heart attack occurred after her second use of the drug. The researcher emphasized that his study supports the fact that cocaine used by the intranasal route can be fatal, even to someone who has presumably used the drug only once or twice.

A recent interview article reported on thirteen patients with a mean age of 34 who died from acute myocardial infarction related to cocaine use. Ten of the patients had pre-existing heart problems. Three were younger than 40 and had no prior history of heart disease. This review shows that cocaine can be fatal to anyone - those with pre-existing heart problems as well as those without.

While the cardiovascular effects of the drug have received more atten-tion in recent times, the potentially fatal central nervous system effects should not be de-emphasized. Cocaine overdose from effects on the central nervous system can typically involve a user with no apparent symptoms who lapses suddenly into grand mal convulsions, followed by respiratory collapse and death. In addition, death from cocaine use can also be caused by rapid elevation in blood pressure resulting in cerebral haemorrhage.

Last year, NIDA funded 23 research projects on cocaine and other stimu-lants. Many of the grants have been useful in developing our current know-ledge of the bio-behavioural consequences of cocaine abuse.

Dr. Roy Wise (1984) has shown that cocaine works directly on the reward centre of the brain and shares the reinforcing properties of the opiates and the amphetamines. He found that, like the rewarding effects of food and water which activate the brain's reward circuitry via physiological proces-ses, cocaine can activate, powerfully and directly, the circuits of goal-directed behaviours. This may explain why the pursuit of this drug may come to dominate the lives of some cocaine users. Dr. Wise and his co-investiga-tors were able to demonstrate that given free access to the drug, laboratory rats will select cocaine over food and water to the point of death.

In another study, Drs. Frank Gawin and Herbert Kleber (1985) evaluated patients entering outpatient treatment programmes regarding the amount and route of their cocaine use, and the psychiatric and other symptoms. This study raises questions regarding some commonly held beliefs about cocaine use. They found:

- Psychiatric diagnoses in over half their sample of thirty men.
- Cocaine freebase smokers used twice as much cocaine weekly as did
    snorters or intravenous users, and had "runs" or "binges" that
    lasted 2½ times longer than IV users.
- The average IV user injected large amounts of cocaine in short time
    periods, using up to 1 gramme per injection.
- Ingestion of cocaine in amounts of approximately 1 gramme, previously
    presumed to be lethal, suggests that tolerance develops, at least
    among some chronic users.
- Daily cocaine use in moderate amounts precedes development of severely
    dysfunctional use, which more likely occurs in extended binges.
- Dangers associated with intranasal cocaine administration may have
    been seriously understated.

During the past four years, there have been significant increases in the numbers of cocaine users seeking help from hospital emergency rooms for various health problems caused by cocaine use. In fact it was one of the major signals that indicated the extensive nature of the cocaine crisis. The number of people admitted to emergency rooms with a cocaine problem more than doubled between 1981 and 1984; a large portion of this increase occurred between 1983 and 1984 when cocaine mentions increased by almost half in the Drug Abuse Warning Report (DAWN) of 1985. According to this study, last year in 26 major metropolitan areas, more than 12,000 people visited emergency rooms for cocaine-related problems, almost a 100% increase since 1983. Of those who experienced adverse reactions to cocaine, 67% were male; 37% were white and 43% were Black; 52% were between the ages of 20 and 29 years old.

Likewise, the number of cocaine-related deaths reported by medical examiners increased almost threefold, from 195 deaths in 1981 to 580 deaths in 1984. Preliminary data for 1985 indicate that cocaine-related deaths were continuing at unprecedented levels. So far, medical examiners have reported 563 cocaine-related deaths for 1985. According to the DAWN survey, most of the cocaine-related deaths were male (80%); most were between 20 and 39 years of age (83%); slightly over half were white (54%); about one-third were Black (34%); and about one-tenth were Hispanic (9%). These are very conservative figures since they do not include data from New York City, the nation's largest city and one of the major areas of reported cocaine use. Also, medical examiners do not always look for cocaine as a cause for a fatal heart attack. The significance of this data is that the recent deaths of several sports figures are not isolated events. Deaths from cocaine have been increasing nationwide over the past several years.

Also, data from 19 states and the District of Columbia reveal that the percentage of clients in treatment who use cocaine several times per week or more, increased between 1977 and 1981 from 47.5% to 58.3%, while those who used cocaine once a week or less decreased.

The most common drugs used in combination with cocaine in DAWN emergency room reports in 1984 included heroin, alcohol, marijuana, and phencyclidine. Although the majority of people still snort or inhale cocaine, those entering emergency rooms are also injecting the drug. Injection of cocaine now reflects 44.2% of emergency room cocaine visits. In addition, smoking (free-basing refers to the cocaine base rather than the salt form) increased substantially from 1.3% of admissions in 1981 to 6.1% in 1984.

This latest practice (smoking freebase) increases the pharmacological effects of the drug, resulting in an enormous craving for the drug. According to Perez-Reyes et al. (1982), the very intense effects reported by cocaine freebasers probably reflect the fact that smoking is a very efficient way of delivering any drug in a very concentrated form to the brain.

Very recently, a NIDA grantee at the University of California reported that cocaine stimulates the body's natural killer (NK) cells and may therefore adversely affect the body's immune system (Jones, 1986). Dr. Reese Jones found that the NK cells function in host defense against certain malignancies and viral infections and may impair the differentiation, proliferation, and activity of B cells, especially their ability to produce antibodies.

TREATMENT

Treatment of cocaine abuse is a major area of research at the NIDA. As with other types of mental health and substance abuse disorders, research has emphasized the heterogeneous and the multi-drug-using nature of this client population. Also, the apparent heterogeneity of cocaine users, in terms of demographic, socioeconomic, cultural, and environmental characteristics, combined with differing drug use patterns and combinations, suggests that a number of different treatment settings and approaches are necessary.

Similarly with other chronic high-frequency abusers of psychoactive drugs, many cocaine abusers demonstrate marked psychopathologies. Whether these pathologies were the primary cause, or contributed to, or were exacerbated by cocaine use, they must be a focus of treatment. It is highly unusual for a chronic abuser not to show social or psychological areas of impaired functioning, even when the substance abuse has been discontinued. A diagnosis of the individual's specific pathology must be performed before a specific treatment approach is developed. It may be that with more moderate abusers, more directive systems and/or cognitively-oriented psychotherapeutic strategies may be sufficient. This may be accomplished through existing behavioural therapies.

The Addiction Research Centre, NIDA's intramural research programme, and a host of other institutions such as the Johns Hopkins School of Medicine, the University of California at San Francisco, the University of Pennsylvania, McLean Hospital in Boston, have been engaged in studying the specific behavioural and environmental factors contributing to the development, maintenance, and elimination of drug-taking behaviour. Several approaches have shown promising results, including psychotherapy, behaviour modification, self-help strategies, and various pharmacotherapies.

For example, a NIDA-supported researcher (Crowley, 1984) has developed a very promising treatment for medical professionals based upon the behavioural modification approach. His particular approach precludes hospitalization and takes advantage of his patients' need to continue their careers without abusing drugs despite their continued exposure to cocaine in the workplace. Specifically, by employing the technique of contingency contracting, Dr. Crowley has reported dramatic and abrupt reductions in his patients' drug use. They were able to continue their careers while unlearning previous compulsive drug-using behaviours.

In addition to behavioural modification techniques, other forms of therapy may be indicated for some patients. For example, Drs. Kleber and Gawin (1984), as previously reported, have made findings which suggest that chronic cocaine abuse may lead to neurophysiological adaptations which require more than psychological intervention. They are studying the use of other drugs with similarities in physiological effect to cocaine as possible replacements from which the cocaine abuser could then be weaned. The results of such studies suggest that the tricyclic antidepressants might be particularly effective in attenuating cocaine-induced psychological changes. In addition, these chemotherapies may help in the treatment of those individuals whose depressive disorders have predisposed them to chronic cocaine use. Other drugs, including methylphenidate and lithium, are also being investigated and may be of special value for treating specific subpopulations of cocaine abusers.

The application of self-help strategies is another promising area of research for the treatment of cocaine abusers. These strategies might be used with clients after they complete traditional inpatient or outpatient programmes. Cocaine Anonymous, modelled closely after Alcoholics Anonymous, is one such aftercare group. Some leading investigators are also actively researching the use of professionally guided self-help groups as a primary intervention for cocaine abusing individuals.

PREVENTION

The National Institute on Drug Abuse supports a broad prevention research programme covering the following areas: (1) research on causes of use and abuse; (2) research on trends in use; (3) research on the abuse potential of drugs; (4) research on the risks and consequences of using various drugs; and (5) research on methods of prevention and early intervention. The objective of the first area is to expand understanding of the factors that contribute or inhibit the risk of drug abuse in later life. Knowledge of these factors, or the etiology of drug abuse, contribute significantly in the formulation of preventive interventions. Research on trends of use helps to identify prevention priorities, and enables prevention programmes to target high risk groups. The third area, research on abuse potential of drugs, provides for the development of appropriate guidelines for the use of drugs and helps to determine the scheduling of certain drugs in order to control their availability. The fourth area, research on the risks and consequences associated with particular drugs, creates the knowledge necessary for the development of accurate information on drug abuse central to so many prevention efforts. The final area, preventive intervention research, examines the effectiveness of different prevention approaches in deterring or delaying the onset of drug abuse behaviours.

Cocaine abuse prevention activities, therefore, should be based upon knowledge provided by the research programme. Emphasis must be placed on the seductive, highly reinforcing, and unpredictable nature of cocaine. Severely dependent cocaine users will sacrifice their health, their families, and eventually their jobs for cocaine. For many people, this message is brand new, and it often contradicts what they observe in their immediate environments. Some individuals appear to use the drug and to suffer no apparent adverse consequences. Indeed, research indicates that many people do not seek treatment until after several years of use. When treatment is finally sought, it is often because the individual feels he or she is out of control, or a profound crisis has occurred. The challenge is to reach people if possible before they begin to use cocaine, but certainly before their addiction has reached a dependent stage.

The workplace has evolved as a key setting for drug abuse preventive/ interventive approaches. We have learned that persons involved with cocaine perceive their job as a major link - and sometimes the last link - to normative behaviour. Therefore, the supervisor and co-worker may be even more influential than the family in moving an individual to treatment.

Through various research efforts there exists knowledge about the prevalence of drug use in the general public, and about the effects of drugs on performance and productivity. However, it has proven to be quite difficult to characterize the general nature and extent of drug use and abuse in the workplace. Accidents, loss of productivity, loss of trained personnel, theft, treatment and security costs contribute to corporate costs overall, but actual in-house statistics of incidents, accidents, and the real dollar cost due to drug abuse have been difficult to obtain.

As both cocaine and marijuana rates have increased in the general population, concerns about drug use in the work environment have become of primary interest to business and industry. As recent as two years ago, business and industry were reluctant to admit that drug abuse had reached the worksite. However, as the impact of drug use on production levels was uncovered, management officials began to address the problem of employee drug abuse.

Recent technological advancements in clinical diagnostic techniques, which have grown out of NIDA's research technology programme, have made widely available assays suitable for the detection of drugs in the body fluids. These tests are being used by the Department of Defense (DOD) in an effort to detect and reduce the incidence of drug use in the Armed Forces. Clearly, the availability of this new technology has contributed to more effective means of addressing the concern about drug use in industry. Drug

screening is increasingly utilized by employers for both deterrent and intervention purposes.

How best to use such assays to screen employees for drug use is a complex legal and ethical question. Principles of public safety, efficient performance, and optimal productivity can be seen as conflicting with individual
rights and civil liberties. In some instances, management and unions may
have considerable difficulty in achieving consensus and agreement on how or
when to utilize this new technology. Such disagreement has recently received
wide public attention in the area of professional sports.

In response to this developing phenomenon, in the spring of 1986, the
National Institute on Drug Abuse sponsored a national forum on "Interdisciplinary Approaches to the Problem of Drug Abuse in the Workplace" (NIDA,
1986). This conference brought together several hundred representatives of
the major businesses and industries in the country. The topic of urine
screening was discussed in detail and a consensus of thought emerged. Among
the several significant consensus statements adopted on this topic were the
following:

> There is ample evidence available that drug abuse is a significant
> problem for the work setting and that any solution to the problem
> demands a multidimensional approach which includes identification,
> education, prevention, and treatment.
> ...The decision to establish a screening programme for drug
> abuse should be based upon conclusions after consideration: (1)
> the awareness of or concern about impaired performance at the
> worksite; (2) the impact of drug abuse upon the health, safety,
> security, and productivity of employees; and (3) supportive or
> alternative means to detect drug use in the workplace...If the
> decision is to introduce drug screening, consideration of the
> usefulness of drug screening in assessing the employee's health
> and/or fitness for duty is important.

As has been described earlier, cocaine users are primarily young adults,
who for the most part are familiar with the drugs of abuse. Most grew up in
the 1960's and early 1970's when a more permissive attitude about drugs and
drug taking existed. In addition, their occasional drug use has not significantly interfered with their life goals in education and their profession,
and they feel invulnerable to the negative consequences of drugs. They perceive cocaine as the new and ultimate high and believe that only losers fall
victim to its effects. They are resistant to traditional prevention messages
and they tend to ignore warnings and reports from the federal government
about the negative consequences of drug use.

How then can these young adults be reached with a credible message?
Dramatic stories of "throwing it all away" for cocaine from former users who
are colleagues or peers of the user is one promising approach to the problem.
It must also be made clear in any prevention effort that this is a drug with
a high potential for abuse, and that no matter how "well-adjusted", or how
competent, physically fit, or successful in other areas, no one can be
immune from cocaine's inherent dangers.

Of concern, also, are the significant others who surround the user.
These people need to know the research-based information about cocaine and
the signs and symptoms of compulsive use to successfully move the user into
treatment or other appropriate intervention.

REFERENCES

Branch, M. N. and Dearing, M. E., 1982, Effects of acute and daily cocaine administration on performance under a delayed-matching-to-sample procedure, Pharmacol Biochem Behav, 16(5):713-718.

Byck, R., 1975,"Cocaine Papers: Sigmund Freud", Stonehill Publishing Co., New York.

Clayton, R. R., and Voss, H. L., 1982, Marijuana and cocaine: the causal nexus. Paper read at meeting, National Association of Drug Abuse Problems, New York.

Clayton, R. R., and Ritter, C. J., 1983, Cigarette, alcohol, and drug use among youth: selected consequences. Paper read at meeting, National Council on Alcoholism and Research Society on Alcoholism, Houston, TX.

Crowley, T. J., 1984, Contingency contracting treatment of drug abusing physicians, nurses, and dentists, in: "Behavioural Intervention Techniques in Drug Abuse Treatment, NIDA Research Monograph No. 46," J. Grabowski, M. L. Stitzer, and J. E. Henningfield, eds., National Institute on Drug Abuse, U.S. Printing Office, Washington, D.C.

Dupont, R. L., 1977, Foreword in "Cocaine, 1977, NIDA Monograph No. 13" R. C. Petersen, and R. C. Stillman, eds., U.S. Government Printing Office, Washington, D.C.

Epstein, P. N., and Altshuler, H. L., 1978, Changes in the effects of cocaine during chronic treatment, Res. Commun. Chem. Pathol. Pharmacol., 22(1):93-105.

Fischman, M. W., Schuster, C. R., Resnekov, L., Shick, J. F. E., Krasnegor, N. A., Fennel, W., and Freedman, D. X., 1976, Cardiovascular and subjective effects of intravenous cocaine administration in humans, Arch. Gen. Psychiatry, 33:983-989.

Fischman, M. W., and Schuster, C. R., 1980, Cocaine effects in sleep-deprived humans, Psychopharmacology, 72(1):1-8.

Fischman, M. W., and Schuster, C. R., 1980, Experimental investigations of the actions of cocaine in humans, in "Cocaine, 1980: Proceedings, InterAmerican Seminar on Medical and Sociological Aspects of Coca and Cocaine," F. R. Jeri, ed., Pacific Press, Lima, Peru.

Fischman, M. W., and Schuster, C. R., 1981, Acute tolerance to cocaine in humans, in "Problems of Drug Dependence, 1980, Vol. 34," L. S. Harris, ed., National Institute on Drug Abuse, Rockville, Md.

Fischman, M. W. and Schuster, C. R., 1982, Cocaine self-administration in humans, Fed. Proc., 41:241-246.

Fischman, M. W., Schuster, C. R., Hatana, Y., 1983, A comparison of the subjective and cardiovascular effects of cocaine and lidocaine in humans, Pharmacol Biochem Behav, 18(1):123-127.

Foltin, R. W., Preston, K. L., Wagner, G. C., Schuster, C. R., 1981, The aversive stimulus properties of repeated infusions of cocaine, Pharmacol Biochem Behav, 15(1):71-74.

Johanson, C. E., 1984, Assessment of the dependence potential of cocaine in animals, in "Cocaine: Pharmacology, Effects, and Treatment of Abuse, NIDA Research Monograph Series 50," U. S. Government Printing Office, Washington, D. C.

Johnson, D., 1973, "Marijuana Users and Drug Subcultures," Wiley Press, New York.

Johnston, L. D., O'Malley, P. T., Bachman, J. G., 1986, "Drug Use Among American High School Students, College Students, and Other Young Adults, National Trends Through 1985," U. S. Printing Office, Washington, D. C.

Jones, R., 1986, Cocaine increases natural killer cell activity, Journal of Clinical Investigation, Inc., 77:1387-1390.

Kandel, D. B., 1975, Stages in adolescent involvement in drug use, Science, 190:912-914.

Kandel, D. B., Murphy, D., Karun, D., 1985, Cocaine use in young adulthood: patterns of use and psychosocial correlates, in "Cocaine use in America Epidemiologic and Clinical Perspectives, NIDA Research Monograph Series

No. 61," N. Kozel and E. Adams, eds., U. S. Government Printing Office, Washington, D. C.

Kleber, H. D., and Gawin, F. H., 1984, Cocaine abuse: a review of current and experimental treatments, in "Cocaine: Pharmacology, Effects, and Treatment of Abuse, NIDA Research Monograph Series 50," J. Grabowski, ed., U.S. Government Printing Office, Washington, D. C.

Kleber, H. D. and Gawin, F. H., 1985, Cocaine use in a treatment population: patterns and diagnostic distinctions, in "Cocaine Use in America: Epidemiologic and Clinical Perspective, NIDA Research Monograph Series 61," N. Koxel and E. Adams, eds., U.S. Government Printing, Office, Washinton, D. C.

Kozel, N. J., 1985, Reports of coca paste smoking field investigations in South Florida and New York, internal report, National Institute on Drug Abuse, Rockville, MD.

Miller, J. D., Cisin, I. H., Gardner-Keaton, H., Harrell, A. V., Wirtz, P. W., Abelson, H. I., and Fishburne, P. M., 1983, "National Survey on Drug Abuse: Main Findings, 1982," U. S. Government Printing Office, Washington, D. C.

Mortimer, W. G., 1901, "Peru History of Coca, The Divine Plant of the Incas, with an Introductory Account of the Incas and of the Andean Indians of Today," J. H. Vail and Co., New York.

Musto, D. F., 1973, "The American Disease: Origins of Narcotic Control," Yale University Press, New Haven.

National Commission on Marijuana and Drug Abuse, 1973, "Drug Use in America: Problem in Perspective, Second Report of the National Commission on Marijuana and Drug Abuse," U. S. Government Printing Offics, Washington, D. C.

National Institute of Drug Abuse, 1984, "Drug Abuse and Drug Abuse Research, The First Triennial Report to Congress," U. S. Government Printing Office, Washington, D. C.

National Institute on Drug Abuse, 1986, "National Drug Abuse Survey, 1985 Preliminary Report," internal report,

Post, R. M., Kopanda, R. T., Black, K. E., 1976, Progressive effects of cocaine on behaviour and central amine metabolism in rhesus monkeys: relationship to kindling and psychoses, Biol Psychiatry, 11:403-419.

Resnick, R. B., Kestenbaum, R. S., Schwartz, L. K., 1977, Acute systemic effects of cocaine in man: A controlled study by intranasal and intravenous routes, Science, 195:696-698.

Siegel, R. K., 1982, Cocaine smoking, J. Psychoactive Drugs, 14(4):271-359.

Strategy Council on Drug Abuse, 1973 "Federal Strategy for Drug Abuse and Drug Traffic Prevention, 1973," U. S. Government Printing Office, Washington, D. C.

Stripling, J. S., and Hendricks, C., 1981, Effect of cocaine and lidocaine on the expression of kindled seizures in the rat, Pharmacol Biochem Behav, 14(3):397-403.

Van Dyke, C., Ungerer, J., Jatlow, P., Barash, P., Byck, R., 1982, Intranasal cocaine: dose relationships of psychological effects and plasma levels, Int J Psychiatry Med, 12(1):1-13.

Van Dyke, C., and Byck, R., 1977, Cocaine: 1884-1974, in "Cocaine and Other Stimulants," E. Ellinwood and M. Kilbey, eds., Plenum Press, New York.

Walsh, M. ed., "Interdisciplinary approaches to problems of drug abuse in the workplace," National Institute on Drug Abuse, U. S. Government Printing Office, Washington, D. C., in press.

Wetli, C. V., and Fishbain, D. A., Cocaine-induced psychosis and sudden death in recreational cocaine users, J. Forensic Sciences, JFSCA, Vol. 30, No. 3, July, 1985.

Wise, R., 1984, Neural mechanisms of the reinforcing action of cocaine, in "Cocaine: Pharmacology, Effects, and Treatment of Abuse, NIDA Research Monograph Series 50," J. Grabowski, ed., U. S. Government Printing Office, Washington, D. C.

COCAINE: DRUG EPIDEMIC OF THE '80's

Arnold M. Washton

800-COCAINE National Hotline
Fair Oaks Hospital
Summit, New Jersey

THE COCAINE EPIDEMIC

Cocaine use in the United States has reached epidemic levels in recent years. Nationwide surveys estimate that over 22 million Americans have already tried cocaine at least once in their lifetime and that as many as two million or more may be severely dependent on the drug (1). The "800-COCAINE" National Hotline, established in 1983, has received over 1.5 million calls in its first three years of existence, with a continuing influx of more than 1,400 calls per day (2,3,4).

At the outset of this epidemic it appeared that cocaine use was especially pervasive among America's middle class (5), but now with lower prices, increased supplies, and more widespread acceptance, cocaine has spread to virtually all socioeconomic groups and to all geographic areas of the U.S. (2,4).

The willingness of many people to at least try cocaine probably stems in part from the drug's long-standing popular image as a chic, non-addictive, and harmless intoxicant. Most who try it believe that they will have no difficulty in controlling their use and will suffer no adverse effects. Contrary to these popular beliefs, recent observations suggest that the potent reinforcing properties of cocaine can rapidly take control of the user's behaviour such that even "occasional" use may pose significant risks. Observations from at least several different sources, such as Hotline callers (2,3,4), treatment applicants (6), and animal experiments (7), have revealed the powerful addictive properties of cocaine. Moreover, apart from initial exposure to the drug and continued access to supplies, it seems impossible to predict exactly who among the total pool of "occasional" users will become addicted to it. A striking feature of the current cocaine epidemic is that so many reasonably mature, well-integrated people with good jobs, a history of good functioning, no previous drug addiction, and no serious psychiatric illness seem to have become full-blown cocaine addicts (4,5).

As cocaine use in the U.S. has become more prevalent, the full range of medical, psychological, and social consequences associated with its use has been revealed. Some of the statistical indicators of this trend include a more than threefold increase in cocaine-related emergency room visits over the first few years of this decade and a more than sixfold increase in requests for treatment of cocaine dependence in government-sponsored programmes (1). Many cocaine abusers who seek help prefer private rather than public treatment settings, and so it is not surprising that private clinicians (psychologists, psychiatrists, and family physicians) and private treatment facilities in many different areas of the U.S. have seen an increasing number of cocaine abusers requesting assistance. The rapid emergence of a nationwide cocaine addiction problem has forced treatment personnel, both public and private, to acquire skills in the proper assessment and treatment of the problem.

EFFECTS OF COCAINE

Cocaine is a potent central nervous system stimulant (8) derived from the erythroxylon coca plant grown primarily in the mountainous regions of Central and South America. Cocaine is extracted from the leaves of the coca plant and processed into a white powder (cocaine hydrochloride), which sells on average for $75 to $150 per gramme on the street.

Cocaine is most commonly self-administered by one of three different methods: (1) Intranasal Use: The white powder is inhaled ("snorted" or "sniffed") into the nostrils, usually through a straw, from a small "coke spoon," or through a rolled-up dollar bill. (2) Freebase Smoking: The white powder can be transformed into cocaine freebase, using either baking soda, ammonia, or ether as the reagent, and then smoked in a pipe or cigarette. The purpose of transforming the powder into freebase is to yield a form of the drug with a vaporization point low enough so that it can be smoked. Smoking cocaine, as compared to snorting it, produces a much more rapid and intense euphoria, similar to intravenous injection. A recent development in the illegal marketing of cocaine has been the direct sale of cocaine freebase in the form of ready-to-smoke tiny pellets or "rocks," known on the street as "crack" (9). The widespread availability of crack has led to a dramatic rise in the incidence of cocaine freebasing problems in many areas of the U.S.. (3) Intravenous Injection: Cocaine hydrochloride powder is readily soluble in water, and thus a solution of the drug can be drawn up into a syringe and injected directly into a vein. Most IV users have previous or current involvement with heroin and sometimes mix the drugs simultaneously in the same injection, known as a "speedball."

The pleasurable mood-altering effects of cocaine include feelings of euphoria, exhilaration, energy, social confidence, and mental alertness. Many users report instantaneous mood elevation, sexual arousal, and decreased inhibitions. The "high" is extremely short-lived, lasting no more than ten to twenty minutes after each dose, and is often followed by an equally unpleasant rebound reaction, the "crash," characterized by feelings of depression, irritability, restlessness, and craving for more cocaine. The intensity of the crash tends to increase with the dosage and chronicity of cocaine use and also tends to be more severe with freebase smoking and IV use. The "high-crash" process is one of the pharmacologic characteristics of cocaine that actively promotes a pattern of escalating and compulsive use. The user is driven to recreate the short-lived "high" and to escape the unpleasant "crash." This phenomenon is exaggerated with smoking or IV administration where the intense euphoria may last only 5 to 10 minutes and the rebound dysphoria is more exaggerated.

Many of the physiological effects of cocaine include those typical of other stimulant drugs, such as amphetamines (8). Cocaine increases the user's heart rate, blood pressure, and dilates the pupils. Unlike other stimulants, cocaine is also a local anaesthetic and a vasoconstrictor. It blocks afferent nerve conduction in sensory pathways causing a numbing or "freezing" of mucous membranes to which it is applied. It also constricts surrounding blood vessels in the applied area thereby prolonging the drug's anaesthetic action. When cocaine is "snorted" it temporarily numbs the user's nasal and throat passages, a side-effect anticipated by all experienced users. Dealers often adulterate cocaine with cheaper local anaesthetics such as procaine, lidocaine, or tetracaine as a way of reducing the purity of street cocaine while retaining the drug's anticipated effects. Other adulterants or "cuts" include various types of white powders (e.g., lactose, mannitol) and stimulants such as methamphetamine.

CONSEQUENCES OF USE

Perhaps the most obvious evidence indicating that a tolerance to cocaine develops with chronic usage is the fact that many users escalate to dose levels that would have been lethal at the beginning stages of use. Moreover, chronic users find that the euphoria and other pleasureable effects diminish rapidly with increasing doses and frequency of use. Chronic users often find themselves caught in a futile, obsessive chase to recapture the original cocaine high, but as dosages and frequency increase, so does the user's tolerance to the euphoric effects.

When cocaine use continues well beyond the point of tolerance to the euphoria, there will usually be a complete reversal of the drug's effects. For example, mood elevation and euphoria are replaced by depression, anxiety, and irritability. Increased alertness and spontaneity are replaced by distractibility, poor concentration, and mental confusion. Increased drive and energy are replaced by apathy and fatigue. Increased sociability, talkativeness, and sexuality are replaced by social withdrawal, non-responsiveness, and a total loss of sexual desire. Continued use can lead to extreme agitation, explosiveness, personal neglect, and in some cases to a severe paranoid psychosis with elaborate delusions and hallucinations (10).

The exact consequences of cocaine use cannot be predicted with certainty for any single user (6), but will vary according to many different factors, including: the amount used, the chronicity and frequency of use, the route of administration, the expectations, mood, and personality of the user, the setting and circumstances under which the drug is taken, and the concomitant use of other drugs. The mood-altering effects of cocaine can range from mildly pleasant to profoundly euphoric, from slightly unpleasant to terrifying. Medical evidence shows quite clearly that cocaine can indeed be fatal (11) even in an "occasional" user and that intranasal administration or snorting, the most popular method of using cocaine, offers no inherent protection against developing an addiction to it or against suffering serious medical and psychological consequences (2,4,12).

COCAINE HOTLINE

In January, 1983, the author established the Cocaine Hotline of Greater New York, the first such hotline for cocaine abusers in the U.S. (13). This hotline revealed, in perhaps the most dramatic way up to that time, that there were large numbers of employed middle-class cocaine abusers who had become severely dependent on the drug and were having great difficulty in finding professional help. They were being told by private clinicians and by drug abuse experts that cocaine was not addictive, that no detoxification was required, and that no specific treatment was either necessary or available. It was clear from the reports of these early hotline callers that neither the public drug abuse treatment programmes which tended to be concerned mainly with hard-core heroin addiction, nor private clinicians who were typically unfamiliar with the treatment of drug addiction, were equipped to deal with what appeared to be a large but previously unrecognized population of cocaine abusers desperately seeking help. The callers sensed that they had become addicted to cocaine, in direct contradiction to what they had been previously told about the drug's apparent harmlessness, but their observations were being refuted by professionals. Many of these callers asked: "How can I be addicted to a drug that is supposedly non-addictive?"

"Is there something especially wrong with me - are others having this problem?"

In addition to uncovering these phenomena, the New York Hotline served as a vehicle for collecting research data. From among the 200-300 calls received per day, a random sample of 55 callers was selected to participate in an anonymous telephone research interview lasting 30-40 minutes. The interview was based on an extensive research questionnaire that was designed to collect information on the demographics of the callers, their history, and current patterns of cocaine use, their use of other drugs, the effects and side effects of cocaine, and any negative consequences of cocaine use on their health and functioning. The results of this survey (13) provided clear-cut and dramatic evidence of compulsive patterns of cocaine use associated with a wide range of medical, psychological, and social problems. The typical respondent in the random sample was a white, middleclass male, 25-35 years old, with no prior history of drug addiction or psychiatric illness. Nearly all said they felt addicted to cocaine and were unable to stop using the drug despite numerous adverse effects.

When 800-COCAINE was subsequently established in May, 1983, it became clear that the cocaine epidemic was not restricted to New York: it was nationwide. The phones began to ring incessantly almost from the very first instant that the Hotline went into operation. At first the calls came mainly from the New York City area and other locations in the northeast since these were the places where the Hotline initially received the most media attention and publicity. As news about the Hotline spread to other parts of the country the volume of calls from those areas started to increase accordingly. Within the first three months the Hotline had received calls from more than 37 different states in the U.S. Forced by a rapidly soaring volume of calls to expand its operating hours to 24 hours per day and to increase its number of incoming phone lines, the Hotline became a dramatic example of how serious and widespread the cocaine problem had become without either the public or the professional community being sufficiently aware of it.

The 800-COCAINE Hotline afforded unprecedented access to large numbers of cocaine abusers who would otherwise not be available for study by traditional methods. Our Hotline studies provide one of the few available sources of information about changes in the cocaine epidemic, i.e., trends or shifts in patterns of cocaine use in the U.S. By comparing surveys taken from nationwide samples at different points in time, the Hotline has helped to detect some of the major trends in cocaine abuse over the past two and one-half years.

HOTLINE SURVEYS

First National Survey, 1983

The first national survey on the Hotline (14) was based on a randon sample of 500 callers to the Hotline during its initial three months of operation, May through July, 1983. The sample included callers from 37 different states across the U.S. with the majority being from New York, New Jersey, California, and Florida, which collectively represented 63% of the entire sample. Each caller voluntarily consented to an anonymous 30-40 minute telephone interview during which the research questionnaire was administered.

Demographic Profile. The average age of the 500 respondents was 30 with most being between 25 and 40 years old at the time of their call to the Hotline. Some were as young as 16 and others as old as 78. Sixty-seven percent were male, 33% were female. The overwhelming majority (85%) were white; 15% were Black or Hispanic. Many were well-educated, on average having completed just over 14 years of schooling. The sample included many college graduates, degreed professionals, and highly skilled business executives and technicians. Forty percent had incomes of over $25,000 per year.

Cocaine and Other Drug Use. The respondents had begun their use of cocaine on average 4.9 years before calling the Hotline, and over 90% had started with intranasal use (snorting). At the time of their call, 61% were taking the drug intranasally, 21% were freebase smoking, and 18% were intravenous users. About half the sample were using cocaine daily, at a street cost of $75 to $150 per gramme. Many reported "binge" patterns of use in which they used the drug continuously for two or three days at a time until their supply of the drug, their money, or their physical energy was totally exhausted. Some said they used the drug only on weekends. Although the range was from 1 to 32 grammes per week, on average they were using about 6 grammes per week. They said they had been spending on average $637 per week for cocaine during the week before they called the Hotline, with a range from about $100 to $3,200. The vast majority of callers (80%) said that when the cocaine high wore off, they felt depressed, irritable, restless, and drained of energy - the rebound dysphoric reaction commonly referred to as the cocaine "crash." The importance of the crash lies in the fact that 68% of the respondents said that in order to alleviate the unpleasant aftereffects of cocaine they were led to abuse other drugs such as alcohol, sleeping pills, tranquilizers, or opiates. This finding revealed that cocaine abuse is likely to promote polydrug abuse and dependence. Many of the callers had become multiply-dependent on a combination of cocaine, alcohol, and tranquilizers.

Addiction and Dependency. The questionnaire included a series of items to probe the issue of drug dependency and addiction. Callers were asked to indicate which, if any, of the items characterized their involvement with cocaine. Their replies provided clear evidence of the dependence-producing ability of the drug and its ability to dominate the user even in the face of extreme negative consequences. Overall, 61% said they felt addicted to cocaine, 83% said they could not turn down the drug when it was available, 73% said they had lost control and could not limit their cocaine use; 67% said they had been unable to stop using cocaine except for brief periods lasting no longer than one month. Over half the respondents said that cocaine had become more important to them than food, sex, recreational activities, social relationships, and job or career. Although it had become clear that cocaine use was impairing their functioning, most said that they feared feeling distressed and were unable to function properly without the drug.

Drug-Related Consequences. Their compulsion to continue using cocaine despite serious adverse effects became even more clearly evident from the questionnaire items of drug-related consequences to health and functioning. This section of the questionnaire included a total of 59 items on physical, psychological, and social problems associated with cocaine use.

Over 90% of the respondents reported five or more adverse effects that they attributed to their cocaine use. The five leading physical problems were chronic insomnia (82%), chronic fatigue (76%), severe headaches (60%), nasal and sinus infections (58%), and disrupted sexual functioning (55%). Other serious physical problems included cocaine-induced brain seizures with loss of consciousness reported by 14% of the sample, and nausea and vomiting reported by 39%. The leading psychological problems were depression, anxiety and irritability, each reported by more than 80% of the sample. Paranoia, loss of interest in non-drug-related activities, loss of non-drug using friends, and difficulty in concentrating were each reported by more than 60%. Nine percent reported a cocaine-induced suicide attempt. Numerous personal and social problems were also reported. For example, 45% said they had stolen money from their employers, family, or friends to support their cocaine habit. Most were in debt having spent all of their monetary assets on cocaine. Some had mortgaged their home, lost their business or profession, squandered their inheritance or trust fund, or sold their valuables for cocaine. Thirty-six percent said they had dealt drugs to support their cocaine habit; 26% reported marital/relationship problems which ended in separation or divorce; 17% had lost a job due to cocaine; 12% had been arrested for a drug-related crime of dealing or possession of cocaine; 11% reported a cocaine-related automobile accident.

Route of Administration. It is commonly believed that people who use cocaine by snorting it are immune to becoming dependent on the drug or from suffering serious adverse consequences. Our clinical experience did not support this view nor did the large numbers of intranasal users who were calling the Hotline reporting serious problems (12). We therefore made a special effort to examine our survey data for a comparison between the different methods of use (14). We found that intranasal users reported patterns and consequences of cocaine use similar to those of freebase and IV users. The incidence and types of consequences reported by each group of users were comparable to one another, with one major difference: the freebase and IV users tended to show greater disruption of psychosocial functioning and were more likely to report nearly all of the most serious ill effects. For example, freebase and IV users reported higher rates of cocaine-related brain seizures, automobile accidents, job loss, and extreme paranoia. This finding may have been at least partly due to the higher dosages of cocaine used by the freebase and IV groups. Intranasal users who reported comparably high levels of use also reported comparably serious adverse effects. The data suggested that possibly another important difference between the different methods of cocaine use is the speed at which the user becomes addicted. Intranasal users typically reported a rather long period of occasional non-problematical use, sometimes 2 to 4 years, before becoming dependent on the drug. Those who started out with freebase or IV use or those who later switched to these methods after an initial period of snorting cocaine, said that their drug use had reached problematic levels almost immediately; relatively few described their freebase or IV use, even in its beginning stages, as "recreational" or non-problematic. Most felt that they were overcome almost immediately by irresistible drug cravings.

National Surveys: 1983 vs 1985

Table 1 compares selected results of the 1983 survey with a similar survey conducted almost two years later in 1985. These data reveal a number of significant shifts in patterns of cocaine use, as outlined below.

Geographic Shifts. The proportion of Hotline calls from southern and midwestern regions of the U.S. has increased significantly, although the absolute number of calls remains highest from the northeast and western regions of the country. This finding indicates that cocaine use has spread to virtually all areas of the U.S., including many small towns and rural areas which in 1983 were thought to be largely exempt from the cocaine epidemic. We continue to receive calls on the Hotline from many sparsely populated areas in the U.S., including small towns in Wyoming, Montana, Mississippi, New Mexico, Alabama, and others. Large cities and adjoining suburbs in the northeast, California, and Florida continue to show the highest absolute volume of calls.

Demographic Shifts. The profile of Hotline callers has changed substantially since 1983 when most callers were white middle/upper class employed males between the ages of 25 and 35. With the continuing spread of cocaine use over the past two years, it appears that a broader cross-section of the American population has become involved in this phenomenon. As a result, no single demographic profile accurately describes the majority of cocaine users. There is no longer a "typical" user. The data in Table 1 indicate increasing

Table 1: HOTLINE SURVEYS: 1983 vs 1985. Each survey is based on a random
          sample of 500 callers during a three-month time period: May-July,
          1983, and January-March, 1985.

	1983	1985
Origin of calls:		
Northeast	47%	32%
Midwest	11%	23%
West	33%	22%
South	9%	23%
Demographics:		
Males	67%	58%
Females	33%	42%
Whites	85%	64%
Black/Hispanic	15%	36%
Average Age	30yr	27yr
Adolescents	1%	7%
Yearly Income:		
$0-25,000	60%	73%
over $25,000	40%	27%
Cocaine Use:		
Consumption	6.5gm/wk	7.2gm/wk
Expenditure	$637	$535
Intranasal	61%	52%
Freebase	21%	30%
Intravenous	18%	18%
Use of Other Drugs to Alleviate Cocaine Side-Effects	68%	87%
Auto Accident on Cocaine	11%	19%
Use of Cocaine at Work	42%	74%

cocaine use among women, minority groups, lower-income groups and adolescents. The major demographic shifts can be summarized as follows:
     WOMEN: In 1983, women cocaine abusers comprised about one-third of the
randomly-sampled callers to the Hotline. In 1985, they comprised nearly one-
half of the callers. Our more detailed surveys of women callers reveal that
the vast majority (87%) are introduced to cocaine by a male companion and
often receive "gifts" of cocaine from men, which indicates how the drug has
become incorporated into social relationships. On average, women report
using less cocaine than men and are less likely to resort to drug dealing as
a way to support their use. However, women are more likely than men to
report extreme depression due to chronic cocaine use and to exchange sexual
"favours" for the drug.

     MINORITY AND LOWER-INCOME GROUPS: The proportion of black and hispanic
callers has more than doubled since 1983. Similarly, more callers now report
earning less than $25,000 per year.

     ADOLESCENTS: The average age of callers has decreased, reflecting the
spreading use of cocaine among younger users including adolescents, college
students and other young adults. For adolescents, aged 13-19, there has
been a sevenfold increase in the percentage of Hotline calls since 1983. A
more detailed study of adolescent callers is described later in this article.

Levels of Cocaine Use. The surveys indicate that levels of cocaine use have increased, as shown by the callers' self-reported estimates of weekly consumption. There has also been an increasing tendency for users to shift from snorting cocaine to freebase smoking. The data further indicate that the price of cocaine on the illicit market has dropped considerably, from an average of approximately $98/gm in 1983 to about $75/gm in 1985.

Concomitant Use of Other Drugs. Our surveys show an increasing problem of polydrug abuse among current cocaine abusers. With increasing levels of cocaine consumption, users say they are more likely to resort to other drugs and alcohol in order to relieve the unpleasant side effects of cocaine. Abuse of alcohol and sedative-hypnotic drugs appears to be the rule rather than the exception among current cocaine abusers.

Automobile Accidents. Cocaine-related automobile accidents reported by Hotline callers have nearly doubled since 1983. In 1985, nearly one-fifth of all callers said they had had at least one automobile accident resulting in personal injury or property damage while under the influence of cocaine or a combination of cocaine and other drugs. The opposing effects of stimulants and depressants that differ with regard to the onset and duration of their actions appears to create an especially dangerous situation. Cocaine's short term stimulant effects temporarily mask the depressant effects of alcohol. The cocaine user therefore is able to consume a large quantity of alcohol and initially not feel the intoxicating effects of the alcohol that might otherwise lead them to refrain from driving. When the cocaine wears off, in only 20-30 minutes, the driver may suddenly become severely intoxicated or even stuperous from the alcohol, resulting in unexpected and gross impairment of driving ability.

Cocaine Use in the Workplace. The percentage of callers who say they use cocaine at work increased sharply from 42% in 1983 to 74% in 1985. A more detailed survey of cocaine use in the workplace is presented later in this article.

Adolescent Survey

Our surveys continue to show dramatic increases in cocaine use by adolescents. In a survey focusing on adolescent cocaine users (15), we interviewed 100 randomly-selected Hotline callers who were between the ages of 13 and 19. The interview included a structured research questionnaire requiring a 30-40 minute telephone interview.

Survey Results. A descriptive profile of the adolescent sample is shown in Table 2. Most were white male high school students in the 11th or 12th grade, with many from urban and suburban middle-class families. The time lag between snorting their first "line" of cocaine and evidence of cocaine-related disruption of functioning which led them to call the Hotline averaged 1.5 years as contrasted with over 4 years in adult survey samples. Most were snorting cocaine although 12% had switched to the more intensified methods of use. Nearly every subject reported multiple, combination drug use. Cocaine was often combined with or followed immediately by use of marijuana, alcohol, and sedative-hypnotic drugs, usually to counteract the unpleasant side effects of cocaine. Most said that they were purchasing drugs from school mates and older users or dealers, often in or around school premises.

A wide range of cocaine-related medical, social, psychiatric, and school problems were reported by the adolescent respondents. School performance was reported to have suffered considerably because of the continuing cocaine use and its resulting problems. Seventy-five percent had missed days of school, 69% said their grades had dropped significantly, 48% had ex-

Table 2: Adolescent Cocaine Abusers   (N=100)

Demographics:		Current Use:	
Males	65%	Consumption	1.4gm/wk
Whites	83%	Expenditure	$95/wk
Average age	16.2yr	Intranasal	88%
Avg. education	11.4yr	Freebase	10%
		Intravenous	2%

First Use:		Use of other drugs to relieve cocaine side effects:	
Time before call	1.5yr	Marijuana	92%
Intranasal	100%	Alcohol	85%
		Sedatives	64%
		Heroin	4%

perienced disciplinary problems due to drug-related disruption of mood and behaviour, and 31% had been expelled for cocaine-related difficulties. To support their escalating drug use, 44% had been selling drugs, 31% were stealing from family, friends, or employer, 62% were using lunch or travel money or income from a part-time job to buy drugs. Among the most serious drug-related consequences were: cocaine-induced brain seizures with loss of consciousness (19%), automobile accidents (13%), suicide attempts (14%), and violent behaviour (27%). Similar to adult users, common complaints included insomnia, fatigue, depression, irritability, short-temper, paranoia, head-aches, nasal and sinus problems, poor appetite, weight loss, memory and con-centration problems, and heart palpitations. In most cases, loss of interest in non-drug using friends, family activities, and sports or hobbies were also reported. Nearly every subject said that the only limit on their cocaine use was money: if they had more money they would use more cocaine. The results of this survey showed that adolescents do have sufficient access to cocaine to become serious abusers and that their vulnerability to the dependence-producing properties of the drug and subsequent disruption of functioning may be greater than that of adults.

Cocaine in the Workplace

The problem of drug use in the workplace is a matter of rapidly escala-ting concern not only to employers but to society at large. The broadening scope of this problem in terms of its negative impact on individual health and safety as well as on the nation's economy are just starting to become recognized. It seemed to us that if most cocaine users are employed, then there must be a great deal of cocaine and other drug use occurring in the workplace. To our knowledge, no previous attempts had been made to study this phenomenon by actually interviewing drug-using employees. We took a random sample of 227 employed cocaine abusers who called the Hotline and who consented to a research interview concerning drug use on the job.

Survey Results. The demographic profile of these drug-using employees was as follows: 70% were male, 61% were white, 53% were 20-29 years old, 40% were 30-39 years old, and 7% were 40 years old or over. Sixty-seven per-cent earned under $25,000 per year, 32% earned $26-$50,000 and 1% earned over $50,000. Their occupations included the following: automobile mechan-ic, attorney, stock broker, legal secretary, salesperson, real estate agent, airline flight attendant, dentist, nurse, optician, pharmacist, physician, laboratory technician, bank executive, prison guard, carpenter, electrician, office clerk, postal employee, public utility worker, security guard, compu-ter programmer, retail store owner, pipe fitter, bus driver, and railway switchman.

Seventy-four percent said they used drugs at work. This included respondents who said that they had self-administered drugs during working hours (or breaks) as well as those who had come to work while already under the influence of drugs. The types of drugs used at work were as follows: cocaine (83%), alcohol (39%), marijuana (33%), sedative-hypnotics (13%), and opiates (10%). (The total of these percentages exceeds 100% because most subjects reported multiple drug use.) Sixty-four percent said that drugs were readily obtainable at their place of work, and 44% said they had dealt drugs to fellow employees. Twenty-six percent reported being fired from at least one previous job because of drug-related problems, 39% feared that a raise in salary would lead to further escalation of their drug use. Eighteen percent said they had stolen money from co-workers in order to buy drugs and 20% reported having at least one drug-related accident on the job.

"CRACK"

Sudden Problem

Smoking cocaine freebase is not a new phenomenon in the U.S., but the selling of ready-to-smoke freebase rather than cocaine powder on the illicit drug market does represent a new problem. Tiny chunks or "rocks" of cocaine freebase weighing approximately 100 mg are now sold on the streets in small plastic vials for $5 to $15 each. The drug is called "crack" and the phenomenon has spread rapidly to many different areas of the U.S. within an extremely short period of time (9). This phenomenon is very similar to the recent experience of the Bahamas (16) where dealers have switched from selling powder to selling ready-to-smoke freebase.

No statistics are as yet available on the prevalence of crack use in the U.S., although marked increases in cocaine smoking problems have already been noted in over 15 major cities. Among callers to the 800-COCAINE National Hotline, nearly one-third of all cocaine users in a recent survey (9) reported addiction to crack. Most of these crack users were young adult males between the ages of 20-35 years, although crack use among teenagers also appears to be increasing.

The Hotline has received calls from crack users in at least 25 different states with most saying that the drug is "readily available" in their area. Only one year ago, not a single Hotline caller had mentioned crack. Similarly, over 75% of the recent cocaine admissions to our treatment programmes at Regent Hospital and Stony Lodge Hospital have been for crack-related problems. Only nine months previously, not a single admission was for crack.

Special Appeal

Drug dealers prefer to sell crack rather than cocaine powder because of its higher profits, ease of handling, and also because the higher addiction potential of crack ensures the dealer a more reliable clientele. Users prefer crack to powder because of its lower unit cost ($10-$20 for a vial of crack vs. $75-$100 for a gramme of powder) and its more potent, instantaneous euphoria or "rush."

Crack is extracted from its parent compound, cocaine hydrochloride powder, in a simple procedure using baking soda (sodium bicarbonate), heat, and water. As this procedure does not involve volatile chemicals such as ether, large quantities of powder can be more safely converted into freebase for sale as crack. Unlike cocaine powder, freebase can be smoked rather than snorted because of its lower vaporization point (17). Freebase is readily volatilized into smoke with only modest heating, using a regular match or lighter. It is usually smoked in a glass waterpipe, using a butane "torch"

54

or lighter as the heat source.   Sometimes crack is smoked in a tobacco or
marijuana cigarette.

Special Dangers

From a medical standpoint, the special dangers of crack stem from the
drug's extremely high addiction potential and its ability to cause serious
medical and psychiatric problems.   In general, the adverse consequences of
freebase smoking are dose-related exaggerations of those typically associated with intranasal use.

The pharmacologic properties of cocaine when smoked as freebase are especially conducive to the development of compulsive use patterns (9).   Upon
inhalation of the smoke, cocaine is absorbed via the large surface area of
the lungs directly into the pulmonary circulation and transmitted to the
brain in less than 10 seconds.   The instantaneous euphoria is intensified
and compressed into a 3-5 minute period of intoxication that is followed immediately by an equally unpleasant rebound dysphoria ("crash") coupled with
intense cravings to return to the euphoric state.

The speed of the addiction to crack can be extremely rapid.   Many users
say they literally "fall in love" with the drug the first few times they use
it.   Others may develop compulsive use patterns over the course of several
weeks or months.   A single deep inhalation of the freebase smoke produces a
sharp, rapid rise in cocaine plasma levels.   Inhalations tend to be repeated
as often as every three to five minutes during prolonged smoking episodes or
"binges."   These binges may extend over 1-3 days of continuous use until the
supplies of cocaine are depleted or the user collapses from exhaustion.   As
much as 10-30 grammes or even several ounces of cocaine (approximately 28
grammes to the ounce) may be consumed during a marathon smoking binge.   Abuse
of alcohol and sedative-hypnotics is often associated with chronic freebase
smoking because of the user's attempt to self-medicate against the unpleasant cocaine side effects.

In addition to rapid addiction, the extremely high plasma levels of cocaine achieved by freebase smoking can cause serious toxic or overdose reactions, including brain seizures, hypertension, and cardiac arrhythmia.   Respiratory problems from chronic cocaine smoking are common.   These include:
chest congestion, wheezing, black phlegm, and chronic cough.   Potentially
irreversible lung damage with impairment of diffusing capacity has also been
reported (18).   Other signs may include: extreme hoarseness, parched lips,
tongue, and throat, and singed eyebrows and eyelashes.

The psychiatric consequences of freebase smoking may include: extreme
irritability and short-temper, acute paranoid reactions sometimes leading to
an increased potential for violent and suicidal behaviour, a full-blown cocaine psychosis with delusions, hallucinations, and extreme agitation, profound dysphoria, depression, and anhedonia, and a complete loss of sexual
desire.   There is often a rapid, marked deterioration in psychosocial functioning as the drug obsession escalates.   An inordinate amount of time may
be devoted to obtaining, using, and recuperating from the drug to the exclusion of other responsibilities.   The chronic freebase smoker often becomes
socially withdrawn and isolated, preferring to use the drug alone, even
though the activity may have started as a social experience.   The psychiatric consequences of crack may be due to the drug's potent effects on brain
neurotransmitters.

Although there is no single diagnostic sign that will reliably tell the
clinician or family member whether someone is using crack, many of the medical and psychiatric symptoms previously mentioned are possible clues.   Urine
testing for cocaine metabolites (19) can aid in the diagnosis, although

Table 3: Cocaine dependence: clinical characteristics

1.   Loss of control over use.

> inability of refuse the drug when offered.
> inability to limit amount of use.
> "binge" patterns of excessive use for 24 hours or longer.
> unsuccessful attemps to stop usage for a significant
>   time period.

2.   Drug compulsion

> persistent or episodic cravings and compulsions
> compulsive use despite the absence of drug-induced
>   euphoria
> compulsions override desire to stop usage.
> fears of being without cocaine
> compulsion to use other drugs in the absence of cocaine

3.   Continued use despite adverse consequences

> medical complications: lethargy, insomnia, nasal or
>   sinus problems, appetite disturbance, loss of sex
>   drive, impotence
> psychiatric complications: depression, irritability,
>   anhedonia, paranoia, suicidal/homicidal ideation or
>   gestures
> social complications: financial, relationship, legal, or
>   job problems; general deterioration in psychosocial
>   functioning

4.   Denial

> denies that the problem exists
> denies or downplays the seriousness of the adverse effects
> acts defensive in response to inquiries about the drug use

these tests cannot differentiate crack from other forms of cocaine.

TREATMENT

Treatment is indicated when the cocaine user is unable to stop taking
the drug in spite of physical, psychological, or social problems resulting
from the drug use. It is important to emphasize that severe cocaine abuse
can develop with any route of administration. Intranasal administration
offers no protection against addiction or medical consequences, although
problematic use may develop more rapidly with freebase smoking and intravenous administration. Some of the clinical characteristics of cocaine dependence are described in Table 3.

Treatment Issues

The usual severity of cocaine freebase problems suggests that proper
treatment of the crack user requires a highly structured drug abuse programme. Moreover, freebase smokers are more likely than snorters to require
initial hospitalization in order to disrupt the compulsive pattern of drug
use and properly treat the more severe medical and psychiatric problems.
However, the patient who is highly motivated and displays no serious medical
or psychiatric problems can be given an initial trial in a highly structured
outpatient treatment programme (20,21,22). The programme must require total
abstinence from all drugs, including marijuana and alcohol. Treatment services for the crack patient should include: daily counselling sessions;

cocaine-specific drug education lectures, urine testing for all drugs at least twice weekly, self-help and professionally-led recovery groups, and treatment contracting for an initial 30-day abstinence period (verified by urine testing) with the stipulation that immediate hospitalization will be required if the initial abstinence period is not successfully completed. Family members and significant others should be involved in the treatment, whenever possible, to assist in forming a "safety net" around the primary patient who will remain extremely prone to sudden relapse for at least the first month or two of outpatient treatment.

Although there is no severe physical withdrawal syndrome following abrupt cessation of chronic freebase smoking, symptoms of depression, anergia, and sleep or appetite disturbance (consisting of prolonged periods of sleep interrupted by short periods of hyperphagia) may persist for several days after last use. During this period, drug craving is usually intense and patients often report dreams and nightmares related to smoking cocaine. Frequent contact with the patient for motivation and support is essential.

Inpatient or Outpatient?

Hospitalization of the cocaine abuser is usually required only in severe cases and so most can be treated as outpatients. Unlike heroin or alcohol, cocaine can be stopped abruptly without medical risk or a dramatic withdrawal syndrome and no substitute drugs or gradual weening from cocaine are needed.

Criteria for Hospitalization. Inpatient treatment of the cocaine abuser is usually indicated for those with the following characteristics:
1. Heavy users whose drug compulsion is uncontrollable, especially heavy freebase and IV users.
2. Those with physical dependency on other drugs or alcohol.
3. Those with severe medical or psychiatric complications.
4. Those with severe psychosocial impairment.
5. Those who have failed in outpatient treatment.

Role of Hospitalization. The major objectives of inpatient treatment should be to break the cycle of compulsive drug use, to address related and co-existing problems, and to strengthen the patient's motivation and skills for maintaining abstinence following hospital discharge. Inpatient treatment should be seen as only the first step in a more comprehensive recovery plan that must include outpatient aftercare treatment. The critical task of recovery is to maintain a drug-free lifestyle without the artificial protection of the hospital environment. Permanent abstinence following inpatient treatment is highly unlikely and relapse rates following hospitalization will remain unacceptably high unless the discharge plan includes intensive aftercare treatment.

Complete Abstinence

From Cocaine. The treatment must require immediate and complete cessation of cocaine use. A treatment goal of returning to "occasional" use is not only unrealistic, but potentially dangerous for anyone who has been dependent on the drug. Attempts to reduce rather than discontinue cocaine use may be temporarily successful in some cases, but usually lead back to heavy use and additional drug-related consequences that could have been avoided by complete abstinence.

From Other Drugs/Alcohol. Abstinence from all drugs of abuse, including marijuana and alcohol, is important in order to maximize the benefits of treatment and to minimize the possibility of relapse. The major goals of treatment must be to develop a reasonably satisfying drug-free lifestyle and to develop ways of coping without resorting to mood-altering chemicals.

Many cocaine abusers resist the idea of giving up alcohol or marijuana stating that they have had no problem with these substances in the past and would like to continue "social" or "recreational" use. The clinician must emphasize that complete abstinence offers the widest margin of safety, as evidenced by the following considerations: (1) while staying away from cocaine one is often more likely to switch to other substances for substitute "highs" even in the absence of previous problems with these substances; (2) substances that have been used in conjunction with cocaine, such as alcohol and marijuana, acquire the capacity through associative conditioning to trigger intense urges and cravings for cocaine; (3) even a single glass of wine or beer or a single marijuana cigarette may reduce one's ability to resist temptation for cocaine due to the "disinhibiting" effect of these substances; and (4) thorough evaluation of the patient's present and past use of various mood-altering substances often reveals more significant abuse patterns than previously recognized - since alcohol and other drugs are often used to self-medicate cocaine side effects, rather than to get "high", the user is often unaware of having acquired a simultaneous dependency on other substances.

Urine Testing

Urine testing is essential to the success of outpatient treatment (20, 21). Throughout the entire course of treatment, the patient's urine should be tested at least two to three times per week (19) for cocaine and all other commonly-abused drugs (e.g., opiates, barbiturates, benzodiazepines, amphetamines, marijuana, PCP, etc.). Despite mutual trust and a strong therapeutic alliance between patient and clinician, urine testing is necessary because of a re-emerging denial and self-deceit that is characteristic of the chemical dependency problem. Clinicians who fail to take the necessary counteractive measures are likely to become enablers of the patient's continuing drug use. It must be emphasized that the purpose of urine testing is NOT to catch the patient in a lie. Rather, urine testing is a valuable treatment tool that helps to break through denial, promote self-control over drug use impulses, and to provide an objective monitor of treatment progress. Consequences for drug-positive urines must be stipulated at the outset of treatment. Although a rare or infrequent "slip" might be expected, any emerging pattern of regular or frequent drug use should lead to revision of the treatment plan, including temporary hospitalization if other efforts fail.

Treatment Contracting

A written treatment contract should be used to clarify and concretize treatment requirements. The contract should specify the following: (1) that the patient immediately discontinue all use of mood-altering substances, (2) that the patient remain in treatment for no less than 6-12 months since this is the minimum amount of time needed to begin a solid recovery and promote long-lasting changes in lifestyle and behaviour, (3) that a urine sample be given whenever requested, (4) that a severe relapse may necessitate immediate hospitalization, and, (5) that designated family members and/or significant others can be contacted in the event of relapse, premature termination from treatment, or for discussion of treatment progress.

Stages of Treatment

Phase 1: Stopping All Drug Use. The first goal of treatment is to stop all drug use and to achieve total abstinence for an initial 30-day period. Patients are seen daily or every two or three days for abstinence training, supportive counselling, urine testing, and drug education meetings involving family members. Joining a self-help group such as Cocaine Anonymous (CA) is strongly encouraged and may be required in some cases. Abrupt cessation

rather than gradual reduction of cocaine use is essential. The likelihood
of early treatment failure is increased if there is a tapering-off period because any involvement with the drug at all only heightens the patient's ambivalence about stopping and saps their motivation to achieve total abstinence. Major tasks of this phase include: breaking through denial, discar-
ding all drug supplies and paraphernalia, establishing a commitment to total
abstinence, breaking off relationships with drug dealers and users, anticipating and handling drug urges, and forming an initial support network.

Phase 2: Relapse Prevention and Lifestyle Change. This phase lasts for
approximately 6-12 months and focuses on counteracting the most common and
predictable factors that may lead to relapse. Patients attend twice-weekly
cocaine recovery groups in conjunction with individual or family sessions at
least once a week. Critical issues in this phase include: recognizing the
earliest warning signs of relapse, combatting "euphoric recall", overcoming
the desire to test personal control over drug use, avoiding "high-risk" situations, preventing "slips" from becoming full-blown relapses, learning how
to cope with stress, how to have a good time without drugs, and forming new
social relationships.

Phase 3: Preparing for Long-Term Abstinence. This phase usually lasts
for an additional 6-12 months and focuses on issues that are crucial to maintaining long-term abstinence. Treatment consists of a once weekly "advanced"
recovery group with other senior members of the programme and/or individual
psychotherapy based on individual clinical needs. Major issues include:
enhancing personal relationships and feelings of self-esteem, counteracting
"flare-up" periods, overconfidence, and renewed denial, addressing issues of
"arrested maturity," strengthening the commitment to drug-free living, rea-
lizing plans, goals, and aspirations.

Phase 4: Follow-Up. When the formal treatment phases have been completed, patients are followed on a reducing schedule of periodic visits at 3-6
month intervals. Returning for more frequent contact during high-stress or
"flare-up" periods is strongly encouraged. Continued participation in CA
or some other self-help recovery network may become increasingly important
during this phase of recovery.

Family Involvement

Close family members and especially the spouse or parents of the cocaine abuser should be involved in the treatment for a number of reasons.
Family members can provide additional information about the patient's drug
use and other behaviour. Well-intentioned family members often function as
enablers by making excuses for the cocaine abuser, providing money for the
drug, or otherwise trying to spare the patient from suffering the consequences of his/her behaviour. Family members need instruction and guidance in
how to deal with the cocaine abuser and how to provide the necessary support
to foster the patient's recovery. They also need an opportunity to deal
with their own feelings of anger, blame, guilt, and victimization so as to
minimize family stress and confusion which could itself lead to the patient's
early relapse and treatment failure.

Self-Help Groups

Participation in a self-help group is essential to long-term recovery
for many patients. These groups should be used in conjunction with a programme of professional treatment and should also be used as a continuing
support system after treatment has ended. Professionals who employ an absti-
nence-oriented treatment approach will find that self-help groups enhance
their therapeutic success and provide an invaluable source of information
and emotional support for their patients. Cocaine Anonymous chapters are

rapidly proliferating in many areas of the U.S. Modeled after Alcoholics
Anonymous (AA) and Narcotics Anonymous (NA), CA utilizes the same 12-steps
of recovery which include admitting one's powerlessness over drugs and the
need for total abstinence to arrest the disease of chemical dependency. CA
provides an intensive peer-support network that is immediately available at
no cost to any newcomer who wants to stop using cocaine. In places where CA
chapters do not exist, patients should be encouraged to attend AA or NA
meetings.

Pharmacologic Treatment

No substitute drugs or gradual detoxification from cocaine is needed.
Moreover, there is no definitive evidence that any known medication can alleviate post-cocaine symptoms, cravings, or block the drug-induced euphoria,
despite earlier claims from uncontrolled clinical trials. A dopamine substitute, bromocryptine, has shown initial promise in reducing post-cocaine symptoms and cravings (22,23) in hospitalized patients, but controlled outpatient trials are needed to determine whether this treatment can actually increase abstinence rates. The potential usefulness of bromocryptine was suggested by evidence (24) that cocaine-induced dopamine depletion may underlie
the post-drug cravings and dysphoria.

Although carefully-researched pharmacologic adjuncts might be helpful in
some cases, there can be no substitute for the behavioural, attitudinal, and
lifestyle changes that are required for successful recovery from cocaine.

Psychiatric Issues

Cocaine-induced psychosis tends to be self-limiting with disappearance
of psychotic symptoms usually within 2-5 days after cessation of drug use.
Hospitalization and short-term use of neuroleptics or sedative-hypnotics may
be needed to facilitate initial management of some patients, but others may
show complete remission of symptoms within 24-48 hours without medication.

Depression is a common side effect of chronic cocaine abuse and a common
complaint during initial cocaine abstinence. Symptoms mimicking bipolar disorders, attention deficit disorders, and anxiety disorders may also be generated by cocaine abuse. Therefore, it is essential to allow a sufficient
post-cocaine recovery period before making a definitive psychiatric diagnosis
or introducing psychotropic medication. In cases where there is a genuine
dual diagnosis of psychiatric illness and chemical dependency, both problems
must be treated. It is nonetheless imperative that the drug abuse problem
be dealt with as a primary disorder and not merely as a symptom of the psychiatric illness. In order to avoid unrealistic or distorted expectations,
patients who receive psychotropic medication should be informed that the
medication cannot be a substitute for the lifestyle change and other treatment efforts that are essential to recovery.

Other Interventions

In addition to formal treatment interventions, a regular schedule of
exercise and planned leisure-time activities is an important feature of many
patients' recovery plan. These activities not only help to reduce stress
but also to instill a feeling of greater control over one's life. Worka-
holism and lack of satisfying social or leisure time are often precursors
to relapse.

Success Rates

Success rates will depend upon a variety of factors including the
severity of abuse, the patient's motivation to be drug free, and the extent

to which the treatment programme meets certain clinical needs. In our
highly structured and intensive RPR outpatient programme, we have found that
over 65% of patients complete the 6-12 month programme and over 75% are still
drug-free at 1-2 year follow-up (20). The highest success rates are found
in those with a strong desire to stop using cocaine, a history of good functioning before cocaine, and a genuine acceptance that the chemical dependency problem exists.

FINAL COMMENT

It is clear that cocaine use has grown from what appeared to be a relatively minor problem in the sixties and seventies to a major public health
problem today. Government reports issued in 1973 (24,25) concluded that
problems associated with cocaine did not appear to be significant and that
few who used the drug actually sought professional help in drug abuse treatment programmes or elsewhere. It was further stated that there had been no
confirmed cases of cocaine overdose deaths. Unfortunately, these reports
and others in the medical literature have fueled the myth that cocaine is
harmless and nonaddictive. The earlier low rates of problematic cocaine use
were probably due mainly to the drug's high price and limited availability
in the U.S. at that time rather than to its presumed low abuse potential.

As seen from our Hotline surveys and other sources, the situation has
changed drastically in the past few years. The evidence attesting to the
addictive patterns of cocaine use and the wide-ranging adverse consequences
is indisputable. The academic debates about whether or not cocaine is truly
addicting will probably continue, but the behavioural evidence is already
clear. Consistent with our Hotline experience, government surveys in more
recent years, show that the prevalence of cocaine use in the U.S. has increased dramatically since the late seventies. Similarly, there have been
significant increases in the medical consequences of cocaine use according
to the government statistics on cocaine-related emergency room episodes,
deaths, and admissions to government-sponsored drug abuse treatment programmes.

In addition to an overall increase in the prevalence of cocaine use and
its health consequences, our Hotline surveys identify other noteworthy
trends. Here too, our findings are consistent with government surveys. The
Hotline data show major demographic shifts in user profiles indicating increased cocaine use among women, minority groups, lower-income groups, and
adolescents. Cocaine use is no longer restricted primarily to white middle
and upper class adult males. A major contributor to the spreading use has
been the increasing availability of cocaine supplies at reduced prices,
making the drug more accessible to a much larger segment of the population.
On average, a gramme of cocaine is now cheaper than an ounce of marijuana in
many places across the U.S. The greater accessibility to cocaine has no
doubt also contributed to more intensified use among current users as reflected in the higher dosages reported by Hotline callers and the increasing
popularity of freebase smoking - a method of administration that almost invariably leads to more compulsive, higher-dose use.

The presumed safety of intranasal cocaine use has been challenged consistently in all of our surveys. Intranasal users continue to account for
over 50% of callers to the Hotline and for the majority of treatment admissions for cocaine problems to our own programmes and elsewhere. The potential dangers of snorting cocaine should not be underestimated. In addition
to problems of addiction and drug-related dysfunction, there have been instances of death from intranasal cocaine use verified by coroner's reports
(11).

One might hope that the current upsurge in cocaine use would be a short-lived, temporary phenomenon - a passing fad that would dissipate as quickly as it seemed to appear. Unfortunately, this seems highly unlikely. The current cocaine epidemic has already become too pervasive and indications are that it is still growing at a rate that would make rapid resolution of the problem all but impossible. The enormous profits of the illicit cocaine industry have resulted in increased production and supplies of the drug and a powerful motivation to continue making the drug available to as large a segment of the U.S. population as possible. Since there is often a time lag of about 4-5 years between the onset of snorting cocaine and its escalation to the point where the user is driven to seek help, the peak effects of the current epidemic in the U.S. may not be seen for several years. Although public education and other prevention efforts may help to discourage future experimentation, at present it appears that if supplies of the drug continue to increase while prices continue to decline, the current epidemic is likely to become even more widespread and intensified.

The recent appearance of crack has confirmed the fears of many who were already seriously concerned about the spreading cocaine menace in America. This highly addictive form of cocaine is leading to further increases in cocaine-related medical, psychiatric, and addiction problems and straining an already overcrowded treatment system stretched to its limits.

Unfortunately, there is no simple "quick-fix" solution to America's drug problem and the belief that there should be such a solution may be part of the problem itself. Our country is now suffering the consequences of a more than 25-year "love affair" with powerful mood-altering drugs that started in the '60s, expanded in the '70s, and became dramatically evident in the '80s. We cannot reverse this trend overnight. We simply cannot eliminate the supply of illegal drugs and so we must put our best efforts into eliminating the demand. This will require nothing short of a major overhaul in our society's attitudes and values about drugs. Through education in the schools at home, in the media, and anywhere else we can transmit the message, drug use must once again become unacceptable, undesirable, and an obstacle to success. Our goal should be to make total avoidance of drug use as desirable to our youngsters as being physically fit and attractive. It should be emphasized that in addition to being unhealthy, drug use is illegal and socially and morally irresponsible.

Prevention of drug use is a massive, long-term undertaking: no single programme or law will eliminate the problem overnight, but we cannot allow ourselves to be frustrated by the lack of an instantaneous solution or have it lead us into taking radical measures (such as legalizing drugs) that are likely to backfire. It is not just drug abuse experts and law enforcement officials who will solve the problem, all segments of society must join the fight!

REFERENCES

1.   E. H. Adams, N. J. Kozel, Cocaine Use in America, in "Cocaine Use in America: epidemiological and clinical perspectives; NIDA Research Monograph No. 61," Government Printing Office, Washington, D.C.(1985)
2.   A. M. Washton, M. S. Gold, Changing patterns of cocaine use in America: a view from the National Hotline, 800-COCAINE, Advances in Alcohol and Substance Abuse, in press.
3.   M. S. Gold, "800-COCAINE", Bantam, New York (1984)
4.   A. M. Washton, The cocaine abuse problem in the U.S., Testimony presented before the Select Committee on Narcotics Abuse and Control, U.S. House of Representatives, Washington, D.C., July 16, 1985.
5.   N. S. Stone, M. Fromme, D. Kagan, "Cocaine: Seduction and Solution", Clarkson N. Potter, New York (1984)
6.   A. M. Washton, N. S. Stone, The human cost of chronic cocaine use,

Medical Aspects of Human Sexuality, 18:122-130 (1984)

7. C. Johanson, Assessment of the dependence potential of cocaine in animals, in "Cocaine, pharmacology, effects, and treatment, NIDA Research Monograph No. 50", J. Grabowski, ed., U. S. Government Printing Office, Washington, D.C. (1984)

8. R. T. Jones, The pharmacology of cocaine, in "Cocaine:pharmacology, effects, and treatment, NIDA Research Monograph No. 50," J. Grabowski, ed., U. S. Government Printing Office, Washington, D.C. (1984)

9. A. M. Washton, Crack: The latest threat in addiction, Medical Aspects of Human Sexuality, in press.

10. R. M. Post, Cocaine psychosis: a continuum model, Am. J. Psychiatry, 132:225-231 (1975)

11. C. V. Wetli, R. K. Wright, Death caused by recreational cocaine use, Journal of the American Medical Association, 241:2519-2522 (1979)

12. A. M. Washton, M. S. Gold, A. C. Pottash, Intranasal cocaine addiction, Lancet, 11:1378 (1983)

13. A. M. Washton, A. Tatarsky, Adverse effects of cocaine abuse, in "Problems of drug dependence, 1983, NIDA Research Monograph No. 44," L. S. Harris, ed., U. S. Government Printing Office, Washington, D. C. (1984)

14. A. M. Washton, M. S. Gold, Chronic cocaine abuse: evidence for adverse effects on health and functioning, Psychiatric Annals, 14:733-743 (1984)

15. A. M. Washton, M. S. Gold, A. C. Pottash, L. Semlitz, Adolescent cocaine abusers, Lancet 11:letter (1984)

16. J. F. Jekel, D. F. Allen, S. Dean-Patterson, H. Podlewski, et. al., Epidemic freebase cocaine abuse: case study from the Bahamas, Lancet, March 1:459-462 (1986)

17. R. K. Siegel, Cocaine smoking, J. of Psychoactive Drugs, 14:271-355, (1982)

18. J. Itkonen, S. Schnoll, J. Glassroth, Pulmonary dysfunction in freebase cocaine users, Archives of Internal Medicine 144:2195-2197 (1984)

19. K. Vereby, D. Martin, M. S. Gold, Drug Abuse: interpretation of laboratory tests, in "Diagnostic and laboratory testing in psychiatry," M. S. Gold, A. C. Pottash, eds., Plenum, New York (1986)

20. A. M. Washton, M. S. Gold, "Cocaine Treatment Today," American Council on Drug Education, Bethesda (1986)

21. A. M. Washton, Cocaine abuse treatment, Psychiatry Letter 3:51-56 (1985)

22. M. S. Gold, A. M. Washton, C. A. Dackis, Cocaine abuse: neurochemistry, phenomenology, and treatment, in "Cocaine Use in America: Epidemiologic and Clinical Perspectives, NIDA Research Monograph No. 61", N. J. Kozel and E. H. Adams, eds., U. S. Government Printing Office, Washington, D.C. (1985)

23. C. A. Dackis, M. S. Gold, Bromocryptine as a treatment for cocaine abuse, Lancet 2:1151-1152 (1985)

24. National Commission on Marijuana and Drug Abuse, "Drug Use in America: Problem in Perspective, Second Report of the National Commission on Marijuana and Drug Abuse," National Institute on Drug Abuse, Washington, D. C. (1973)

25. Strategy Council on Drug Abuse, "Federal Strategy for Drug Abuse and Drug Traffic Prevention, 1973," U. S. Government Printing Office, Washington, D.C. (1973)

This paper was based on a speech presented at the First International Drug Symposium, Nassau, Bahamas, November, 1985. Portions of this manuscript are adapted from previous articles by the author published in Advances in Alcohol and Substance Abuse, Volume 6, 1985; and reprinted with the permission of Haworth Press, New York, N.Y.

THE SELF-MEDICATION HYPOTHESIS OF ADDICTIVE DISORDERS: FOCUS ON HEROIN

AND COCAINE DEPENDENCE

Edward J. Khantzian

Cambridge Hospital
Harvard Medical School
Cambridge, Mass.

PREFACE

The article which follows this introduction was originally published as a Special (Cover) Article in the American Journal of Psychiatry in the November, 1985 issue, the same month in which the First International Drug Symposium, sponsored by The Bahamas Ministry of Health and The Embassy of the United States of America, was convened to discuss the rock-cocaine epidemic in the Bahamas and other Caribbean Islands. Based on my article, I was invited to participate in the Symposium and to speak about some of my views on the psychological predispositions for drug dependence in general, and in particular, on the psychological predisposition for cocaine dependence. At first, I did not grasp the seriousness and scope of the cocaine problem, but I accepted the invitation, believing I might make a contribution to the Symposium. I was not long in attendance at the Symposium before I realized that the Bahamian citizens, professionals, and health care leaders were facing a major crisis as a consequence of the cocaine epidemic.

With this realization, I began to experience some trepidation about whether my ideas on the self-medication motive for drug dependence could or should be considered at the Symposium. Later, I likened the situation to a friend by saying it would be comparable to being in Rome and discussing the finer architectural details of the buildings while Rome was burning. While I believe in the validity and relevance of "the self-medication hypothesis" as a significant factor in compulsive drug use, I thought that there were factors of availability, potency and the population involved in the Bahamian situation that were unique and to some extent outside the scope of my clinical experience. The situation certainly highlighted the often held contention that heavy use of substances caused psychopathology at least as much as psychopathology causes substance dependency. Given the intensely disruptive effects of inhaled freebase cocaine on neurotransmitters and the likely possibility that this effect can powerfully fuel craving mechanisms as much as psychological factors can, I at first thought that self-medication factors probably played a more modest role. However, upon further reflection and somewhat removed from the crisis atmosphere, I believe that the hypothesis advanced here does have relevance.

Addiction to drugs, including cocaine addiction, is intimately related to human suffering. Clearly, substances of abuse, and particularly freebase cocaine abuse, cause much physical and psychological distress. However, not everyone that experiments with or uses addictive drugs advances to abuse or addiction. The article which follows traces the development of a perspective and presents findings which suggest that an important factor accounting for heavy reliance on drugs is the result of painful emotional states involving dysphoric affects (or feelings) and related, co-existent psychiatric disorders. The drugs that addicts select are not chosen randomly. Their drug of choice is the result of an interaction between the psychopharmacologic action of the drug and the dominant painful feelings with which they struggle. Narcotic addicts prefer opiates because of their powerful muting action on disorganizing and threatening affects of rage and aggression. Cocaine has its appeal because of its ability to relieve distress associated with depression, hypomania and hyperactivity.

The Bahamian experience with rock cocaine (or "crack") is an extreme and tragic one. It remains to be seen, with the passage of time and abatement of the crisis climate, whether we will discover that the distress associated with heavy exposure to cocaine has antecedents related to those described in this chapter, or, whether they principally reflect how much a drug like cocaine can alter our brain chemistry and emotions and produce the suffering which fuels addictive behaviour. However, given that all the answers are not yet in on the cause of substance dependence, and it is unlikely we will reduce our understanding to a single etiology, I believe it is important to consider the self-medication hypothesis as an important causative factor in drug dependency.

HISTORICAL PERSPECTIVES

Developments in psychoanalysis and psychiatry over the past 50 years have provided enabling new insights and approaches in understanding mental life and in treating its aberrations. In psychoanalysis, there has been a shift from a focus on drives and conflict to a greater emphasis on the importance of ego and self structures in regulating emotions, self-esteem, behaviour, and adaptation to reality. In psychiatry, we have witnessed the advent of psychotropic medications, a more precise understanding of the neurobiology of the brain, and the development of standardized diagnostic approaches for identifying and classifying psychiatric disorders. Such developments have had implications for understanding and treating addictions, especially given the recent dramatic rise in drug abuse in all sectors of our society and our growing inclination to treat our drug-dependent patients through private practice, in community mental health centres, and in methadone-maintenance or self-help programmes, in close proximity to the surroundings in which their addictions evolved.

Popular or simplistic formulations in the early 1970s emphasized peer group pressure, escape, euphoria, or self-destructive themes to explain the compelling nature of drug dependency. In contrast, the work of a number of psychoanalysts in the 1960s and 1970s has led to observations, theoretical formulations, and subsequent studies representing a significant departure from these previous approaches and explanations. On the basis of a modern psychodynamic perspective, these analysts succeeded in better identifying the nature of the psychological vulnerabilities, disturbances, and pain that predispose certain individuals to drug dependence. This perspective, which has spawned a series of diagnostic studies over the past decade, emphasizes that heavy reliance on and continuous use of illicit drugs (i.e., individuals who become and remain addicted) are associated with severe and significant psychopathology. Moreover, the drug of choice that individuals come upon is not a random phenomenon.

On the basis of recent psychodynamic and psychiatric perspectives and findings, I will elaborate on a self-medication hypothesis of addictive disorders, emphasizing problems with heroin and cocaine dependence. This point

of view suggests that the specific psychotropic effects of these drugs interact with psychiatric disturbances and painful affect states to make them compelling in susceptible individuals.

PSYCHODYNAMIC FINDINGS

An extensive review of the psychoanalytic literature on addiction goes
beyond the scope of this paper. I have reviewed elsewhere the early psychoanalytic literature on addiction, which principally emphasized the pleasurable aspects of drug use (1,2). Psychoanalytic reports that pertain most to
this thesis date back to the work of Chein et al. (3) and the earlier related
work of Gerard and Kornetsky (4,5), who were among the first groups to study
addicts in the community (inner city, New York addicts) and attempt to understand the psychological effects of opiates and how they interacted with addicts' ego, superego, narcissistic, and other psychopathology. They emphasized that individuals use drugs adaptively to cope with overwhelming (adolescent) anxiety in anticipation of adult roles in the absence of adequate
preparation, models, and prospects. Because they did not have the benefit of
a modern psychopharmacologic perspective, they referred to the general "tran-
quilizing or ataractic" properties of opiates and did not consider that the
appeal of narcotics might be based on a specific effect or action of opiates.
In addition, their studies were limited to narcotic addicts, thus providing
little basis to compare them to addicts dependent on other drugs.

Around 1970, a number of psychoanalysts began to report findings based
on their work with addicts, who were coming in increasing numbers to their
practices and to a variety of community treatment settings. In contrast to
the early emphasis in psychoanalysis on a drive and topographical psychology,
their work paralleled developments in contemporary psychoanalysis and placed
greater emphasis on structural factors, ego states, and self and object relations in exploring the disturbances of addicts and understanding their suffering. More particularly, this literature highlighted how painful affects
associated with disturbances in psychological structures and object relations
interacted with the psychopharmacologic action of addictive drugs to make
them compelling.

Despite a superficial resemblance to earlier formulations that stressed
regressive pleasurable use of drugs, work by Wieder and Kaplan (6) represented an important advance and elaboration of trends set in motion by Gerard
and Kornetsky in the 1950s. Wieder and Kaplan used recent developments in
ego theory, which enabled them to appreciate that individuals self-select
different drugs on the basis of personality organization and ego impairments.
Their emphasis on the use of drugs as a "prosthetic," and their focus on developmental considerations, adaptation, and the ego clearly sets their work
apart from earlier simplistic formulations based on an id psychology.

On the basis of this and other recent work that considers ego and adap-
tational problems of addicts, and following lines pursued by Wieder and Kaplan, Milkman and Frosch (7) empirically tested the hypothesis that self-selection of specific drugs is related to preferred defensive style. Using the
Bellak and Hurvich Interview and Rating Scale for Ego Functioning, they compared heroin and amphetamine addicts in drugged and nondrugged conditions.
Their preliminary findings supported their hypothesis that heroin addicts
preferred the calming and dampening effects of opiates and seemed to use
this action of the drug to shore up tenuous defenses and reinforce a tendency
toward withdrawal and isolation, while amphetamine addicts used the stimulating action of amphetamines to support an inflated sense of self-worth and a
defensive style involving active confrontation with their environment.

The work of Wurmser (8,9) and Khantzian (1,10-12) suggested that the
excessive emphasis on the regressive effects of narcotics in previous studies

was unwarranted and that, in fact, the specific psychopharmacologic action of opiates has an opposite, "progressive" effect whereby regressed states may be reversed. Wurmser believed that narcotics are used adaptively by narcotic addicts to compensate for defects in affect defense, particularly against feelings of "rage, hurt, shame and loneliness." Khantzian stressed drive defense and believed narcotics act to reverse regressed states by the direct antiaggression action of opiates, counteracting disorganizing influences of rage and aggression on the ego. Both these formulations proposed that the pharmacologic effects of the drug could substitute for defective or nonexistent ego mechanisms of defense. As with previously mentioned recent investigators, Wurmser and Khantzian also considered developmental impairments, severe predisposing psychopathology, and problems in adaptation to be central issues in understanding addiction. Radford et al. (13) reported detailed case material which supported the findings of Wurmser and Khantzian that opiates could have an antiaggression and antiregression action or effect. They further observed that opiate use cannot be exclusively correlated with any particular patterns of internal conflict or phase-specific developmental impairment.

Krystal and Raskin (14) were less precise about the specific effects of different drugs but allowed that they may be used either to permit or prevent regression. However, their work focused more precisely on the relationship between pain, depression, and anxiety and drug and placebo effects. They explored and greatly clarified addicts' difficulties in recognizing and tolerating painful affects. They proposed that the tendency for depression and anxiety to remain somatized, unverbalized, and undifferentiated in addicts resulted in a defective stimulus barrier and thus left such individuals ill-equipped to deal with their feelings and predisposed them to drug use. Their work also focused in greater depth on the major problems that addicts have in relation to positive and negative feelings about themselves and in relation to other people. Krystal and Raskin believed that addicts have major difficulties in being good to themselves and in dealing with their positive and negative feelings toward others because of rigid and massive defenses such as splitting and denial. They maintained that drug users take drugs not only to assist in defending against their feelings but also briefly and therefore "safely" to enable the experience of feelings like fusion (oneness) with loved objects, which are normally prevented by the rigid defenses against agression.

DIAGNOSTIC AND TREATMENT STUDIES

Partly as an extension and outgrowth of the psychodynamic studies and partly as the result of the development of standardized diagnostic approaches for classifying and describing mental illness, a number of reports over the past decade have documented the coexistence of psychopathology in drug-dependent individuals. Some of these investigators have also reported on the results of conventional psychiatric treatment, including psychotherapy, with drug-dependent individuals. Although the results of such studies to date have been inconsistently successful or inconclusive, they are suggestive enough to further support the concept that drug dependence is related to and associated with coexistent psychopathology.

In a placebo-controlled study, Woody et al. (15) treated a series of narcotic addicts with the antidepressant doxepin and documented significant symptom reduction. Their study suggested that this group of patients suffered with an anxious depression, and as the depression lifted with treatment, there was a corresponding reduction in misuse and abuse of drugs and an improvement in overall adaptation. More recently, this group of investigators (16) reported longitudinal data on individuals dependent on psychostimulants, sedative-hypnotics, or opiates which suggested that addicts might be medicating themselves for underlying psychopathology. Their study suggested that such individuals might respond to the administration of appropriate psychopharmacologic agents for target symptoms of phobia and depression.

Dorus and Senay (17) and Weissman et al. (18-20) evaluated large samples of narcotic addicts and, using standardized diagnostic approaches, documented a significant incidence of major depressive disorder, alcoholism, and antisocial personality. Rounsaville et al. (19,20) concluded that their findings were consistent with the clinical theories of Wurmser and Khantzian, i.e., that depressed addicts used opiates as an attempt at self-treatment for unbearable dysphoric feelings.

At a conference sponsored by the New York Academy of Sciences on opioids in mental illness, Khantzian, Wurmser, McKenna, Berken, Millman, Vereby, and others presented clinical findings and theoretical observations that support a self-medication hypothesis cf addictions (21). The sponsors of the conference and the participants reviewed the role of exogenous opiates as well as endorphins in regulating emotions. One of the conclusions drawn from these findings was that the long-acting opiate methadone might be an effective psycotropic agent in the treatment of severe psychoses, especially cases refractory to conventional drugs and in instances associated with violence and rage.

Treece and Nicholson (22), using diagnostic criteria from DSM-III, published findings indicating a strong relationship between certain types of personality disorder and methadone dose required for stabilization. They studied this same sample and compared "high drug" and "low drug" users (i.e., in addition to their prescribed methadone dose) and were able to show that the high drug users were significantly more impaired in the quality of their object relations than the low drug users (23). In a recent report, Khantzian and Treece (24) studied 133 narcotic addicts from three subject samples (i.e., a methadone programme sample, a residential setting sample, and a street sample) and, using DSM-III, documented depression in over 60% and a range of personality disorders (that included but was not limited to antisocial disorder) in over 65%. We also explored the possible relationships between the disturbed/disturbing behaviour of addicts as reflected by the personality diagnosis, and the painful affects with which addicts suffer as reflected by the diagnosis of depression.

Recently, Blatt et al. (25) used the Loevinger Sentence Completion, the Bellak Ego Function Interview, and the Rorschach to extensively study 99 opiate addicts and compare them to normal subjects. Their findings provide further evidence that opiate addicts suffer significantly in their interpersonal relations and in affect modulation. The authors indicate that addicts use drugs in the service of isolation and withdrawal.

In two carefully executed studies (26,27) based on the assumption that opiate dependence is associated with psychopathology, the effectiveness of psychotherapy on the psychopathology and presumed related drug dependence of 72 and 110 narcotic addicts, respectively, was tested. Rounsaville et al. (26) found no evidence that psychotherapy appreciably influenced treatment outcome. Woody et al. (27) demonstrated that the addicts receiving psychotherapy had greater improvement than the addicts who received only drug counselling and that the psychotherapy subjects required less methadone and used fewer psychotropic drugs.

Finally, carefully executed studies dating back to the early 1970s document that in selected cases and samples of substance-dependent individuals, target symptoms and psychopathology have been identified and successfully treated with psychotropic drugs (15,28-31).

CLINICAL OBSERVATIONS - NARCOTIC AND COCAINE DEPENDENCE

Clinical work with narcotic and cocaine addicts has provided us with compelling evidence that the drug an individual comes to rely on is not a random choice. Although addicts experiment with multiple substances, most prefer one drug. Wieder and Kaplan (6) referred to this process as "the drug-of-choice phenomena," Milkman and Frosch (7) described it as the "preferential use of drugs," and I (32) have called it the "self-selection" process. I believe that narcotic and cocaine addicts' accounts of their subjective experiences with and responses to these drugs are particularly instructive. They teach us about how addicts suffer with certain overwhelming affects, relationships, and behavioural disturbances and how the short-term use of their drug of choice helps them to combat these disturbances.

Narcotic Addiction

Although narcotics may be used to overcome and cope with a range of human problems including pain, stress, and dysphoria (33), I have been impressed that the antiaggression and antirage action of opiates is one of the most compelling reasons for its appeal. I base this conclusion on observations of over 200 addicts whose histories reveal lifelong difficulties with rage and violent behaviour predating their addiction, often linked to intense and unusual exposure to extreme aggression and violence in their early family life and the environment outside their homes. These experiences included being both the subject and the perpetrator of physical abuse, brutality, violent fights, and sadism. In the course of their evaluation and treatment these patients repeatedly described how opiates helped them to feel normal, calm, mellow, soothed, and relaxed. I have also observed addicts in group treatment whose restlessness and aggressiveness, especially manifested in their abusive and assaultive use of obscenities, subsided as they stabilized on methadone (1,10-12). I was also impressed that many narcotic addicts discovered the antirage action of opiates in a context of violent feelings, often of murderous proportion, being released in them by sedatives and alcohol or being manifested as a consequence of amphetamine and cocaine use (34).

Clinical vignette. A 29-year old ex-felon, admitted to a closed psychiatric ward because of increasing inability to control his alcohol and cocaine use, demonstrated dramatically this special relationship between violence and drug use and why opiates were his drug of choice. I saw this patient in consultation on a day when he had become very agitated and intimidating as he witnessed a very disturbed female patient being placed in four-point restraints. An alert attendant, who was a felon himself on work release, ascertained from the patient that this scene triggered panic and violent reactions similar to those he had experienced when he had been attacked by guards in prison because he had been threatening or assaultive. When I met him I was surprised by his diminutive stature and reticence. I told him I wanted to understand his drug-alcohol use and determine whether our unit was okay (i.e., safe) for him. He immediately apologized for overreacting to the restrained patient and explained how much it reminded him of his prison experience. He quickly launched into his worry that his alcohol and cocaine use was causing increasingly uncontrolled outbursts of verbally assaultive behaviour and, as a consequence, an increasing tendency to use opiates to quell his violent reactions. He openly admitted to past overt assaultive behaviour, most frequently involving knifings, when he felt threatened, provoked, or intimidated. He kept returning to the confrontation and restraint of the patient, apologizing for his reaction but also explaining how disorganizing and threatening it was for him. He said there had been a time when attack would have been a reflex in such situations, but he wanted to reassure me and the staff that he really understood why we were doing what we were. He seemed to be begging to stay and said he wanted help with his alcohol and drug use. An inquiry about his drug use and its effect on him revealed that he preferred opiates, so much so that he knew he had to avoid them. (He explained that he knew too many people who had become hopelessly dependent on or had died because of opiates.) Whereas alcohol or stimulants could cause violent eruptions, he explained that opiates-and he named them all correctly-countered or controlled such reactions. He said that the only person he had to rely on was his mother but that she was very ill and in the hospital. He complained that he had suffered as a consequence of his father's alcoholism, the associated violence, and his premature death (alcoholic complications) when the patient was in his early teens. His father's unavailability and early death had left him without supervision or guidance. He bitterly lamented that a brother 5 years older had been "useless" in providing any guidance on how to control his drug alcohol use or his impulsive and aggressive behaviour ("he didn't help me to smarten up").

The patient's description of his violent side before prison and once in prison, confirmed by the mental health aide who corroborated his story, was chilling and convincing. He was equally graphic in describing the many attacks he suffered at the hands of sadistic correction officers and other inmates. What was clear was that, whether he was the perpetrator or the victim, the violence was a recurrent, regular, and repetitious part of his adult life. I have concluded (11,12) that such individuals welcome the effects of opiates because they mute uncontrolled aggression and counter the threat of both internal psychological disorganization and external counter-aggression from others, fears that are not uncommon with people who struggle with rage and violent impulses. The discovery that opiates can relieve and reverse the disorganizing and fragmenting effects of rage and aggression is not limited to individuals who come from deprived, extreme, and overly violent backgrounds.

Clinical vignette. A successful 35-year-old physician described how defensive and disdainful he had become since his early adulthood as a consequence of his mother's insensitivity and his father's cruel and depriving attitude toward the patient and his family, despite their significant affluence. He said he became dependent on opiates when his defense of self-sufficiency began to fail him in a context of disappointing relationships with women and much distress and frustration working with severely ill patients. More than anything else, he became aware of the calming effects of these drugs on his bitter resentment and mounting rage. He stressed how this effect of the drugs helped him to feel better about himself and, paradoxically, helped him to remain energized and active in his work.

I have described (12,34) similar patients from privileged backgrounds in which sadistic or unresponsive parents fueled a predisposition to angry and violent feelings toward self and others.

Cocaine Addiction

From a psychodynamic perspective, a number of investigators have speculated on the appeal of stimulants and, in particular, cocaine. For some, the energizing properties of these drugs are compelling because they help to overcome fatigue and depletion states associated with depression (32). In other cases the use of stimulants leads to increased feelings of assertiveness, self-esteem, and frustration tolerance (6) and the riddance of feelings of boredom and emptiness (9). I have proposed that certain individuals use cocaine to "augment a hyperactive, restless lifestyle and an exaggerated need for self-sufficiency" (34,p.100). Spotts and Shontz (35) extensively studied the characteristics of nine representative cocaine addicts and documented findings that are largely consistent with the psychodynamic descriptions of people who are addicted to cocaine.

More recently, we have considered from a psychiatric/diagnostic perspective a number of factors that might predispose an individual to become and remain dependent on cocaine (36,37): 1) pre-existent chronic depression; 2) cocaine abstinence depression; 3) hyperactive, restless syndrome or attention deficit disorder; and 4) cyclothymic or bipolar illness. Unfortunately, studies of representative larger, aggregate samples of cocaine addicts do not yet exist to substantiate these possibilities.

Clinical vignette. A 30-year-old man with a 10-year history of multiple drug use described the singularly uplifting effect of cocaine, which he came to use preferentially over all the other drugs. In contrast to a persistent sense of feeling unattractive and socially and physically awkward dating back to adolescence, he discovered "that (snorting) it gave me power-and made me happy. It was pleasant-euphoric; I could talk-and feel erotic." Subsequently injecting it intensified these feelings, but more than anything else, the cocaine helped him to not worry what people thought about him.

I have been repeatedly impressed how this energizing and activating property of cocaine helps such people, who have been chronically depressed, overcome their anergia, complete tasks, and better relate to others, and, as a consequence, experience a temporary boost in their self-esteem (37).

Clinical vignette. In contrast, a 40-year-old accountant described an opposite, paradoxical effect from snorting cocaine. Originally, when I evaluated this man, I thought he was using the stimulating properties of the drug as an augmentor for his usual hyperactive, expansive manner of relating. He finally convinced me to the contrary when he carefully mimicked how he put down several lines in the morning, snorted it, and breathed a sigh of relaxation and then described how he could sit still, focus on his backlog of paper work, and complete it.

This man's story, a recent dramatic and extreme case (38), and two other related reports (36,37) suggest that cocaine addicts might be medicating themselves for mood disorders and behavioural disturbances, including a pre-existing or resulting attention deficit/hyperactive-type disorder. The extreme case responded dramatically to methylphenidate treatment. I have successfully treated several other patients with methylphenidate. The patients I have treated with methylphenidate provide further evidence to support a self-medication hypothesis of drug dependency. At this point it would be premature to conclude precisely what the disorder or disorders are for which cocaine addicts are medicating themselves. However, the pilot cases and my previous clinical experiences suggest several possibilities. The patients share in common lifelong difficulties with impulsive behaviour, emotional lability, acute and chronic dysphoria (including acute depressions), and self-esteem disturbances that preceded cocaine use. All of the patients experienced a relief of dysphoria and improved self-esteem on cocaine; they also experienced improved attention leading to improved interpersonal relations, more purposeful, focused activity, and improved capacity for work. The substitution of the more stable, long-acting stimulant drug methylphenidate provided an opportunity for me to observe these patients clinically and to confirm the stabilizing effect of stimulants on them.

COMMENT

Clearly, there are other determinants of addiction, but I believe a self-medication motive is one of the more compelling reasons for overuse of and dependency on drugs. Clinical findings based on psychoanalytic formulations have been consistent with and complemented by diagnostic and treatment studies that support this perspective, which, I believe, will enable researchers and clinicians to further understand and treat addictive behaviour. Rather than simply seeking escape, euphoria, or self-destruction, addicts are attempting to medicate themselves for a range of psychiatric problems and painful emotional states. Although most such efforts at self-treatment are eventually doomed, given the hazards and complications of long-term, unstable drug use patterns, addicts discover that the short-term effects of their drugs of choice help them to cope with distressful subjective states and an external reality otherwise experienced as unmanageable or overwhelming. I believe that the perspective provided by the self-medication hypothesis has enabled me and others to understand better the nature of compulsive drug use and that it has provided a useful rationale in considering treatment alternatives. The heuristic value of this hypothesis might also help us to more effectively understand and treat the most recent elusive addiction, cocaine dependence.

REFERENCES

1. E.J.Khantzian, Opiate addiction: a critique of theory and some implications for treatment, Am. J. Psychother., 28:59-70, (1974).
2. E.J.Khantzian, C. Treece, Psychodynamics of drug dependence: an overview, in: "Psychodynamics of Drug Dependence: NIDA Research Monograph 12," J.D. Blaine, D.A. Julius, eds., National Institute on Drug Abuse, Rockville, Md.,(1977).
3. I. Chein, D.L.Gerard, R.S.Lee, et al.,"The Road to H: Narcotics, Delinquency, and Social Policy," Basic Books, New York, (1964).
4. D.L.Gerard, C. Kornetsky, Adolescent opiate addiction: a case study, Psychiatr. Q., 28:367-380, (1954).
5. D.L.Gerard, C. Kornetsky, Adolescent opiate addiction: a study of control and addict subjects, Psychiatr.Q., 29:457-486, (1955).
6. H. Wieder, E.H. Kaplan, Drug use in adolescents: psychodynamic meaning and phamacogenic effect, Pschoanal. Study Child, 24:399-431, (1969).
7. H. Milkman, W.A. Frosch, On the preferential abuse of heroin and amphetamine, J. Nerv. Ment. Dis., 156:242-248, (1973).
8. L. Wurmser, Methadone and the craving for narcotics: observations of patients on methadone maintenance in psychotherapy, in "Proceedings of the Fourth National Methadone Conference, San Francisco, 1972," National Association for the Prevention of Addiction to Narcotics, New York, (1972).
9. L. Wurmser, Psychoanalytic considerations of the etiology of compulsive drug use, J. Am. Psychoanal. Assoc., 22:820-843, (1974).
10. E. J. Khantzian, A preliminary dynamic formulation of the psychopharmacologic action of methadone, in "Proceedings of the Fourth National Methadone Conference, San Francisco, 1972," National Association for the Prevention of Addiction to Narcotics," (1972).
11. E.J. Khantzian, An ego-self theory of substance dependence: a contemporary psychoanalytic perspective, in "Theories on Drug Abuse: NIDA Research Monograph 30," D.J. Lettieri, M. Sayers, H.W. Pearson, eds. National Institute on Drug Abuse, Rockville,Md., (1980).
12. E. J. Khantzian, Psychological (structural) vulnerabilities and the specific appeal of narcotics, Ann. N.Y. Acad. Sci., 398:24-32 (1982).
13. P. Radford, S. Wiseberg, C. Yorke, A study of "main line" heroin addiction, Psychoanal. Study. Child., 27:156-180 (1972).
14. H. Krystal, H. A. Raskin, "Drug Dependence: Aspects of Ego Functions," Wayne State University Press, Detroit (1970).
15. G. E. Woody, C. P. O'Brien, K. Rickels, Depression and anxiety in heroin addicts: a placebo-controlled study of doxepin in combination with methadone, Am. J. Psychiatry, 132:447-450 (1975).
16. A. T. McLellan, G. E. Woody, C. P. O'Brien, Development of psychiatric illness in drug abusers, N. Engl. J. Med., 201:1310-1314 (1979).
17. W. Dorus, E. C. Senay, Depression, demographic dimension, and drug abuse, Am. J. Psychiatry, 137:699-704 (1980).
18. M. M. Weissman, F. Slobetz, B. Prusoff, et al., Clinical depression among narcotic addicts maintained on methadone in the community, Am. J. Psychiatry, 133:1434-1438 (1976).

19. B. J. Rounsaville, M. M. Weissman, H. Kleber, et al., Heterogeneity of psychiatric diagnosis in treated opiate addicts, Arch. Gen. Psychiatry, 39:161-166 (1982).

20. B. J. Rounsaville, M. M. Weissman, K. Crits-Cristoph, et al., Diagnosis and symptoms of depression in opiate addicts: course and relationship to treatment outcome, Arch. Gen. Psychiatry, 39:151-156 (1982).

21. K. Vereby, ed., Opiods in Mental Illness: Theories, Clinical Observations, and Treatment Possibilities, Ann. N. Y. Acad. Sci., 398:1-512 (1982).

22. C. Treece, B. Nicholson, DSM-III personality type and dose levels in methadone maintenance patients, J. Nerv. Ment. Dis., 168:621-628 (1980).

23. B. Nicholson, C. Treece, Object relations and differential treatment response to methadone maintenance, J. Nerv. Ment. Dis., 169:424-429 (1981).

24. E. J. Khantzian, C. Treece, DSM-III psychiatric diagnosis of narcotic addicts: recent findings, Arch. Gen. Psychiatry (in press).

25. S. J. Blatt, W. Berman, S. Bloom-Feshback, et al., Psychological assessment of psychopathology in opiate addicts, J. Nerv. Ment. Dis., 156-165 (1984).

26. B. J. Rounsaville, W. Glazer, C. H. Wilber, et al., Short-term interpersonal psychotherapy in methadone-maintained opiate addicts, Arch. Gen. Psychiatry, 40:629-636 (1983).

27. G. E. Woody, L. Luborsky, A. T. McLellan, et al, Psychotherapy for opiate addicts, Arch. Gen. Psychiatry, 40:639-645 (1983).

28. A. T. Butterworth, Depression associated with alcohol withdrawal: imipramine therapy compared with placebo, Q. J. Stud. Alcohol., 32:343-348 (1971).

29. J. E. Overall, D. Brown, J. D. Williams, et al., Drug treatment of anxiety and depression in detoxified alcoholic patients, Arch. Gen. Psychiatry, 29:218-221 (1973).

30. F. M. Quitkin, A. Rifkin, J. Kaplan, et al., Phobic anxiety syndrome complicated by drug dependence and addiction, Arch. Gen. Psychiatry, 27:159-162 (1972).

31. F. H. Gawin, H. D. Kleber, Cocaine abuse treatment, Arch. Gen. Psychiatry, 41:903-908 (1984).

32. E. J. Khantzian, Self-selection and progression in drug dependence, Psychiatry Digest, 10:19-22 (1975).

33. E. J. Khantzian, J. E. Mack, A. F. Schatzberg, Heroin use as an attempt to cope: clinical observations, Am. J. Psychiatry, 131:160-164 (1974).

34. E. J. Khantzian, Impulse problems in addiction: cause and effect relationships, in "Working With the Impulsive Person," H. Wishnie, ed., Plenum, New York (1979).

35. J. V. Spotts, F. C. Shontz, "The Life Styles of Nine American Cocaine Users," National Institute on Drug Abuse, Washington, D.C. (1977).

36. E. J. Khantzian, F. Gawin, H. D. Kleber, et al., Methylphenidate treatment of cocaine dependence - a preliminary report, J. Substance Abuse Treatment, 1:107-112 (1984).

37. E. J. Khantzian, N. J. Khantzian, Cocaine addiction: is there a psychological predisposition?, Psychiatric Annals, 14(10):753-759 (1984).

38. E. J. Khantzian, An extreme case of cocaine dependence and marked improvement with methylphenidate treatment, Am. J. Psychiatry, 140:784-785 (1983).

PSYCHOTHERAPY FOR COCAINE ABUSERS

Kathleen M. Carroll, Daniel S. Keller, Lisa R. Fenton, and Frank Gawin

Yale University
New Haven, Conn.

In spite of the burgeoning demand for treatment of cocaine abuse and the explosive growth of treatment centres, there is as yet no concensus on optimal treatment strategies for cocaine abuse. The reason for this lack of concensus is twofold: First, while a variety of promising treatments for cocaine abuse have recently been proposed (Anker and Crowley, 1982; Rounsaville, Gawin and Kleber, 1985; Gawin and Kleber, 1984; Gold, Washton, and Dackis, 1985; Tennant and Rawson, 1982; Siegel, 1982; Khantzian, 1983), none has as yet been systematically evaluated. Second, treatment-seeking cocaine abusers appear to be heterogeneous on a variety of dimensions, particularly the level of the severity of use and the psychiatric diagnoses (Helfrich, Crowley, Atkinson and Post, 1983; Schnoll, Karrigan, Kitchen, Daghestani, and Hansen, 1985; Kleber and Gawin, 1986; Weiss, Mirin, Michel and Sollugub, 1983). Given such heterogeneity, it follows that no single treatment will be identified as superior for all cocaine abusers.

There is one consistency across current treatments. Psychotherapy, in some form, is provided in most existing cocaine treatment programmes and is generally recognized as an essential component of treatment (Resnick and Resnick, 1984; Siegel, 1982; Wesson and Smith, 1985; Kleber and Gawin,1984). Furthermore, many clinicians contend (Rawson, Obert, McCann and Mann, 1986; Siegel, 1982; Wesson and Smith, 1985; Anker and Crowley, 1982; Gold, et al., 1985; Resnick and Resnick, 1984) that psychotherapeutic treatment alone may be sufficient for many cocaine abusers, in contrast to opiate or alcohol abusers, where purely psychotherapeutic efforts have been perceived as insufficient (Rounsaville, Glazer, Wilbur, Weissman and Kleber, 1983; Woody, Luborsky, McLellan, O'Brien, Beck, Blaine, Herman and Hale, 1983).

Many of the psychotherapies used in the treatment of cocaine abuse have been borrowed without alteration from the treatment of opiate or alcohol dependence. While there are many treatment similarities across types of substance abuse, there are many psychotherapeutic issues specific to cocaine abusers that prior treatments have not addressed. This chapter will describe general principles for psychotherapeutic treatment of cocaine abusers, describe the theory and technique of the dominant models of psychotherapy now applied to cocaine abuse, as well as propose an integrated approach where different models and techniques are incorporated at various stages of treatment in accordance with the specific needs of the patient. The treatments described below are based on current practice and developed over the past four years at the Yale Cocaine Abuse Treatment Programmes. Systematic evaluation of the approaches described has recently begun, thus further research should shortly clarify for the first time the short- and long-term efficacy of these approaches in cocaine abusers as well as to identify variables associated with optimal patient-treatment matching.

GENERAL TREATMENT CONSIDERATIONS

Defining the Need for Treatment

Cocaine abuse has been considered a psychological phenomenon; thus a variety of psychotherapeutic models for its treatment have been offered. Most of these treatments described below, share three fundamental goals. These include (a) helping the abuser to recognize the deleterious effects of cocaine abuse and to accept the need to stop; (b) helping the abuser to manage cocaine use and impulsive behaviour in general; and (c) helping the abuser to recognize the function that cocaine has served in his/her life and to develop strategies for meeting these needs without drugs (Rounsaville, et al. 1985).

We define as needing treatment any cocaine abuser who finds that he or she cannot curb cocaine use despite problems arising from that behaviour. This definition is broader than DSM-III criteria for cocaine abuse, which include a pattern of pathological use, impairment in social functioning due to cocaine use, and duration of the disturbance of at least one month (APA, 1980). Our definition is more consistent with the proposed DSM-III-R criteria for cocaine dependence which include: (1) repeated effort or persistent desire to cut down or control substance use; (2) frequent intoxication when the individual is expected to fulfill other obligations or when substance use is hazardous; (3) tolerance; (4) withdrawal; (5) frequent preoccupation with taking the substance; (6) forsaking other important social or occupational activity in order to use the substance; (7) unintentional, unplanned, excessive use of the substance; and (8) continued use of the substance despite problems arising from its use (Rounsaville, Spitzer and Williams, 1986).

Inpatient versus Outpatient Treatment

There is still controversy concerning both the relative efficacy and the indications for inpatient versus outpatient treatment for cocaine abuse. Unlike alcohol and opiate dependence, acute cocaine withdrawal syndromes do not require inpatient hospitalization or present physiological danger. Some clinicians (Anker and Crowley, 1982; Gawin and Kleber, 1984; Rawson et al., 1986; Resnick and Resnick, 1984) believe that outpatient treatment may be adequate for the majority of cocaine abusers. However, Siegel (1982) and many others (Washton et al., 1985) strongly favour hospitalization for the initial "detoxification."

No convincing evidence supporting either side exists thus far. However in one uncontrolled study, Rawson et al.(1986) followed 83 cocaine users who self-selected either outpatient, inpatient, or no formal treatment. At eight month follow-up, significantly fewer subjects in outpatient treatment returned to at least monthly cocaine abuse, compared to those who chose inpatient or no formal treatment (13% vs 43% and 47%, respectively). Despite methodological shortcomings, this study offers preliminary evidence that outpatient treatment for cocaine abusers is not only a viable form of treatment, but also may be superior in preventing relapse. This is consistent with many outcome studies of abusers of other substances, which have not demonstrated the superiority of inpatient or more "intensive" treatments (Miller and Hester, 1986). Furthermore, the lack of demonstrated evidence for the superiority of inpatient treatment and the increased accessibility of outpatient care warrant continued emphasis on more cost-effective outpatient programmes for the increasing numbers of cocaine abusers seeking treatment.

Cocaine's ever increasing supply and availability dictate that treatment focus on making the drug psychologically unavailable, as prolonged physical unavailability is almost impossible to achieve. A particular advantage of inpatient treatment is its temporary effectiveness in removing the cocaine abuser from ready access to the drug, particularly in cases of very severe abuse. However, it is well known that relapse following hospitalization is quite high. Treating abusers exclusively as inpatients may not prepare them adequately for when they return to their usual setting. Outpatient treatments enable assessment and treatment of the abuser in the context of cocaine availability as well as amidst the conditioned psychological, social, and environmental cues that maintain or precipitate cocaine craving and abuse. It is our impression that outpatient treatment facilitates gradual exposure to these cues and stressors while supporting abstinence. This is akin to a period of "extinction" in animal conditioning studies, and may be a necessary prerequisite for long-term abstinence. Hence, we feel outpatient care is a requirement for cocaine abusers, whether or not inpatient treatment occurs. Hospitalization may, in cases where abstinence could have been initiated as an outpatient, simply delay confrontation with fundamental issues.

Thus far, the only widely accepted factors indicating need for inpatient treatment are (a) severe depression, suicidal ideation, or psychotic symptoms persisting beyond the cocaine "crash" and (b) repeated outpatient failures (Kleber and Gawin, 1986). Other clinical indicators for the consideration of inpatient treatment have been described by Gold et al. (1985) and include: (1) chronic freebase or intravenous use; (2) severe impairment of psychosocial functioning and/or lack of family or social supports; (3) medical or psychiatric complications; (4) concurrent physical dependency on other drugs; and (5) inability or unwillingness to stop using cocaine as an outpatient, or lack of motivation. Cohen (1981) has suggested that another indicator is ready access to large amounts of cocaine, as is the case with most users who are also dealers. It is our impression that most users of ready-to-smoke freebase cocaine (crack) may require a period of hospitalization before they can meaningfully participate in outpatient treatment due to the intense craving produced by freely available, ready-to-smoke freebase cocaine (crack).

Pretreatment Assessment

A comprehensive, multidimensional pretreatment assessment is essential for all abusers entering treatment. The assessment should facilitate appropriate placement and enable the therapist to tailor a treatment plan for the patient, establishing a hierarchy of problems to be addressed in treatment. The assessment should include the following: (1) current level of severity of cocaine use, with a detailed description of route, the history of cocaine use, changes in the pattern of use as well as the development of dyscontrol. The clinician should keep in mind that route of administration is not necessarily associated with level of severity (Helfrich et al., 1983; Gawin and Kleber, 1984). (2) Presence and frequency of use of other psychoactive substances. The clinician should be aware that multisubstance use is the rule, not the exception in cocaine abuse (Chitwood, 1985; Schnoll et al., 1985; Schuster and Fischman, 1985). CNS depressants, particularly alcohol, opiates, and barbiturates, frequently accompany cocaine use to medicate the dysphoric side effects. Physiological dependence on another substance dictates inpatient treatment. (3) A thorough description of previous periods of abstinence, including circumstances, dates, length, and precipitants of relapse. (4) History of previous treatment attempts. (5) A mental status exam and detailed psychiatric and family history. Assessment of concurrent psychopathology should be made carefully, attempting to distinguish between transient, acute cocaine intoxication or "crash"-related abstinence symptomatology which closely mimic several Axis I disorders (Gawin and Kleber, 1986), and more enduring symptomatology. The most commonly diagnosed Axis I disorders in cocaine abusers are affective disorders: Major Depressive, Dysthymic, Bipolar and Cyclothymic disorders (Gawin and Kleber, 1986; Helfrich, et al., 1983; Weiss, et al., 1983). Personality disorders are also frequently seen in cocaine abusers, most commonly borderline, narcissistic and antisocial types (Gawin and Kleber, 1986). Accurate diagnoses dictate appropriate treatment consideration, and family history can often facilitate detection of underlying Axis I psychiatric disorders. Also to be assessed are: (6) Existence and level of family support. (7) Educational and occupational history, including current level of stress from cocaine-related indebtedness. (8) Legal history, probationary status and pending legal stressors. (9) The user's determination of precipitants and motivations for treatment, as well as his/her expectations from treatment, and (10) A comprehensive physical examination, including EEG.

The Addiction Severity Index (ASI) (McLellan, Luborsky, Woody and O'Brien, 1980) is the briefest comprehensive evaluative instrument for the assessment of cocaine abusers. The ASI is a brief, structured clinical interview that uses objective information to make subjective ratings in six areas commonly affected by substance use including substance use, psychological, medical, family/social, employment/support, and legal problems. Thus far, no superior quantitative measure of drug-induced dysfunction has emerged.

Prognostic Factors

Prognostic factors associated with outcome of psychotherapeutic treatment of cocaine abusers have so far not been systematically evaluated. Predictor studies have been carried out in other areas of addictive behaviours (Miller and Hester, 1986) and given apparent similarities across such behaviours (Brownell, Marlett, Lichtenstein, Nilson, 1986; Miller and Hester, 1986), limited generalization to cocaine abusers seems appropriate at this time.

Our experience is that several factors appear to predict better outcome of psychotherapy in outpatient cocaine abusers. These are, however, only clinical impressions that must be substantiated with further research. These factors include: (1) Internal, rather than external motivation for treatment. Self-referred abusers who acknowledge their inability to control their cocaine use, and experience some psychic distress or guilt, in most cases do better in psychotherapeutic treatment than those who deny or minimize the extent of their abuse, and come into treatment only under pressure or coercion by their families or the courts. (2) In general, abusers with less severe, less frequent, or less chronic cocaine and other substance abuse do better in outpatient psychotherapy than more severely impaired abusers. (3) Previous ability to maintain abstinence of at least several weeks' duration is a positive prognostic sign. (4) As with other types of substance abuse (McLellan, Luborsky, Woody, O'Brien and Druley, 1983; Rounsaville, Tierney, Crits-Christoph, Weissman and Kleber, 1982; Vaillant, 1973), absence of concurrent psychopathology is generally a positive prognostic sign. However, the type of co-existent psychiatric disorder may be more important prognostically than its presence alone. When cocaine abuse develops as a self-medication for Axis I psychiatric disorders, appropriate pharmacotherapy often ends cocaine abuse (Gawin and Kleber, 1984). The desire for relief from depressive symptomatology may motivate some abusers to stay in treatment and such patients may thus have a better outcome than patients with no Axis I diagnosis. (5) Other positive prognostic signs include presence of family support, and good premorbid psychosocial functioning, particularly a stable employment history (Rounsaville et al., 1982).

(6) "YAVIS"-like qualities which are good prognostic signs for participation in psychotherapy in general, bode well for psychotherapy with cocaine abusers. In particular, patients who seek treatment in order to bolster self-control over cocaine abuse, who realize treatment can be a long and difficult process, who can tolerate ambiguity and setbacks, and who have the ability to detach themselves and act as observers of their own behaviour, do better in psychotherapy than abusers who do not assume responsibility for the control of their own behaviour, but depend on the therapist to provide them with a quick "cure" or who cannot readily achieve a self-observative stance.

When evaluating an abuser for treatment, a preponderance of negative prognostic factors should lead the therapist to consider more intensive structured approaches such as inpatient or even long-term residential treatment. However, patients with a preponderance of poor prognostic factors may do poorly regardless of the type of treatment they receive. For example, a chronic freebase smoker with an antisocial personality disorder and a long legal history may not respond to any treatments. More work is needed on developing effective treatments of difficult subpopulations of cocaine abusers who may represent a substantial proportion of treatment failures.

Attaining Abstinence

In general, early psychotherapy with cocaine abusers is highly structured. It is presented to patients as open ended, but focused primarily on the attainment and maintenance of abstinence. In our experience, it is useful to devote the first 12 to 16 weeks of treatment almost exclusively to the control of cocaine use. Other issues invariably emerge, but these are treated as secondary contributors to the primary problem of cocaine abuse. By the time most abusers present for treatment, their lives are dominated by the procurement and ritualized use of cocaine, with repetitive cycles of cocaine binges and crashes. Thus, it is unlikely that meaningful work can be done on these secondary issues until the cocaine abuse is brought under control.

Some abusers find it relatively easy to attain short periods of abstinence through application of simple self-control and avoidance strategies, such as destroying their cocaine paraphernalia, breaking off contacts with dealers, etc. Such individuals often present for treatment having been abstinent for several weeks. With such individuals, Relapse Prevention, as described below, and exploration of other issues can begin immediately.

Other abusers find it difficult to break the unceasing use or binge-crash cycles to attain abstinence even for short periods of time. While "abstinence" of 4 or 5 days, or even 10 days is common in cocaine abusers, abstinence of over two weeks is uncommon, so we employ this period (of greater than 14 days abstinence) to determine if initial abstinence has been attained. With actively abusing individuals, it is necessary to institute interventions to establish abstinence as quickly as possible. "Common sense" measures should be used to make cocaine as inaccessible as possible. Therefore, all remaining cocaine caches should be thrown away (not sold), paraphernalia destroyed, and cocaine contacts broken. Drastic lifestyle changes may be necessary; the abuser should avoid drug-using friends, change employment in fields where cocaine is routinely used and sometimes separate from a spouse who refuses to stop using cocaine.

While self-control of cocaine use is the eventual goal of treatment, it is often helpful with such individuals to establish temporary external controls. This may take the form of having significant others "police" the abuser's behaviour, through taking control of their financial affairs, demanding the abuser account for time not spent at home or at work, etc. The therapist can also impose such external control through frequent contacts with the abuser during the early phases of the treatment, daily urine monitoring, and the use of contingency contracts (Anker and Crowley, 1982; Boudin, 1972) which are described in detail below.

Coping with Ambivalence

Despite the negative consequences of continued cocaine use, cocaine is an extremely powerful reinforcer (Aigner-Balster, 1978; Goldberg, 1973; Griffiths, Findley, Brady, Dolan-Gutcher, Robinson, 1975; Schuster and Johanson, 1980). It is therefore not surprising that many abusers are ambivalent about giving up cocaine-induced euphoria and completely terminating all cocaine use. Some abusers express such ambivalence through leaving "their foot in the door" to cocaine abuse by not breaking contacts with dealers, by continuing to hide their cocaine abuse from their families, by retaining cocaine paraphernalia, etc. Such ambivalence can sabotage treatment and must be acknowledged and dealt with directly by the therapist. The therapist should acknowledge that giving up cocaine is difficult, but necessary. Reiterating the negative consequences of continued cocaine use and having the abuser generate a list of reasons for giving up cocaine can be useful. (Later, reference to this list by the patient when confronted with a craving or temptation can also be helpful in preventing a slip.) Often by the time treatment is sought, the abuser views cocaine as the only reinforcing or pleasurable activity in which he/she can engage. Other reinforcing activities (such as work, exercise, hobbies, sex, or social contact) may not be immediately available to the user by virtue of his/her chronic use (as in the abuser's spouse who has left in disgust, or the abuser who has been fired due to repeated absences), or these activities may no longer be perceived as being pleasurable. The therapist can remind the abuser that other activities were enjoyed before the onset of cocaine abuse, and with stable abstinence, pleasurable activities can and will be rediscovered in time.

Ambivalence can also be expressed as the belief that he/she will be able to return to "controlled" or "recreational" cocaine use after abuse has been brought under control. It has been our experience that this is rarely if ever, possible once an individual has crossed the line from controlled to abusive cocaine use, even if only episodically. This should be made explicit in the early phases of treatment. Similarly, usage of psychoactive substances, particularly alcohol, have frequently been implicated as precipitants of relapse (Kleber and Gawin, 1986: Rawson et al., 1986). Abstinence from other psychoactive substances also precludes the possibility of sympton substitution.

Preventing Premature Dropouts

Treatment dropouts are high across the addictions and treatment of cocaine is no exception. While duration and frequency affects for outpatient treatment in cocaine abuse have not yet been systematically evaluated, it is advisable to retain patients in treatment as long as possible.

New patients who have never before been in psychotherapy frequently have unrealistic expectations for length of therapy, often expecting the entire course of treatment to last only a matter of weeks (Geller and Schwartz, 1972). If dramatic results are not achieved immediately, the uninformed patient may become discouraged. This may in some cases precipitate premature termination or relapse. Thus, at the outset of therapy the patient is told that treatment, like any change process, can be lengthy and at times difficult. It is unrealistic for a disorder such as cocaine abuse, which may have persisted for years before the abuser sought treatment, to be resolved in a matter of weeks.

At the beginning of treatment abusers often attain two or three weeks of abstinence as a result of great enthusiasm and conviction. Assuming they are "cured" and therefore have no further need of treatment, they announce their intention to leave treatment. The patient should be told that such periods of abstinence are but a good start towards prolonged abstinence. It is essential to remind them of the lessons of their personal history, as most abusers have previously experienced brief periods of abstinence followed by relapse, or if they have had no prior abstinence period, of the collective clinical experience. Longer treatment is necessary to help the abuser successfully negotiate several episodes of exposure to conditioned cues, high cravings, etc., without resorting to cocaine use. In fact, a patient who relapses during treatment but learns to cope effectively in the "safe" environment of treatment may be better prepared for the difficulties inherent in long-term abstinence than the abuser who has become isolated, and never experienced slips or high cravings during treatment.

Education

The unfortunate history of misinformation about cocaine by street culture and by the scientific community alike resulted in the popular perception of cocaine as a safe, non-addicting drug which persisted until recently. Studies done during the 1960's and 1970's which employed animal models to demonstrate cocaine's unusual strength as a reinforcer (Aigner-Balster, 1978; Goldberg, 1978; Griffiths et al., 1975; Pickens and Thompson, 1968) and warning of its compelling abuse liability were given only passing mention in recent descriptions of human abuse patterns (Grinspoon and Bakalar, 1976). Despite the change in these beliefs through the recent dissemination of more accurate information, patients often carry misconceptions that may hinder treatment. These misconceptions should be explored and corrected by the therapist if necessary. Simply providing accurate information on the neurochemical actions of cocaine, its acute effects and after effects as well as the polyphasic consequences of chronic cocaine use can do much to allay unwarranted fears, and alter magical thinking and the misattribution of mythological powers to cocaine. Observations of cocaine abusers suggest that craving may be the most intense in the period of anhedonia following resolution of the crash (Gawin and Kleber, 1986), when the abuser may feel him/herself "out of danger." Describing and explaining the psychological and somatic symptomatology associated with the various stages of the cocaine abstinence syndrome, from "crash" to euthymia to anhedonia (Gawin and Kleber, 1986) and including the different phases of relapse liability associated with these stages will provide a context for coping effectively with the symptoms as they emerge and thus may prevent relapse.[1]

INDIVIDUAL PSYCHOTHERAPY APPROACHES

Cognitive-Behavioural Models

Many aspects of substance abuse can be readily conceptualized in learning theory terms, relating reinforcement strength to learned behaviour. Such conceptualizations seem particularly appropriate for cocaine abuse. There are many animal studies that demonstrate: (a) cocaine's unusually high strength as a reinforcer (Aigner-Balster, 1978; Goldberg, 1973; Griffiths,

---

1.   Subtypes of cocaine abusers may also benefit from specific types of medical information. For example, intravenous users, if not already aware, should be told about the risks associated with this route of administration, such as AIDS, hepatitis, thromboses, endocarditis and cellutitis, and be evaluated accordingly. Chronic freebase smoking has been associated with pulmonary dysfunction (Weiss, et al., 1981). Female abusers, most of whom are in childbearing years, should be made aware of recent reports of neonatal risk (Chasnoff, Burns, Schnoll, and Burns, 1986).

et al., 1975; Schuster and Johanson, 1980); (b) its similarity to other, more traditional reinforcers, such as food and water, in shaping and maintaining operant behaviours (Dougherty and Pickens, 1973; Pickens and Thompson, 1968; Woods, 1977); and (c) cocaine self-administration is a behaviour under stimulus control (Stewart, 1983). Such studies have also provided an animal model of relapse by demonstrating that once cocaine self-administration has been extinguished in animals, it can be reinstated through a single forced injection of the drug, or "prime" (Gerber and Stretch, 1975; Pickens and Harris, 1968; Pickens and Thompson, 1971) or , more importantly, through presentation of a conditioned stimulus previously paired with cocaine reinforcement, such as an auditory tone (DeWitt and Stewart, 1981).

While there is ample evidence supporting behavioural and learning models of cocaine abuse and relapse, such data have rarely been incorporated into comprehensive treatments for cocaine abusers. In the one published account of such a treatment, Anker and Crowley (1982) used contingency contracting in an attempt to make cocaine less reinforcing to the abuser by focusing on its formidable adverse consequences. In that study, subjects were asked to sign treatment contracts that stipulated "severely punishing" adverse consequences contingent on a cocaine-positive urine report. Such consequences were individualized and usually involved notification of the subject's employer, licensing board, or the Drug Enforcement Agency of the subject's cocaine use. The success of this approach was limited: only 50% of the original sample agreed to sign the contracts, perhaps selecting for less impaired abusers. Although 81% of those entering into a contingency contract were abstinent during the three-month contract, 50% relapsed after completion of the contract (Crowley, 1982). While we recognize the potential value of treatment focusing on the adverse consequences of cocaine use, we believe that this particular behavioural approach is too limited and unlikely to be successful with the general population of cocaine abusers for several reasons. First, such programmes are unattractive to many potential clients, as evidenced by the refusal of half of Anker and Crowley's sample to enter into contracts. It is likely that the patients who refused to sign treatment contracts were heavier abusers and aware of their difficulty in maintaining complete abstinence. Second, this approach may be unnecessarily punitive: in the likely event of a slip during treatment, loss of employment, licensure, etc. may result, thereby further reducing needed patient resources and severely damaging the helping alliance with the therapist. Third, the ethical considerations raised by these punitive procedures may make them unwarranted, particularly in the light of their low efficacy (only 20% of the original group was abstinent at follow-up). Finally, while the Crowley study is important because it represents the first application of behavioural treatment principles to the treatment of cocaine abusers, self-selection effects may have accounted for differences between comparison groups because the study lacked random assignment.

We believe that positive contingencies may be more useful in encouraging adaptive behaviour and terminating cocaine abuse. Positive contingencies may involve, for example, having the abuser deposit a sum of money controlled by a significant other, at the beginning of treatment. A portion of that sum is then returned to the patient at pre-set intervals as long as urinalysis indicates abstinence is sustained.

Relapse Prevention

Background. Marlatt and colleagues (Cummings, Gordon and Marlatt, 1980; Marlatt and Gordon, 1980) have extended cognitive-behavioural principles to address the significant clinical problem of relapse after treatment. Although Marlatt's approach was originally formulated for the treatment of alcohol abuse, it is clinically applicable across most addictive behaviours (Marlatt and Gordon, 1985), including cocaine abuse.

A relapse prevention approach to the treatment of substance abusers is attractive for several reasons. First, many existing treatments focus mainly on reducing cocaine use, but offer little to facilitate enduring change or to decrease susceptibility to abuse. Relapse prevention approaches offer skills and general coping strategies designed not only to prevent relapse, but to create a general approach to cope with the problems and stressors that are inevitable in the abuser's life. Second, other psychotherapeutic approaches often tend to stress abstinence and to ignore relapse. Conversely, relapse prevention strategies (a) focus specifically on relapse, (b) teach skills that will help the individual identify and cope with high-risk situations that may precipitate relapse, and (c) recognize that occasional slips do occur and provide an alternative means of conceptualizing such slips, i.e., relapses represent a failure to cope effectively with a situation rather than a weakness or loss of control in the individual. Third, rather than relying on external controls, relapse prevention strategies are based on the assumption that the individual can gain control over and has responsibility for his/her own behaviour and substance use.

Related to this concept is the idea that individuals can be taught to be their own therapists and are not necessarily dependent on indefinite programme attendance or group membership to maintain abstinence. Fourth, these programmes promote education and awareness of drug effects, the role of substance use in the individual's life, the processes underlying relapse or continued substance use. Fifth, recognizing that no single treatment may be best for all individuals, such an approach can be individualized, focusing on the specific situations and problems likely to be troublesome to the individual and helping the individual devise specific strategies for coping with them. Finally, cognitive-behavioural approaches are flexible, incorporating and integrating the best elements of several approaches. For example, Relapse Prevention integrates (a) behavioural principles that substance abuse is an overlearned behaviour amenable to extinction procedures, (b) awareness of the role of neuropharmacological factors in continued substance use, and (c) the role of maladaptive cognitive processes in relapse (Pechacek and Danaher, 1979).

While not yet supported by a wealth of clinical data, Relapse Prevention techniques have been adopted to be used with cocaine abusers (Carroll, 1985; Gold, 1985) and appear quite promising with this group (Gold et al., 1985; Rawson et al., 1986). In addition to the advantages listed above, this approach may be particularly effective in the treatment of cocaine abusers because (a) cocaine hydrochloride is a drug with a limited, relatively mild withdrawal syndrome, (b) a self-control, coping-skills rationale is likely to be attractive to many cocaine abusers who tend to be better educated, middle-class, goal-oriented and cognitively intact in comparison with other groups of substance abusers, (c) it is a flexible approach, adaptable to the needs of different cocaine abusers, who tend to be a heterogeneous group and (d) as animal studies indicate, conditioning effects may be especially important in sustaining stimulant use, and relapse prevention strategies include specific interventions designed to extinguish associations between high-risk situations, conditioned craving, and drug use.

Description. Marlatt's theory extends cognitive-behavioural theories of drug abuse to address the significant clinical problem of relapse to substance use following treatment. The relapse curve (Hunt and Matarazzo, 1970) illustrates that within the first three months after treatment, approximately 60-70% of all individuals who achieve abstinence by the end of treatment will relapse. Although no data are available yet on the process of relapse after treatment of cocaine abuse, the striking similarity of the relapse curve across types of substance abuse suggests that cautious generalization is appropriate at this time (Brown et al., 1986; Marlatt and Gordon, 1985).

83

Briefly, Marlatt's theory states that during a period of self-imposed abstinence from an abused substance, the individual will feel a sense of control, or self-efficacy (Bandura, 1977) over his/her behaviour. However, the individual may be confronted at any time with a number of high-risk situations likely to precipitate a relapse. These high-risk situations are discriminative cues for reinforced responding and can elicit craving. Typical high-risk situations include exposure to drug paraphernalia or drug-using friends, high levels of interpersonal stress, and negative emotional states. When confronted with such a situation, if the individual executes an effective coping response (e.g., leaving the situation, repeating supportive self-statements), a relapse is unlikely to occur. Each time a high-risk situation is successfully negotiated, self-efficacy will increase, making future relapses less likely. Alternatively, if the individual does not have the appropriate coping response in his/her behavioural repertoire or is unable to execute a coping response due to some inhibiting factor, the individual will experience a reduced sense of self-efficacy. That, coupled with positive expectations for drug effects, may lead to a slip.

Immediately after a slip, the individual may experience the Abstinence Violation Effect (AVE), an emotional reaction that can lead to a full relapse after a single slip. The Abstinence Violation Effect occurs because the slip is in direct conflict with the individual's self-image as an abstinent individual. When experiencing the AVE, the user attributes the slip to a lack of willpower or some other personal shortcoming rather than to not having made the appropriate coping response. The individual will then feel he or she has no control over the situation and a slip may thus become a full-blown relapse. Relapse Prevention theory has been described in further detail elsewhere (Marlatt and Gordon, 1985).

Techniques. In general, there are few differences between Relapse Prevention for cocaine abusers and other cognitive-behavioural treatments in terms of theoretical orientation, general approaches to treatment and characteristics of the patient/therapist relationship (cf Beck and Emery, 1977). The treatment is structured and goal-directed with large problems and projects being divided into a series of smaller tasks. The therapist is active and directive, clearly elucidating the rationale, structure, methods, and goals of the therapy to the patient throughout treatment. Definitions and explanations of "cognitions," "high risk situations," and other key terms are presented in the first session. Relapse prevention theory and classical conditioning of cues and craving are explained in detail. The therapist makes treatment as unambiguous as possible, explaining why specific tasks and homework assignments are given, with the explicit goal of teaching the patient to function as his or her own "therapist" outside the treatment sessions. A treatment plan is generated by the patient and therapist and updated regularly during treatment. The plan includes a prioritized list of problems to be targeted during treatment and short- and long-term goals for each of the problems.

This approach to treatment assumes that cocaine abuse is a behaviour disorder potentially under voluntary control by the individual. Every instance of cocaine abuse represents a decision on the abuser's part, albeit often automatic or difficult to identify. The therapist helps the abuser identify and control this decision-making process and the maladaptive cognitive processes that accompany it. The patient is taught skills that enable him or her to reduce and eliminate cocaine use. Recognizing that many cocaine abusers experience a multiplicity of problems, these skills are taught as general coping and problem solving strategies that can be applied to a variety of symptoms in addition to cocaine abuse.

Relapse prevention for cocaine abusers (Carroll, 1985) can be roughly divided into behavioural and cognitive strategies. The behavioural techniques are designed to (a) help the abuser alter the parameters of stimulus control, (b) help the abuser identify and manage the overlearned behaviours associated with cocaine abuse, and (c) develop and master adaptive coping behaviours. Cognitive strategies attempt (a) to alter the maladaptive cognitions that may be associated with cocaine abuse, and (b) to help the abuser manage the conditioned craving and relapse during treatment.

An important behavioural technique is the identification of high risk situations. This is done primarily through self-monitoring, where the abuser keeps an hourly written record of his activities, thoughts, feelings, and subjective ratings of cocaine "craving." The patient is encouraged to fill in the form hourly, and in particular when he or she is experiencing "craving." Such a record provides a wealth of data concerning the specific high risk situations likely to be most troublesome to the individual, be they thoughts, specific affective states, interpersonal situations, or instances of temptation. This procedure encourages the abuser to attend to internal cognitive and emotional processes as well as identify his/her individual, idiosyncratic pattern of cocaine craving and use. The self-monitoring forms are reviewed each week in depth during sessions. Specific high risk situations, patterns of cravings, activity level, coping response, and thoughts and feelings are identified and explored in detail. The individual's current repertoire of coping behaviours can also be identified through self-monitoring.

It is likely that by the time the abuser has reached treatment, through repeated use of cocaine in a variety of situations, cocaine use has become the single, overgeneralized coping response of the individual, extinguishing other, more adaptive responses or solutions (Khantzian, 1975). The acquisition of coping and problem solving skills (a) provides a behavioural alternative to cocaine use, (b) strengthens and increases the abuser's self-efficacy, and (c) provides a repertoire of skills that may be used to cope with several problem areas in addition to drug use. Thus, once the individual's high risk situations are identified, the abuser is taught to treat them as "red flags" or warnings for relapse and execute an appropriate coping strategy.

Adaptive coping mechanisms and problem-solving skills are initially taught and modelled by the therapist, rehearsed and mastered during treatment through role-playing and imagery. Patients are taught the components of problem-solving and the major coping modes. At the end of each session, the patient is asked to anticipate any high-risk situation that may be encountered during the next week (e.g., going to a party, being home alone) and to plan in advance what coping strategy will be used to avoid a relapse. Such advance planning often defuses a high risk situation before it occurs. One abuser, anticipating going to a bachelor party, knew cocaine would be used there. He realized it would be exceedingly difficult for him to attend the party and not use cocaine, and opted not to go to the party at all. Such advance planning helps control impulsivity as well. One young cocaine abuser who anticipated attending a concert, planned in advance to be the "designated driver" taking responsibility for driving his friends to and from the concert safely. He not only stayed abstinent, but observed the negative actions of his friends when they were high. Role playing, relaxation training, imagery techniques and lifestyle intervention (Marlatt and Gordon, 1985) may also be used depending on the needs of the individual.

Cognitive Restructuring

The cognitive restructuring procedures first used in the psychotherapy of depression (Beck, Rush, Shaw and Emery, 1979) can be applied to the maladaptive cognitions which may accompany cocaine abuse. These cognitive errors and distortions are identified during sessions as well as through self-monitoring and may include: overgeneralizing, selective abstraction, excessive responsibility, assuming temporal causality, self-reference, catastrophizing, dichotomous thinking, and absolute will power breakdown (Beck et al., 1979; Marlatt and Gordon, 1985). These cognitions are identified, challenged, and altered during treatment using techniques that have been well described elsewhere (Beck and Emery, 1977; Beck et al., 1979). There are two additional special cases of cognitive restructuring that are particularly important to address in the treatment of cocaine abusers: craving and relapse.

Craving. Craving has been identified as an important consideration in the treatment of cocaine abusers and has been implicated as a frequent precipitant of relapse (Gold, et al., 1985; Kleber and Gawin, 1986; Marlatt and Gordon, 1980; Wesson and Smith, 1985). Because of the relatively limited physiological withdrawal syndrome associated with the termination of cocaine abuse, withdrawal-based craving for cocaine would seem an unlikely cause for relapse beyond the first three to four days after the inception of abstinence. Episodes of intense subjective craving for cocaine are often reported weeks and even months beyond the termination of cocaine use. Such craving is often a classically conditioned phenomenon. For example, one IV user reported intense craving for cocaine while seeing boxes of cotton balls in a grocery store. Abusers can be educated about the role of classical conditioning in craving using the well-known example of Pavlov's dog as an illustration of classical conditioning and equating salivation to craving, food to cocaine, and the classically conditioning bell to factors paired with that patient's cocaine use. Abusers can be taught to use self-monitoring to associate the experience of craving with specific cues, treat these cues and cravings as "danger signals" and react with an appropriate coping strategy.

For environmental, externally imposed cues (such as exposure to objects associated with cocaine abuse, accidental contact with a dealer, being offered cocaine at a party, use of other psychoactive substances, etc.) coping strategies can include physically removing oneself from the situation, rehearsing the reasons why abstinence is necessary, and employing urge control techniques (Marlatt and Gordon, 1985). Urge control techniques include allowing the craving to occur without giving in to it, in the knowledge that the craving will decrease with time, becoming less frequent and less intense if the individual does not give in to it. More direct cue extinction procedures, where craving is elicited during sessions (through imagery or exposure to other cue-inducing stimuli such as tetracaine) and paired with relaxation or other states which are incompatible with use, are being explored in our clinics and in others (Gold et al., 1985; Resnick and Resnick, 1984).

For internally-based cues, such as subjective feelings of cravings elicited by negative emotional states or interpersonal conflict, cognitive relabeling of craving may be more appropriate. Through consistent repetitive pairings with cocaine use, certain internal states (depression, anxiety, rage reactions, etc.) or interpersonal situations (arguments with one's spouse) may be misinterpreted as pure craving. Identification and more accurate relabeling of these emotional or interpersonal states enables the abuser to apply a coping strategy more appropriate and more effective than cocaine abuse.

Management of Relapse. The other special case where cognitive restructuring can be particularly helpful in the treatment of cocaine abusers is that of relapse during treatment. As mentioned earlier, slips can elicit the Abstinence Violation Effect, where the slip is attributed by the user to some personal shortcoming. Many users believe that once he/she slips, a full-blown relapse is inevitable. Relapse prevention theory states that how the slip is perceived by the user is an important determinant of whether further cocaine use will ensue. Effective reframing and reattribution of slips may prevent them from becoming protracted relapses. Slips that occur during treatment are viewed as mistakes that occur during the learning process; as specific, unique events that can be reattributed to external controllable factors (Marlatt and Gordon, 1985).

It is our impression that one or more "slips" or brief episodes of relapse, are the rule rather than the exception in the outpatient treatment of cocaine abusers. If the patient is not adequately prepared for the occurrance of such a slip, it can be a devastating experience. The unprepared patient may assume the slip implies he/she is a failure as a patient, or devalue the treatment or the therapist, and leave treatment. In order to prevent slips from terminating the treatment or precipitating full relapses, the possibility of a slip should be discussed at the beginning of treatment. The therapist, while preparing the patient for the possibility of a relapse must be careful that he/she is not giving permission for a slip. Preparing for a slip can be compared to a fire drill (Marlatt and Gordon, 1985) where, although fires are rare, preparedness can prevent disaster in the event a fire occurs. The patient is instructed that a "slip" during treatment should not be viewed as a failure of the patient or the treatment, but as a mistake which can be used as an invaluable opportunity for learning and understanding the behaviour of the patient. While isolated slips can be invaluable in vivo learning opportunities, repeated slips during treatment should be viewed with caution, possibly indicating a need for revision of the treatment plan and escalation of treatment interventions. Clinically, we become concerned when abstinence intervals do not at least double between slips.

In the case of a slip or even a period of intense craving, abusers are taught to do immediate self-monitoring, identifying the events, cognitions, or feelings that led up to the slip, and to identify and institute a more effective coping mechanism. However, given both the illicitness and the danger of cocaine use, a programmed relapse, in which a relapse is planned in advance to occur at a specific time and place or relapse rehearsal as described by Marlatt and Gordon (1985) would be neither an ethical nor a practical intervention in this population.

Supportive-Expressive Treatment of Cocaine Abusers

Recent psychodynamic thinking regarding substance abuse has stressed that drugs are used to self-medicate painful affect states (Khantzian, 1985; Krystal and Raskin, 1970; Weider and Kaplan, 1969; Wurmser, 1978). In supportive-expressive treatment of cocaine abusers, we attempt to explore such affect states and their cognitive sequelae and relate these to interpersonal and intrapsychic conflicts.

In essence, the exploratory psychotherapeutic approach with which we treat cocaine abusers is based upon aspects derived from Luborsky's (1984) recent supportive-expressive model as well as the interpersonal model for brief psychotherapy with cocaine abusers (Rounsaville, Gawin and Kleber, 1985). In supportive-expressive treatment, the therapist attempts to support certain aspects of the patient's personality while exploring other areas in more depth. The question arises: which areas need support and which need expression? In our treatment of cocaine abusers, we focus on two overall goals: (1) achieving and maintaining abstinence, and (2) exploration of and insight into interpersonal and intrapsychic conflicts underlying the addiction. We therefore tend to support those attitudes, efforts and defenses utilized by the patient to become and remain abstinent, such as, for example, non-drug efforts to consolidate self-esteem. Here the efforts of the therapist tend to be mildly directive, suggestive, and where warranted, encouraging. We tend to explore, however, various emotion-laden areas

of the patient's life, such as issues of separation-individuation, depression, dependency, relations to significant others, etc. Here our efforts are
more non-directive, technically neutral and interpretive.

Since therapy is time-limited, the depth and breadth of our explorations
are necessarily more circumscribed and directed than with longer-term insight
oriented approaches. To minimize transference and to express core conflicts
in the here and now is our focus. Transference and resistance are explored but mainly as they relate to conflicts surrounding drug addiction.

We will now describe in greater detail our supportive-expressive interventions. Our generalized supportive efforts in achieving and maintaining
abstinence are not different from the techniques described for relapse prevention. Thus, in the following section we shall focus on the exploration
of core conflicts in terms of: (1) indications for supportive-expressive
treatment (2) the nature of treatment including early phases and initial
resistance, transference reactions, and termination, and (3) cocaine abuse
treatment of severe character disorders.

Exploration of Core Conflicts

Patients presenting with cocaine abuse often manifest a number of difficulties and conflicts related to their abuse, of which they are partially
or totally unaware. Among the more prominent difficulties are disturbances
in self-esteem, regulation of painful affects, and various problems with
respect to significant others. Usually such difficulties are highly interrelated.

All psychotherapeutic interventions will touch upon and illuminate
these problems in varying degrees. What differentiates supportive-expressive treatment from other approaches is that it focuses on unconscious factors contributing to these conflicts and ultimately to the cocaine abuse
itself. Furthermore, it attempts to understand and resolve such conflicts
by interpreting them as they are expressed in the therapeutic relationship
as opposed to providing strategies and other "rational" measures to manage
conflict.

It is important to note, however, that conflict is explored only partially, usually as it manifests itself at higher and surface levels. Our
treatment approach is geared mainly to bringing the cocaine abuse under the
patient's control through the understanding of its manifold functions, not
through alteration of character structure. This does not preclude further
and more in-depth psychotherapeutic work after our interventions, should
the patients desire it.

General Indications

As with other forms of psychodynamic therapy, those patients who make
the best candidates tend to be verbal, articulate, psychologically-minded
and highly motivated (Strupp, 1981). They have a fairly well differentiated
sense of self and others, and possess the capacity to internalize. They
frequently present with a sense that their problems with cocaine are related to other personal difficulties which they find troubling but which they
do not fully understand. Furthermore, they suspect that these difficulties
reside in themselves, to some extent, and they have a capacity to reflect
about them. In short, the patients tend to localize toward the neurotic
end of the psychopathological spectrum.

These qualities and attributes are easy to detect in an initial interview. Such patients will elaborate on questions asked by the therapist,
will make connections between their substance abuse and their psychological

life and will be able to tolerate the various contradictions and ambigui-
ties that arise or are pointed out to them by the therapist. As therapist,
one has the distinct sense that such queries and interventions produce tol-
erable intrapsychic distress.

Psychiatric diagnosis can frequently help to determine to what degree
supportive-expressive treatment is indicated. For instance, many patients
will present with a depression for which cocaine has served as a self-medication (Khantzian, 1985; Gawin and Kleber, 1986). In addition to acting as
an additional motivational spur, these depressions frequently are of an introjective, guilt-ridden type (Blatt, 1974, 1981), reflecting a greater
capacity for object-relatedness and internalization which, in turn, are
prognostically good indicators for the beneficial effect of supportive-expressive interventions. On the other hand, with severe psychopathology such
as schizophrenia and other psychotic disorders, cocaine abuse is better
treated as secondary to the psychosis.

Regarding the more severe character disorders, supportive-expressive
treatment may or may not be indicated depending on the available treatment
resources and the type of disorder. We have included a section below on
the character disorders typically seen in our unit.

Nature of Treatment

When one has made the determination to utilize some supportive-expressive approaches, the nature and extent of treatment need to be explained to
the patient. This is crucial in developing a healthy working alliance
(Greenson, 1967). For many patients presenting with cocaine abuse, even
those with considerable psychological talents, this is often their first experience with a psychological treatment. We explain to the patient the
reasons for choosing the treatment, its timing in relation to other conjunc-
tive treatment efforts and especially the open-ended and somewhat ambiguous
nature of exploratory treatment. Anticipating the development of resistances
which are likely to develop (such as not having anything to say) and the
development of transferential feelings toward the therapist, we emphasize
that such developments are normal and that our task is to understand their
meaning.

Early Phases and Initial Resistances

To facilitate moving in this direction, it is often helpful to refer
back to various issues and conflicts the patient has revealed in his history
and to inquire more deeply into these. In this manner, core areas of conflict can be further delineated, such as thoughts and feelings about one's
self or significant others and their relationship to the onset of and con-
tinuation of cocaine abuse patterns.

With certain patients, eliciting material in this fashion tends to be
relatively uncomplicated, especially at the outset of treatment. Nevertheless, sooner or later, resistances arise. One typical resistance is the
patient's statement that cocaine abuse, especially relapse or slips, has
nothing to do with his/her feelings or behaviour, but occurs accidentally.
One particularly beneficial way to penetrate such a reference is to focus
on how affective conditioned cues may have led to a slip. In our experience
slipping is often preceded by some painful affective state which if uncovered, can help the patient re-establish a connection between cocaine abuse
and him or herself.

For example, one patient described slipping. For some reason, unknown
to him, he had taken an alternate route home from work. On this route, he
stopped for a beer, where he happened to run into a friend, who happened to

have a gramme...etc. We examined what was going through his mind prior to
leaving work. He related in a distinctly bored tone of voice that he wasn't
going to do anything that night; he would merely go home, watch football,
then go to sleep. Drawing his attention to this tone of voice, the patient
realized that he had felt bored at work, that he frequently felt bored prior
to using cocaine, and that he felt a perpetual inner emptiness. In fact, one
of the things he liked about his friend was that the latter wasn't boring.
We were then able to utilize this knowledge not only as a "red flag" for the
onset of the desire to use, but also as a core emotion related to cocaine
onset to be explored further in therapy.

     Another typical resistance is to talk only about cocaine abuse, after
abstinence has been clearly attained, as a means to avoid talking about
other conflictual material. Here, with the proper preparations, one can in-
terpret that just as the patient has filled up his life prior to treatment
with cocaine, he or she now seems to be "filling up" the sessions with "cocaine."

     Of course, as in other psychodynamic work (Freud, 1912), resistances
arise throughout the course of treatment. In supportive-expressive treatment of cocaine abusers, it is important to focus on the patient's tendency
to resist in a particular manner -- to divorce himself (his conflicts- relationships, and feelings) from his use of the drug. Such resistances need
to be consistently interpreted.

Transference Reactions

     In the course of delineating core conflictual material and interpreting
the patient's resistances to exploring it, such conflicts and resistances
begin to penetrate the therapeutic relationship; that it, the patient
develops a transference to the therapist. Several words must now be addressed to the nature of the transference and its interpretation in this
approach to cocaine abuse treatment.

     We believe that solely intellectual understanding of affective-cognitive
and interpersonal conflicts in the patient's extra-therapeutic life will be
of little lasting value to him in his battle to overcome difficulties with
cocaine. Although it is true that cocaine has been used to cover over such
conflicts, lasting insight can be generated in the emotional here and now,
either in external relationships or in the affectively-charged relationship
to the therapist. If there is also a healthy working alliance with which
to attempt to understand such transferences, we believe that the patient
can achieve mastery and control over his previously unconscious motivations
for using cocaine, so in this treatment modality, transference reactions
are not minimized.

     The transference in supportive-expressive treatment is explored and in-
terpreted only in certain aspects and at certain levels of depth. Nevertheless, even within this limited sphere, affective, cognitive, interpersonal
and characterological aspects are highly interrelated and touched upon in
varying degrees, as illustrated with the following case example:

     A 32 year old married salesman with a two year history of intranasal
cocaine abuse presented for treatment in our unit. His cocaine abuse had
gradually increased over this period and he was experiencing severe financial difficulties despite the quite good living he made between his salary
and commissions and his wife's salary. They were now having difficulty
making payments on their "dream house."

     The patient further continued that when he told his wife of his problem
she had been deeply hurt. He had feared she would leave him, but was both

90

surprised and relieved to discover that, despite her hurt, she would support
him in his efforts to get well. He bitterly castigated himself for his
"thoughtless behaviour": How could he have been so disrespectful of his wife;
how could he now subject her to the financial burdens brought on by his
problem; and how good, understanding, selfless and supportive she was about
it all despite his disgraceful actions. It puzzled him that he could act
this way, as they had had a "great" marriage.

The patient had started using cocaine two years earlier to enable him
to work longer hours. He and his wife had just bought their new home at
that time. Actually, they had considered two houses, but had opted to buy
the significantly more expensive one. Initially, the patient maintained
that this was his idea, that he had liked the house when in actuality he had
had doubts about their ability to afford it. By this time, his wife also
wanted the more expensive home and they decided to buy it. However, unless
things changed financially, now they might even have to sell it.

As treatment unfolded, it became apparent that this was not an isolated
incident. At key points in their marriage, the patient had acceded to his
wife's desires for financial improvement. For example, shortly after they
were married, the patient had given up his desire to go to medical school in
order to make good money immediately. Again, this was his idea, although,
he added, his wife certainly appreciated their better financial circumstances.

The patient had a compulsive personality with rigid reaction formations
against the expression of anger. He could not tolerate conflict with his
wife. He would please her even if it meant sacrificing his desires. His
cocaine abuse expressed his unconscious hostility toward her in that it ultimately led to the possibility of denying her the expensive house, as well
as "hurting" her. At the same time, the cocaine abuse was a passive-aggressive assertion of his own desires in that he and his wife might have to
settle for a less expensive house after all.

Exploration of the issue of the house appeared to indicate other mean-
ings as well. The patient spoke of his father with awe, describing him as a
man who, despite his humble upbringing and poor education, now "owned his
own successful business." There were indications that to the patient, "owning" a house meant competing with his father. In fact, the patient had fan-
tasized from time to time about owning his own business and actually had had
several opportunities, but whenever it got around to doing something actively
he would lose interest. Parenthetically, his wife was always opposed to his
opening a business.

In exploring these issues with the patient, he had difficulty in accepting, even considering, that he ought to be angry with his wife. It was
only after he had unconsciously precipitated a "fight" with the therapist
that he was able to understand his feelings and behaviours toward her and
their relation to cocaine abuse. The patient had a particular payment plan
for treatment. He would pay a small percentage weekly, his insurance company making up the difference. His insurance company would pay the medical
costs directly to him and not to the clinic. As his outstanding balance
continued to mount, he claimed he had still not received the payments. In
highly defended ways, he would joke about the therapist being only an appen-
dage of the collection department, wondering whether the clinic was really
concerned about patients apart from their money. At one point, he enter-
tained thoughts of quitting because "treatment" did not seem to be going
anywhere. Another month rolled by with disguised hostile expressions.
These mounted as he explored problems in his marriage. Again asked about
the payments, he became openly hostile, and accused the therapist of not believing him. At this point, it was noted that this was the very first time

he had gotten angry with the therapist openly. The therapist wondered aloud
if it might not be due to something else in addition to being asked about
the money. He admitted then that he felt that the therapy was pushing him
too hard in the exploration of his marriage and he resented it. He feared
losing his relationship with his wife; now he admitted that his marriage had
been suffering for some time. Despite his being abstinent from cocaine for
over four months, he and his wife had not made love more than several times
during that period. In fact, their sex life had been poor prior to his using
cocaine. Maybe their marriage has not been so "great" after all. He was
then reminded that he had also gotten angry in the context of financial
matters. The patient then began to wonder if there was a similar process
in his buying the expensive house, his ensuing use of cocaine, and the ultimate financial problems. In the next sessions, he began to explore his fear
of non-acceptance, and the role of money and self-worth as well as his fear
of his own anger, how he used cocaine to control his fears and to simultaneously express his anger toward his wife. (Also during this period, he had
received a payment cheque which he "forgot" to bring in.)

Several features of this clinical vignette should be abstracted. First,
the promotion of insight occurred in relation to an affectively charged interaction with the therapist that parallelled a similar mode of relating in
the patient's life. Second, our focus was confined to the linking of cocaine
abuse to the patient's experience with anger and how this ramified into interpersonal conflicts. We believe that many other aspects of the patient's
personality could have been investigated - this man felt that any assertion
of his individuality was dangerous, as an expression of oedipal hostility
toward his father. Thus, becoming a doctor, owning a house, owning a business (like his father) had dangerous, aggressive meaning. However, these
and other aspects were not interpreted. Our focus remained on the here and
now conflicts as they contributed to cocaine use.

Termination

As in similar treatments, termination is a crucial part of the treatment
process. Cocaine abusers in supportive-expressive treatment have in certain
respects shifted their dependency on cocaine to dependency on the therapist,
including his/her advice, encouragement, and especially support. Simply
coming repeatedly to the treatment setting may have endowed it with deep,
personal meaning of which the patient may not be aware. The special feelings of cocaine have often been supplanted by the special feeling of succeeding in recovery which diminish with detachment from both the therapist
and the recovery programme. Now the patient must take a further step and
come to depend upon himself. To some extent, exploration of previous sepa-
rations, such as therapist vacations, will have paved the way for termination. Nevertheless, terminating is anxiety-inducing and needs to be
explored in detail. In our experience, the therapist must actively pursue
the issue of termination. There seems to be a tendency with our population
to deny its significance. Often, deeply ambivalent feelings lie just below
the surface. There may be a "recrudescence of symptoms," including slipping.
All such issues need to be examined in the light of the meaning of termination to the patient.

We have attempted to indicate the main features of supportive-expressive
treatment with cocaine abusers. Such treatment attempts to explore core
affective-cognitive and interpersonal conflicts at the same time it attempts
to support and sustain efforts at achieving and maintaining abstinence.
Core conflicts are interpreted within the affectively charged here and now
relationship to the therapist. Other, especially deeper, aspects of the
patient's personality and relationship to the therapist are not explored,
as our main focus is upon cocaine abuse. Supportive-expressive treatment
is most effective with higher functioning personalities using experienced

therapists. Although it is sometimes useful with certain intermediate and lower level character disorders if the staff resources and social support systems in the patient's extratherapeutic life are available. When chosen correctly, supportive-expressive treatment for cocaine abuse can result in post-abuse functioning superior to that preceding abuse.

Cocaine Abuse Treatment of Severe Character Disorders

As noted above, although we do not attempt to alter character structure with out supportive-expressive treatment to any significant degree, an understanding of the patient's character defense is quite helpful.

The types of patients we believe to be best suited for supportive-expressive treatment tend to cluster at higher levels of character pathology with compulsives, hysterics, and depressives predominating. However, we also treat a significant percentage of intermediate and lower level characters (Kernberg, 1976), most typically severe narcissistic, borderline, and antisocial. Inasmuch as specific psychodynamic treatment of borderline and narcissistic conditions has advanced in recent years, we will address the treatment of these characteristics in cocaine abuse populations.

The Role of Narcissism in Cocaine Abuse

As is well known, the narcissistic personality has received much attention by psychoanalytic thinkers in recent years (Kernberg, 1975; Kohut,1971; 1977). Due to severe disturbances in early object relations and ego development, narcissistic personalities have acquired a style of not depending on others, profound needs for self-sufficiency (Modell, 1976), grandiose feelings of self-importance, and manipulative-exploitative use of others to feed themselves with tributes. Although the need for the adoration of others often appears at first to be dependency, psychoanalytic work with such patients reveals that others are regarded as extensions of the narcissist's self-concept and are discarded the moment they no longer provide the needed tributes, or "narcissistic supplies" (Kernberg, 1975).

Such patients appear particularly predisposed to cocaine abuse in that the effect of the drug often produces feelings of grandiosity, omnipotence, and self-sufficiency while at the same time, covering over various forms of inferiority and low self-esteem. Additionally, cocaine may often help to regulate experiences of boredom and emptiness (Khantzian, 1985; Wurmser, 1978) which have been described in narcissistic patients (Kernberg, 1975).

At present, no systematic studies have been undertaken to examine the relationship between narcissism and cocaine. Furthermore, while it is our impression that we treat a number of narcissistic personalities in our unit for cocaine abuse, we also treat many other personality disorders. It may be cocaine appeals to specific aspects of various personality structures. Moreover, it should not be forgotten that all character structures possess narcissistic defenses to regulate self-esteem (Kernberg, 1975). Identifying narcissistic issues is not equivalent to diagnosis of narcissistic personality disorder. It may not only be that cocaine appeals to a variety of patients whose narcissistic defenses have been injured but that cocaine abuse and its associated dyscontrol and dysfunction injure narcissistic defenses.

Narcissistic Personalities

We have noted above that narcissistic personalities may be particularly vulnerable to cocaine abuse. Treatment of these patients by supportive-expressive means may or may not be indicated. In part, one has to evaluate time and resources available, particularly if one is working within a drug rehabilitation setting. Often these patients require a lengthy treatment when psychodynamic interventions are utilized. Modell (1975) has observed that such patients may require as long as a year within a therapeutic "holding environment" before they can begin to depend on the therapist and his or her interventions. Additionally, Kernberg (1975) suggests that for certain types of narcissistic personality disorder, full-scale unmodified psychoanalysis is the treatment of choice since supportive-expressive measures fail to mobilize their conflicts in the transference, due to their extremely rigid character defenses. Kernberg also considers that some narcissistic personalities, especially those whose surface functioning is at a fairly high level and who present with an acute crisis, may best be treated with supportive-reeducative procedures, such as those typical of substance abuse treatment, psychoanalysis being an option at some later time.

Some narcissistic patients presenting with an acute crisis of cocaine abuse do best by "learning" relapse prevention strategies from the therapist. When these "supplies" are no longer forthcoming, such narcissists will begin to devalue the treatment and the therapist, and feeling they no longer "need" help, will lose interest and "discard" the treatment. Kernberg (1975) has formulated the appeal of such learning as the narcissist's "unconscious effort to 'take away' from (the therapist) and make 'his own' what he would otherwise envy in (the therapist)" (p.337). Effective treatment for the problem of cocaine abuse need not make this process conscious, as opposed to effective treatment of character disorders.

For other narcissistic personalities, especially those functioning at an overtly borderline level, supportive-expressive treatment may be valuable providing that the requisite therapeutic resources are available. If such resources are limited, supportive-reeducative measures are the only option.

Borderline Personalities

Certain borderlines can respond well to supportive-expressive interventions for cocaine abuse, but again, a number of factors predispose to favourable outcome. First, available resources in time and effort are crucial: if limited, it is often best to treat such patients in a primary psychotherapy, with adjunctive cocaine abuse treatment; second, such patients should be verbal, articulate and motivated; third, it often helps if social support networks are available for managing this patient's propensity for acting out.

These patients will tend to make an intense attachment to the therapist, alternating with intense periods of devaluing him or her. Thus, the treatment will have a stormy course and will be unable to avoid a focus on this highly volatile and unstable object relationship in the transference. The principal difficulty involves micropsychotic episodes, during which the borderline will devalue the treatment and engage in cocaine binges. Often, gains in relapse prevention as well as other eductional progress appear to evaporate. At such times, for example, attempting to explain "rationally" the importance and value of conditioned cue strategies is of extremely limited value. Such affect storms, in which the patient resorts to drugs, are temporary. They are the result of highly disentegrated and fragmented self and object representations which need to be integrated within a long-term therapeutic relationship. If resources are not available for such treatment, Relapse Prevention strategies may still be partially effective, providing they are adjunctive to other psychotherapeutic treatment which focuses on personality integration to at least remove cocaine abuse from the patient's symptom repertoire.

Antisocial Personality

This personality disorder does not respond well to any psychological treatment, supportive-expressive therapy being no exception. Fortunately, the prevalence of antisocial personality appears decreased in cocaine as opposed to opiate abuse treatment populations (Weiss, 1983). These patients are probably best treated with highly structured, limit setting interventions. In our experience, contingency contract models probably stand the best chance of success, but even here one cannot afford to be overly optimistic.

GROUP APPROACHES

Group Therapy

Group therapy has proved to be an invaluable component of drug abuse treatment. Cocaine abusers present a unique set of problems for therapeutic interaction in a group. The group strategy presented here was designed to meet two important needs in treating cocaine abusers which exist in most settings:

(a) Clients must be able to enter at any phase of their treatment and continue until completion. The groups are always open to new referrals and ideally other group members provide models of behaviour for every stage of the treatment process.

(b) The group must be able to accommodate abusers of divergent backgrounds, skills, and ambitions, who vary in the extent and manner of drug use, the importance of addiction in their lives, and reasons for seeking assistance. Patients who are professionals often initially deny the deterioration in lifestyle they share with street addicts, while freebasers, IV users, and intranasal users may initially see themselves as independent groups; however, strong commonalities inevitably surface as the effects of cocaine abuse on their lives are explored.

Selection criteria for the group is based on ability to benefit from and contribute to the group task. Individuals starting treatment while still using cocaine are often allowed to participate in group only as an adjunct to individual therapy. Patients who enter treatment with the group as their only source of therapy have usually gained some initial control over their use and are less likely to require the day to day careful analyses and coaching around drug seeking behaviour. Others were always low rate users or came for support following hospitalization.

There is also a number of abusers who seek treatment in response to social pressures or as a condition of probation. Group attendance for this population is contingent upon sobriety, or a substantial movement in that direction, as assessed by the individual therapist. Participants who have not freely chosen to attend and who continue to abuse cocaine may readily sabotage the progress of others in the group and are not allowed to remain in the group.

Individual experience with cocaine abusers is essential for a therapist forming a group. The therapist should be familiar with cocaine's biochemical functions and psychosocial sequelae, providing the group with an authority who can comfortably impart knowledge when it is appropriate. Co-therapy is strongly recommended to ensure stability (a two week vacation on the part of a singular therapist has been known to disband the group completely) and for support in remaining on task (crave-inducing "drug rap" can be a favourite diversion for participants) which may initially require the vigilance of more than one non-user.

Curative Factors in Group Psychotherapy

Irvin Yalom (1975) provides a model from which to address the question of how group therapy aids in recovery from cocaine abuse. His central organizing principle delineating an interchange of curative factors is a foundation from which we approach our group. Although these factors remain essentially the same across groups, cocaine abusers enter treatment with unique needs to be addressed. The relevant factors as described by Yalom and as expressed in cocaine abuse groups are delineated as follows:

Instillation of Hope. Cocaine abusers have weathered an extraordinary amount of failure prior to entering treatment. Ultimate acts of resolve, such as breaking the freebase pipe, destroying remaining needles and snorting the last of the stash, are common, repeated conclusions to cocaine runs. Social associations have often become restricted to abusers, who, like themselves, have declared their commitment to stop innumerable times. "Has anyone ever stopped using cocaine for good?" is an inevitable question frequently carried into group treatment; and answer is provided by those further along in treatment. The instillation of hope, therefore, is particularly crucial in group treatment of cocaine abusers.

Exposure to others who are at different points along the dependence-abstinence continuum is both inevitable and helpful in an ongoing group. Individuals can follow the progress of those who, following abstinence, have begun to restore social functioning (e.g., improve work performance, use their money for more constructive purposes). They also have the opportunity to see others at their own level begin to employ effective strategies to avoid drug use. Patients will often inquire about the progress of graduated patients again seeking hope for their own situation. In the face of high turnover, which sometimes occurs in cocaine abuse treatment, the therapist must maintain the conviction that patients can be helped and can help each other. The therapist may also need to fill "gaps" in the spectrum of abusers present with vignettes from therapeutic experience.

Universality. Many cocaine abusers harbour fears that they are unique in losing control over their behaviour. They are ashamed at their compulsion to use drugs in the face of deteriorating relationships, volatile financial or legal situations and aggravated work environment. Upon intake, the extent of their self-defined "crazy" behaviour is discussed, often for the first time. Identifying with others who share identical experiences is a great source of comfort. As one begins to identify with others and realize that he/she is not alone, increased self-disclosure follows and facilitates further progress in treatment.

Imparting Information. Loss of control over one's behaviour can be one of the most confusing and threatening aspects of cocaine addiction. Patients described waking up at 3:00 a.m., hounded by a compulsion to leave the house in stocking feet to find freebase or feeling as if their car takes itself off a highway exit to a dealer's neighbourhood. Compulsive drug use continues, although the high becomes negligible. Therapists and group members can counter the anxiety such dyscontrol brings. Didactic instruction about the biochemical mechanisms of cocaine can reduce the anxiety of clients mystified by their experience. Understanding the brain's demand to restore neurotransmission can assist some patients to resist a cocaine craving. Behavioural models of conditioning and cues are discussed and are applied to cocaine taking behaviour, promoting the patient's understanding of high risk times for relapse and relapse prevention strategies. Many cocaine abusers report increased craving as a result of joining with other users and talking about cocaine. This can and must be addressed in an educational format for every new member to prevent them using the group as a stimulus for getting high. The elicited craving is contextualized as an important

exercise in weakening the association of craving and cocaine -- group members often analogize this to a muscle one must repeatedly exercise that will
grow stronger with consistency. New members are assured by other patients
that the craving will subside quickly if it is not followed by drug use.
Such instruction entails rehearsing after-group plans such as taking an alternate route home to avoid streets where cocaine can be obtained. As new
individuals enter the group, often ongoing members will assume the role of
educator in explaining the way cocaine works, problem solving around cravings or modelling effective use of that knowledge.

   Altruism. The large number of recovering addicts working in the field
of substance abuse treatment supports the curative properties of altruism,
demonstrating that helping another is a powerful way to help oneself. Group
therapy will allow someone who has often lost significant self-esteem to recover self-worth by being helpful to others for the first time since they
have begun abusing drugs. An interest is developed in the situations of
others and patients become eager to share themselves in an attempt to exemplify solutions for others. A sense of power results from transcending
one's own problems to support another. In addition, an accountability to
other group members is frequently described. Early in the group, people
develop a protectiveness and express concern at the absence of a group member. Drug use and participation develops different meaning in this context.
Group members often transcend difficult periods because of the meaning their
relapse would have for others they are attempting to help.

   Social Learning and Imitative Behaviour. Impairment in social functioning is one of the most significant fallouts of continuous cocaine abuse
and among the most difficult and important from which to recover. Individuals often have disassociated themselves from non-drug abusing friends and
fear social incompetence when interacting without the use of cocaine. Some
report the cocaine high has allowed them to become gregarious and uninhibi-
ted, while others report increased isolation from others. A 30 year old
lawyer described herself as "superwoman" -- entertaining guests and family
elaborately in the evenings and weekends while practicing law full time
during the day. Although her functioning began to deteriorate, the price
of abstinence was to test facing a husband, guests, and the judge while
consumed with depression. Unilaterally, cocaine abusers report decreased
recreational activities and difficulty recalling what they did interperson-
ally in the pre-cocaine past. A 20 year old college student, when he discovered his therapist did not use cocaine or frequent bars, queried, "What
do you do for fun? Just mope around the house?" In addition, heterosexual
interaction has been strongly paired with cocaine use; many clients report
their last sexual contact or date was prior to the onset of abstinence.
Cocaine is an effective lure for the opposite sex and allows people to be
less discriminating in satisfying needs for intimacy.

   Group therapy addresses these problems perhaps more effectively than
individual treatment. Above all, group therapy enables intimate interaction
to take place without the use of drugs. Patients are allowed to test out
social skills in the protected structure of a weekly group and receive feedback from others who are sensitive to their issues. The psychological sequelae of the cocaine crash often has allowed people to remain distant and
isolated, deflecting the natural fear of human contact. This insight will
formulate from earnest inquiry into "What can I do with myself if I don't
get high?" The difficulties of sitting down to dinner with a spouse or
facing a party without drugs are rooted in these fears. Patients discover
that, in group, for at least one hour a week, they have a safe setting in
which to learn that they are able to sustain non-drugged intimacy without
negative consequences.

   Initial isolation from many social situations is encouraged in early stages of treatment to avoid cues in one's environment. After initial abstinence is achieved, engaging in pursuit of gratifying alternative behaviours may be equally essential to prevent relapse. Activities that provide a high level of stimulation are recommended. Although there may never be a substitution for a cocaine "rush", patients must be reassured that pleasure will ultimately return if they are disciplined in pursuit of recreation. Again, senior group members are more likely to have reaped the benefits of resumed social activity, and can serve as models for those still suffering from the depression and boredom during the first several months of abstinence. Physical exercise is an ideal activity but sufficient motivation often takes time to develop. Senior group members often provide models for hope and patience.

The full scope of possibilities offered by group therapy is even broader than those just described. A short term open-ended cocaine group as described can be moulded to serve unique functions. Our groups have been predominantly supportive, educational, or dynamic, depending upon the stability of the members, their needs and stage of treatment. The choice of orientation is based on several factors. Patients, at times, must be trusted and respected as experts in knowledge of cocaine use and should be prompted to articulate their opinions and current needs. Assessing future directions in the group dictate obtaining an empirical data-base from which to assess individual and group progress over time. Self-report questionnaires administered in the first five minutes of group while patients arrive are used in our programme to provide data on frequency and quantity of use, as well as to draw the patients' attention to the previous week's activities. Otherwise, it is too easy to lose track of individual progress, or lack thereof, and allow weeks to go by without disclosure of continued problems, if group therapy is the sole therapeutic modality employed. Data collected over time on all individuals entering group can be used to generate guidelines for future structure of the treatment package as a whole.

Multifamily Therapy

Clinicians have noted that families of substance abusers are often rife with family pathology (Stanton et al., 1982; Kosten, Hagan, Jalali, Steidl, and Kleber, 1986; Minuchin and Fishman, 1980). In particular, pathology in families of alcoholics and opiate abusers have received detailed attention. In our experience, this relationship holds for cocaine abusers as well. Limitations in staff resources often preclude individualized family treatment, especially within drug rehabilitation settings. Despite this, recent literature in family therapy with drug abusers has described a powerful therapeutic modality, Multiple Family Therapy (MFT), in which a group of families meets regularly with co-therapists. In addition to its cost-effectiveness, we believe that MFT carries with it certain advantages in its own right, similar to the advantages of group psychotherapy. First, families have the opportunity to observe similar, often identical, pathological patterns in other families. This not only helps the family to overcome the belief that they are alone in their problems, but it can also serve as a springboard in converting ego-syntonic attitudes into ego-dystonic ones. Second, within the MFT context, families often feel better understood, especially if co-therapists are not themselves former substance abusers. Finally, within the MFT setting, the modalities of family and group therapy frequently combine synergistically (denial is not so easy in the family context), resulting in a powerful therapeutic impact rarely observed in either modality taken alone.

We will delineate the format of MFT we have developed for families of cocaine abusers. We shall focus on structure, content, process, and interactions with concurrent individual treatment.

Structure

The group meets weekly for 90 minute sessions. It consists of two or more families. In our experience, three or four families appears optimal. Typically, families are bi-generational with parents and siblings of the identified patient participating. However, we encourage the inclusion of grandparents or other extended family members; especially if, as is often the case, they contain another substance abuser among their number.

We have found it essential to include a minimum of two therapists. In addition to the abundance of material, the sheer intensity of combining group and family dynamics with a drug abuse focus makes more than one therapist necessary. It is important that co-therapists work well together and meet outside the group, to clarify their thoughts and feelings regarding the group and each other and to enhance awareness of developing enmeshment with families. Frequently conflicts emerging between co-therapists can reveal and illuminate conflicts within or between families. Such understanding is then utilized in subsequent sessions.

However, we do not consider it necessary for co-therapists to adhere to the same therapeutic orientation. Most recently, we have had three co-therapists who had received primary training in psychodynamic, behavioural, and AA orientations respectively. Indeed, this diversity of perspectives facilitated rather than hindered the unfolding understanding of the group.

Content

Content of the MFT group will vary greatly. In the initial stages, families are eager to learn about cocaine and its effects and therefore a good portion of group time can be allocated for such pharmaco- and psycho-educational purposes. Often, however, families will seek out such information not only for the purpose of understanding their cocaine abusing member, but also with the view towards "proving" how destructive, detrimental, and "bad" the latter is in his or her drug abuse, as well as to deflect attention away from family conflict. Ironically, it is precisely such efforts that provide an opening for therapists to observe family conflict and to comment on it. A continuum exists ranging from relatively conflict-free psychoeducation on the one hand to family conflict on the other, allowing therapists to oscillate with tact and sensitivity between these poles depending on the total emotional context of the group. Thus, psychoeducation can be quite beneficial in detailing the effects of cocaine, in explaining the functions of conditioned cues, in developing strategies to help prevent relapse, and in modulating functioning in the group as well as to promote group cohesion. At times, we focus on patterns of family and group interactions such as when families use learning to avoid dealing with painful conflicts, or when family conflicts present themselves spontaneously.

Unconscious facilitation of another's substance abuse, or enabling, is another issue which receives abundant attention, and one that must be broached sensitively. Families of substance abusers readily facilitate the identified patient's substance abuse with an astounding lack of perception. For example, the spouse of a 34 year old male cocaine abuser would buy him various gold chains and other expensive jewelry as a "reward" for having been abstinent. The patient readily converted these "gifts" into money with which to buy cocaine. In another case, the mother of a 29 year old female freebase abusing patient left large sums of money in a cookie jar at home "in case someone was short of cash."

While it is important to elucidate how such enabling behaviours contribute to maintaining the identified patient's substance abuse, it is necessary to go further and elucidate the multiple functions such behaviours subserve, including the family's vested interest in maintaining their member's substance use, which they consciously abhor, to deflect attention from other family problems. In this respect, MFT can become quite a powerful mode of intervention: families who can "see" enabling in others are more likely to observe and accept its existence in themselves. Additionally, families who have already observed enabling in themselves can confront another family about it with some authority and be more readily understood and accepted by the other families.

Process

Co-therapists of MFT groups are often confronted with the question: should we approach this as family or group therapy? Should we focus on a particular systems issue or on the group process? Flexibility is essential. Therapists must be prepared to use one approach or the other or both as the situation dictates; at times it will be necessary to focus directly on a particular family's crisis to the relative exclusion of the other families. At other times, the focus will be on group themes, e.g., enabling amongst all families. At still other times, an individual family's crisis can be utilized to reflect a group theme with more attention devoted to that family with other families engaging more passively. At certain times, a group transference can be discerned and interpreted.

Interactions with Individual Treatment

In our unit, MFT is widely utilized as an adjunct to individual treatment of cocaine abusers where family issues are prevalent. Used as such, the two treatments will interact with one another and it is important for therapists to anticipate and deal with issues arising from this interaction. A frequently encountered problem involves disclosing material obtained in an individual session to the family in the context of MFT. We offer the following guidelines: At the outset of treatment, disclosure is described as a potential issue. All family members, including the patient, are asked to be as honest and as forthcoming as possible. However, we acknowledge that individual treatment requires an atmosphere of safety in which the patient can reveal himself to the therapist without the threat of disclosure to the others. Therefore we state that we will not divulge material which the patient asks us to withhold without first discussing the issue with the patient. If after the discussion the patient still insists that the information be withheld, we abide by that decision.

Many other interactions can occur. For example, one patient well into both individual and MFT treatments began to discuss only family issues in her individual sessions, issues that she feared to bring up in the MFT group. This coincided with her recent development of positive transferential feelings toward the therapist, which she also feared. By confronting her resistance to the development of her positive transference toward the therapist, she was able to see that as her positive feelings did not overwhelm her in the individual sessions, she no longer needed the protection of family issues in individual sessions. She was then able to more readily discuss the fearful family issues she had avoided in MFT sessions with the family.

Towards an Integrated Approach

Due to a lack of systematic evaluation, neither the treatment models described above nor other models (pharmacotherapy, self-help groups) can claim superiority in the treatment of cocaine abuse at this time. Similarly while it is recognized that cocaine abusers presenting for treatment are heterogeneous, the variables for matching specific patient groups with the most appropriate type of treatment have not yet been identified. Until such information is available, it may be most prudent to employ an integrated approach which utilizes different models and techniques at different points during treatment according to the specific needs of the patient. An integrated approach uses clinical judgment and patient variables, in particular severity of cocaine use and concurrent psychiatric symptomatology, to tailor a flexible programme for each patient. Such an approach probably reflects actual practice among most existing programmes.

It is essential for the patient entering treatment to become abstinent without delay. It is our experience that due to (a) the pervasiveness and dangers associated with chronic cocaine abuse, and (b) the abuser's preoccupation with the procurement and use of cocaine, meaningful psychotherapeutic work cannot be attempted until a period of stable abstinence is attained. Whether or not cocaine abuse is viewed as a symptom of another underlying disorder or as a primary disorder in its own right, cocaine use must be controlled before other work is possible. While in severe cases, inpatient or pharmacotherapy may be useful in achieving abstinence, it is our clinical impression that a purely psychotherapeutic model can be effective for rapidly achieving abstinence. A progression from abstinence initiation strategies to relapse prevention strategies that focus exclusively on the control of cocaine use appears most successful. Certain other approaches (supportive-expressive, family, individual, or self-help groups) are considered supplemental and are used for most abusers we treat, depending on their needs and progress.

Relapse prevention can also serve as a useful and sometimes necessary foundation for other approaches. Many cocaine abusers, at the time they present for treatment, no longer have the ability to manage affect, tolerate ambiguity, or cope with family conflicts without resorting to cocaine abuse. Without an established repertoire of coping responses ans strategies for managing slips and craving, these other forms of treatment may be less successful. For example, as we have noted, group treatment can elicit intense cravings in group members. Background in the nature of conditioned cues may help the abusers to manage such craving without relapsing, perhaps preventing premature termination of treatment.

Once cocaine abuse is brought under control, other problems can be accurately assessed, concurrent psychiatric disorder can be diagnosed, and appropriate treatment plans can be developed. Here the treatment becomes more flexible and open-ended. The presence of an underlying psychiatric disorder dictates appropriate treatment of that disorder. For example, the diagnosis of a current depressive disorder may be treated in any number of ways: the therapy may consist of cognitive-behavioural or dynamically-oriented individual treatment, with or without pharmacotherapy. An abuser who lives at home and has significant conflicts with his parents may be treated in a multi-family group. Social isolation and need for social support often indicate need for group therapy. Attendance at self-help groups, such as Cocaine Anonymous, can bolster social support. At this point, relapse prevention techniques can be employed on a continuing, as-needed basis. A slip during the middle to latter stages of treatment indicates that a return to focusing on cocaine use is needed until control is regained and stabilized.

Alternatively, it has been our experience that some abusers, particularly those with more severe psychiatric disorders, intense interpersonal disturbances, or severe psychosocial stressors such as homelessness may be unable to participate effectively in treatment until those overwhelming issues are somewhat resolved. Thus another intervention (pharmacotherapy, brief couples treatment, intervention by contact with social service organizations) may be made with the goal of preparing the individual to participate in psychotherapy. For example, a woman whose husband continues to use and deal cocaine may not be able to make use of relapse prevention strategies until supportive-expressive, contingency, or multi-family work enables her to leave him or to force him to seek treatment as well.

Simply stated, there is no room for inflexibility or theoretical dogma in the treatment of cocaine abuse. The therapist must be able to identify and to address a variety of issues as they arise at different points in the treatment. The achievement of abstinence by the abuser, often for the first time in several years, may uncover myriad underlying issues. Here the therapist must be exquisitely sensitive to the needs and abilities of the patient. While structured treatment focusing on the maintenance of abstinence often reveals indications for long-term psychotherapy, the patient's willingness or ability to make this shift must be considered carefully. While many introspective, articulate, relatively stable abusers see the need for and eagerly participate in exploration of deeper issues, many abusers enter treatment with the goal of cocaine cessation, and control of symptoms only. Such patients may see the shift from symptom control to exploration of other issues as threatening. If made prematurely, such a shift can be overwhelming to the patient, terminate the treatment, and pave the way to relapse. For cocaine abuse psychotherapy, as in other psychotherapies, there is no greater attribute than clinical experience and flexibility.

REFERENCES

Aigner, T. G., Balster, R. L., (1978), Choice behaviour in rhesus monkeys: cocaine vs food, Science, 201:534-535.

American Psychiatric Association, (1980), "Diagnostic and Statistical Manual of Mental Disorders, Third Edition", American Psychiatric Association, Washington, D. C.

Anker, A. L., and Crowley, T. J. (1982), Use of contingency contracts in speciality clinics for cocaine abuse, in "Problems of Drug Dependence, NIDA Research Monograph Series No. 41", National INstitue on Drug Abuse, Rockville, Md., ed L. S. Harris.

Bandura, A. (1977), Self-efficacy: toward a unifying theory of behavioural change, Psychological Review, 84:191-215.

Beck, A. T. and Emery, G. (1977) "Cognitive Therapy of Substance Abuse," unpublished manuscript.

Beck, A. T., Rush, A. J., Shaw, B. F., Emery, G., (1979) "Cognitive Theory of Depression," The Guilford Press, New York.

Blatt, S. (1974) Levels of object representation in anaclitic and introjective depression, Psychoanalytic Study of the Child, 29:107-157.

Blatt, S. and Shichman, S. (1983), Two primary configurations psychopathology, Psychoanalysis and Contemporary Thought, 6:187-254.

Boudin, H. M., (1972), Contingency contracting as a therapeutic tool in the deceleration of amphetamine use, Behav Ther, 3:604-608.

Brownell, K. D., Marlatt, G. A., Lichtenstein, E., and Wilson, G. T., (1986), Understanding and preventing relapse, American Psychologist, 41:765-782.

Carroll, K. (1985) "Manual for Relapse Prevention in the Treatment of Cocaine Abuse," unpublished manuscript.

Chasnoff, I. J., Burns, E. J., Schnoll, S. H., and Burns, K. A. (1986), Effects of cocaine on pregnancy outcome, in "Problems of Drug Dependence, 1985, NIDA Research Monograph Series No. 67," L. S. Harris, ed. National INstitue on Drug Abuse, Rockville, Md.

Chitwood, D. D. (1985), Patterns and consequences of cocaine use, in "Cocaine Use in America: Epidemiological and clinical perspectives, NIDA Research Monograph Series No. 61," E. H. Adams and N. J. Kozel, eds., National Institute on Drug Abuse, Rockville, Md.

Cohen, S., (1981) "Cocaine Today", American Council on Drug Education, New York.

Crowley, T. Quoted in "Reinforcing drug-free lifestyles," ADAMHA News, p.3, August 27, 1982.

Cummings, C., Gordon, J. R., and Marlatt, G. A. (1980), Relapse: prevention and prediction, in "The Addictive Behaviours", W. R. Miller, ed., Pergamon Press, New York.

DeWit, H. and Stewart, J. (1981) Reinstatement of cocaine-reinforced responding in the rat, Psychopharmacology, 75:134-143.

Dougherty, J., and Pickens, R. W. (1973), Fixed-interval schedules of intravenous cocaine presentation in rats, Journal of the Experimental Analysis of Behaviour, 20:111-118.

Freud, S., (1912), "The Dynamics of Transference, Standard Edition, Vol. 12" Hogarth Press, London.

Gawin, F. H., and Kleber, H. D. (1984), Cocaine abuse treatment: An open trial with lithium and desipramine, Archives of General Psychiatry, 41:903-910.

Gawin, F. H. and Kleber, H. D. (1986), Abstinence symptomatology and psychiatric diagnosis in chronic cocaine abusers, Arch Gen Psychiat, 43: 107-113.

Geller, J. D. and Schwartz, M. D., "An Introduction to Psychotherapy: A Manual for the Training of Clinicians," unpublished manuscript, (1972).

Gerber, G. J. and Stretch, R. (1975), Drug-induced reinstatement of extinguished self-administration behaviour in monkeys, Pharmacology, Biochemistry and Behaviour, 3:1055-1061.

Gold, M. S. Washton, A. M., Dackis, C. A., (1985) Cocaine Abuse: Neurochemistry Phenomerology and Treatment, in "Cocaine Use in America: Epidemiology and Clinical Perspectives, in NIDA Research Monograph Series No. 61," Adams, E. H. and Kozel, N. J., eds., National Institute on Drug Abuse, Rockville, MD.

Goldberg, S. R. (1973), Comparable behaviour maintained under fixed-ratio and second-order schedules of food presentation, cocaine injection, or d-amphetamine injection in the squirrel monkey, Journal of Pharmacology and Experimental Therapeutics, 186:18-30.

Greenson, R. R., (1967) "The technique and practice of psychoanalysis, Vol. I.", International Universities Press, New York.

Griffiths, R. R., Findley, J. D., Brady, J. V., Dolan-Gutcher, K., and Robinson, W. W. (1975), Comparison of progressive-ratio performance maintained by cocaine, methylphenidate, and secobarbitol, Psychopharmacologia (Berlin), 43:81-83.

Grinspoon, L. and Bakalar, J. "Cocaine: A Drug and its Social Evolution," Basic Books, New York, (1976).

Helfrich, A., Crowley, T., Atkinson, C., and Post, R., (1983), A clinical profile of 136 cocaine abusers, in "Problems of Drug Dependence, 1983, NIDA Research Monograph Series No. 43," L. S. Harris, ed., National Institute on Drug Abuse, Rockville, Md.

Hunt, W. A. and Matarazzo, J. D., (1970), Habit mechanisms in smoking, in "Learning Mechanisms in Smoking," W. A. Hunt, ed., Aldine, Chicago, Ill.

Kernberg, O. (1975), "Borderline Conditions and Pathological Narcissism," Jason Aaronson, New York.

Kernberg, O. (1976) "Object Relations Theory and Clinical Psychoanalysis," Jason Aaronson, New York.

Khantzian, E. J., (1975), Self-selection and progression in drug dependence, Psychiatry Digest, 10:19-22.

Khantzian, E. J. (1983), An extreme case of cocaine dependence and marked improvement with methylphenidate treatment, American Journal of Psychiatry, 140:784-785.

Khantzian, E. J. (1985), The self-medication hypothesis of addictive disorders: Focus on heroin and cocaine dependence, American Journal of Psychiatry, 142:11

Kleber, H. D., and Gawin, F. H., (1984), Cocaine abuse: A review of current and experimental treatments, in "Cocaine: Pharmacology, Effects, and Treatment of Abuse, NIDA Research Monograph Series No. 50," J. Grabowski, ed., National Institute on Drug Abuse, Rockville, Md.

Kohut, H. (1971), "The Analysis of the Self," International Universities Press, New York.

Kohut, H. (1977), "The Restoration of the Self," International Universities Press, New York.

Kosten, T. R., Hogan, I., Jalali, B., Steidl, J., Kleber, H. D., The effect of multiple family therapy on addict family functioning: A pilot study, Adv Alcohol Subs Abuse, 5:51-62.

Krystal, H. and Raskin, H. A., (1970), "Drug Dependence: Aspects of Ego Functions," Wayne State University, Detroit.

Luborsky, L. (1984), "Principles of psychoanalytic therapy: A manual for supportive-expressive treatment," Basic Books, New York.

Marlatt, G. A., and Gordon, J. R. (1980), Determinants of relapse: Implications for the maintenance of behaviour change, in "Behavioural Medicine: Changing Health Lifestyles", P. O. Davidson and S. M. Davidson, eds., Brunner/Mazel, New York.

Marlatt, G. A. and Gordon, J. R., (1985), "Relapse Prevention: Maintenance Strategies in Addictive Behaviour Change," Guilford Press, New York.

McLellan, A. T., Luborsky, L., Woody, G. E., O'Brien, C. P., (1980), An improved diagnostic instrument for substance abuse patients: The Addiction Severity Index, Journal of Nervous Mental Diseases, 168:26-33.

McLellan, A. T., Luborsky, L., Woody, G. E., O'Brien, C. P., and Druley, K. A., (1983), Predicting response to alcohol and drug abuse treatment, Archives of General Psychiatry, 40:620-625.

Miller, W. R., Hester, R. K. (1986), Inpatient alcoholism treatment: Who benefits? American Psychologist, 41:794-805.

Minuchin, S. and Fishman, H. C., (1981), "Family Therapy Techniques," Harvard University Press, Cambridge, Mass.

Modell, A. (1976), "The holding environment" and the therapeutic action of psychoanalysis, Journal of the American Psychoanalytic Association, 24:285-308.

Pechacek, T. F. and Danaher, B. G. (1979), How and why people quit smoking: A cognitive-behavioural analysis, in "Cognitive-Behavioural Interventions: Theory, Research, and Procedures," P. C. Kendall and S. D. Hollon, eds., Academic Press, New York.

Pickens, R. W. and Harris, W. C., (1968), Self-administration of d-amphetamine by rats, Psychopharmacologia (Berlin), 12:158-163.

Pickens, R. W. and Thompson, T. (1968), Cocaine-reinforced behaviour in rats: Effects of reinforcement magnitude and fixed ratio size, Journal of Pharmacology and Experimental Therapeutics, 161:122-129.

Pickens, R. W. and Thompson, T., (1971), Characteristics of stimulant drug reinforcement, in "Stimulus Properties of Drugs" T. Thompson and R. W. Pickens, eds., Appleton-Century-Crofts, New York.

Rawson, R. A., Obert, J. L., McCAnn, M. J., Mann, A. J., (1986), Cocaine treatment outcome: Cocaine use following impatient, Outpitient, and no treatment, in "Problems in Drug Dependence (1985), NIDA Research Monograph No. 67," L. S. Harris ed., National Institute on Drug Abuse, Rockville, M.D.

Resnick, R. B., and Resnic, E. B. (1984), Cocaine abuse and its treatment Psychiatric Clinics of North America, Vol.7 pp. 713-728.

Rounsaville, B. J., Gawin, F., Kleber, H. (1985), Interpersonal psychotherapy adapted for ambulatory cocaine abusers, American Journal of Drug and Alcohol Abuse, 11:171-191.

Rounsaville, B. J., Glazer, W., Wilbur, E. H., Weissman, M. M., and Kleber, H. D., (1983), Short-term interpersonal psychotherapy in methadone maintained opiate addicts, Archives of General Psychiatry, 40:629-636.

Rounsaville, B. J., Spitzer, R. L., Williams, J. B. W., (1986), Proposed changes in DSM-III substance use disorders: description and rationale, American Journal of Psychiatry, 143:463-468.

Rounsaville, B. J., Tierney, T., Crits-Christopher, K., Weissman, M. M., and Kleber, H. D., (1982), Prediction of outcome in treatment of opiate addicts: Evidence for the multidimensional nature of addicts' problems, Comprehensive Psychiatry, 23:462-478.

Schnoll, S. H., Karrigan, J., Kitchen, S. B., Daghestani, A., and Hansen, T. (1985), Characteristics of cocaine abusers presenting for treatment, in "Cocaine Use in America: Epidemiology and Clinical Perspectives, NIDA Research Monograph No. 61", E. H. Adams and N. J. Kozel, eds., U. S. Government Printing Office, Washington, D. C.

Schuster, C. R., and Fischman, M. W., Characteristics of Humans Volunteering for a Cocaine Research Project, in "Cocaine Use In America, Epidemiologic and Clinical Perspectives, NIDA Research Monograph Series No. 61," E. H. Adams and N. J. Kozel, eds., U. S. Government Printing Office, Washington, D. C., (1985).

Schuster, C. R. and Johanson, C. E. (1980), The evaluation of cocaine using an animal model of drug abuse, in "Cocaine, 1980: Proceedings of the interamerican seminar on medical and sociological aspects of coca and cocaine," F. R. Jeri, ed., Pacific Press, Lima, Peru.

Siegel, R. K. (1982), Cocaine smoking, Journal of Psychoactive Drugs, 14: 271-355.

Stanton, M. D., and Todd, T. C., (1982), "Family therapy of drug abuse and addiction," Guilford Press, New York.

Stewart, J., (1983), Conditioned and unconditioned drug effects in relapse to opiate and stimulant drug self-administration, Progress in Neuro-psychobiology, 7:591-597.

Stropp, H., (1981), Toward the refinement of time-limited dynamic psycho-therapy, in "Forms of Brief Therapy," S. H. Budman, ed., Guilford Press, New York.

Tennant, F. S., and Rawson, R. A. (1982), Cocaine and amphetamine dependence treated with desipramine, in "Problems of Drug Dependence, 1982, NIDA Research Monograph Series No 43," L. S. Harris, ed., National Institute on Drug Abuse, Rockville, Md.

Vaillant, G. E., (1973), A twenty-year follow-up of New York narcotic addicts, Archives of General Psychiatry, 29:237-241.

Weiss, R. D., Mirin, S. M., and Michael, J. L. and Sollogub, A., (1983) Psychopathology in chronic cocaine abusers. Paper presented at the 136th Annual Meeting of the American Psychiatric Association, New York, New York.

Wieder, H. and Kaplan, E. H., (1969), Drug use in adolescents: Psychodynamic meaning and pharmacogenic effect, Psychoanalytic Study of the Child, 24:399-431.

Wesson, D. R. and Smith, D. E. (1985), Cocaine: treatment perspectives, in "Cocaine Use in America: Epidemiologic and Clinical Perspectives, NIDA Research Monograph Series No. 61," E. H. Adams and N. J. Kozel, eds., U. S. Government Printing Office, Washington, D. C.

Woods, J. (1977), Behavioural effects of cocaine in animals, in "Cocaine: 1977, NIDA Research Monograph Series No. 13," R. C. Peterson and R. C. Stillman, eds., National Institute on Drug Abuse, Rockville, Md.

Woody, G. E., Luborsky, L. McLellan, A. T., O'Brien, C. P., Beck, A. T., Blaine, J., Herman, I., and Hale, A. (1983), Psychotherapy for opiate addicts: Does it help? Archives of General Psychiatry, 40:639-645.

Wurmser, L. (1978), "The Hidden Dimension," Jason Aaronson, New York.

Yalom, I. D. (1975), "The Theory and Practice of Group Psychotherapy, 2nd Edition," Basic Books, New York.

PUBLIC HEALTH APPROACHES TO THE COCAINE PROBLEM: LESSONS FROM THE BAHAMAS

James F. Jekel

Yale University School of Medicine
New Haven, Conn.

We are witnessing the unprecedented spread of a relatively new form of cocaine called freebase or "crack" cocaine, which is smoked or inhaled rather than being snorted or injected, as is the powdered cocaine hydrochloride. For most North Americans, the first awareness of the new form of cocaine came with Richard Pryor's burning himself in 1980, allegedly while making freebase cocaine from powder using ether. General awareness of the availability of "crack" became news on the east and west coasts of the U.S. during 1985, although many drug users in North America had been making their own freebase cocaine from powder for as long as a decade. Most Americans only began to appreciate the danger of cocaine in 1986, due to intense media coverage and the sudden deaths of two young athletes from the use of cocaine in June, 1986. The cocaine hotline started by Gold and Washton (1-800-COCAINE) has been receiving a rapidly increasing proportion of calls about the freebase form of cocaine. (1)

The industrialized nations have long experienced the use of illegal and addicting drugs. (2-5) The choice of drugs, and the amount of their use, has changed over time according to the availability, cost, legality, and styles of drug use. Concern about drug use, especially about heroin and hallucinogens such as LSD, increased in the 1960's. In the 1970's, new drugs, including speed, PCP (Angel Dust), and MDMA (Ecstacy), became popular, and marijuana became the most popular illicit drug used by North American young people (except for ethanol and nicotine in states where these were illegal for teenagers). Cocaine has been a part of the Western drug scene for more than a century, although until recently its high cost restricted its use primarily to persons with considerable financial resources. Part of the reason cocaine gradually has grown in popularity in industrial nations may be the (incorrect) perception that it is not as dangerous as heroin.

Today, everyone associated with the field of drug abuse problems in the Americas acknowledges that cocaine use is increasing at a very rapid rate and is becoming a serious problem. It is probable that cocaine use now is increasing at a more rapid rate than that for any other illicit drug in history. However, Musto has reminded us that the U.S. experienced heavy cocaine use in all forms from approximately 1885, a century ago, to the 1920's i.e., during a period of approximately 40 years. (6) This era, during most of which time cocaine was legal and was heavily marketed, was brought to a close by a growing popular acceptance of the danger of cocaine and by its being made illegal. We are now seeing a rapid spread of the use of illegal cocaine, with potential consequences for the western hemisphere as serious as those of AIDS.

Why has there been this sudden increase in cocaine use and how should we think about the problem in order to control it? We will illustrate the nature of the problem by describing an epidemiologic case study from the Bahamas and then suggest possible fruitful areas for intervention by considering three public health models.

A NATIONWIDE EPIDEMIC OF FREEBASE COCAINE USE: THE BAHAMAS

The Bahamas are ideally located to be a transshipment area for illicit drugs from South America en route to North America and Europe. Among the 700 islands comprising the Bahamas, the majority of which are uninhabited, there are numerous areas for the easy transfer of drugs from boats and planes coming from South America (especially from Colombia) to other boats and planes as well as to individuals bound for the east and south coasts of the U.S. or for Canada or Europe.

It is inevitable that a nation involved in drug transshipment, however unwillingly, will have abundant access to those drugs. In the 1980's, producers and traffickers alike found that cocaine was much more profitable and far safer to produce and ship than was marijuana. It commanded a much higher price per ounce and was easier to ship as a powder than, for example, marijuana leaves. If law enforcement agents were approaching, cocaine was also much easier to dispose of (e.g., into the ocean or down a toilet). As a result, the production and shipment of cocaine increased, and there was a sharp increase in supply with a consequent drop in price in the Bahamas in late 1982 and early 1983. Addicts affirmed that in this relatively short space of time, the street price of cocaine per gramme in Nassau dropped to about one-fifth of its previous price!

The drop in street price of cocaine presented a problem to the Bahamian drug pushers because initially there was not a corresponding increase in the demand for powder cocaine for snorting to match the increase in supply. The result was that selling cocaine became less profitable. Although a few cocaine users knew how to make the freebase from the powder and had been using freebase cocaine, most cocaine users in the Bahamas either did not know how or did not care to make the freebase form. The pushers, however, were quite aware that freebase cocaine was far more addictive than the snorted powder, and their sales would increase if users switched to using freebase cocaine. The pushers then made a marketing decision to sell only the freebase form of cocaine on the streets. To accomplish this, beginning in early 1983, the pushers produced their own "rocks" of freebase cocaine by home kitchen chemistry and generally sold only the freebase form of the drug in Nassau and other areas of the Bahamas. (7) They were, in effect, using some of the principles of the fast food chains in the U.S. by selling the drug in an attractive form so that it could be purchased and used quickly, easily, and at minimum cost. (8)

The freebase form of cocaine is highly addictive, and with heavy use, symptoms, signs, and behaviour changes develop in the users within weeks or months rather than after months or years of use by snorting users. Not only was the freebase form far more addictive than the hydrochloride powder, but also there is an as yet unproven possibility that the cocaine was more pure than previously. The combination of exclusive and easy availability, low cost, high addictive potential, rapid complications, and, possibly, greater purity, led to the rather sudden appearance of large numbers of cocaine abusers in the few psychiatric facilities in the Bahamas.

The following data are reported from the Community Psychiatry Clinic

Table 1. Trends in first admissions for drug abuse, cocaine vs. all other
drugs, Community Psychiatry Clinic and Sandilands Rehabilitation
Centre, Bahamas, 1980-1985.

	Community Psychiatry Clinic						Sandilands Rehabilitation Centre					
Year	Cocaine		Other Drugs		Total		Cocaine		Other Drugs		Total	
	No.	%	No.	%	No.	%	No.	%	No.	%	No.	%
1980	-		-		-		1	0	64	7	920	100
1981	-		-		-		2	0	40	4	921	100
1982	0	0	4	1	360	100	0	0	59	6	1065	100
1983	37	6	18	3	641	100	32	3	47	4	1123	100
1984	299	34	43	5	883	100	224	17	86	8	1337	100
1985	215	27	43	5	805	100	-		-		-	

(CPC) in Nassau, which provides a large proportion of the non-private, outpatient psychiatric care in the Bahamas, and from Sandilands Rehabilitation
Centre (SRC), the only psychiatric hospital in the Bahamas. The data are
from the monthly summary reports by diagnosis from the CPC and from the admission logs of the SRC. Table 1 reports new (incident) cases of cocaine
abuse to either facility. (There was almost no overlap in data between the
two facilities.) Although many of the drug abusers were polydrug users, it
was usually clear which drug was the primary source of the patient's problems. Private psychiatrists seldom treated cocaine abuse privately in the
Bahamas. Wealthy private patients may have gone to the U.S. for treatment,
but their numbers would not be large compared to those reported in Table 1,
which lists only new admissions for Bahamian citizens. The emergency room
of the acute general hospital in Nassau, Princess Margaret Hospital (PMH),
usually referred cocaine abuse cases to the CPC or the SRC.

Table 1 shows that there were very few cases of cocaine abuse reporting
to psychiatric and medical facilities before 1983, but, starting in mid-1983,
the number increased rapidly to a peak in 1984. Since then, the numbers of
new cases at the CPC have dropped somewhat, but they still remain high, and
the drop in numbers may only reflect that a high proportion of those persons
most at risk for cocaine use have already become symptomatic abusers and,
therefore, are no longer at risk for being new (incident) cases. Treatment
of diagnosed abusers has become an overwhelming burden for the psychiatric
facilities. It should be noted that the CPC was started in 1981, and its
statistics did not distinguish between new and repeat users until 1982, so
the CPC statistics are only reported from 1982. On the other hand, the admission logs for the SRC carefully distinguished between new and repeat admissions until the end of 1984, after which, perhaps due to the marked increase in admission load, this distinction is not possible to make from the
admission logs.

These data present the incidence of medically-seen abuse, not necessarily the incidence of cocaine use. But the severity of medical, psychiatric,
and social problems that result from cocaine addiction are such that a high
proportion of freebase cocaine users will probably come to the attention of
the medical or legal system sooner or later. The new patients with cocaine
abuse were usually either self-referred (for symptoms that worried the patient, such as hallucinations or social problems), by the family (who were
worried by the patient's deviant behaviour), by an employer (who was worried
by the behaviour of the patient, including theft), or by the legal system
(usually following arrest).

From an epidemiological perspective, the fast rise in incident cases, combined with a change in type of drug use, indicated that this was a true epidemic. Many mistakenly equate the concept of an "epidemic" with the idea of a "serious problem." Many serious problems, even those that are steadily increasing, do not fit the criteria for an epidemic. An epidemic is an unusual incidence of a disease or other health problem occurring so rapidly as to indicate a rather sudden change in the equilibrium of forces for and against that problem.

The clear epidemic character of this outbreak of cocaine abuse in the Bahamas led us to be persistent in searching for the change or changes that could produce such an event. The rather rapid drop in the street price of cocaine, due to the increase in supply, was certainly part of the explanation but it did not appear to be sufficient to explain such a sudden outbreak so soon after the drop in price. Therefore, we persisted in the questioning of addicts until one of them mentioned that in early 1983 the marketing strategy of the pushers switched from vending cocaine powder to selling "rocks" of freebase cocaine. When questioned, other addicts confirmed this switch. The new patients in 1983 and 1984 were almost all using freebase cocaine rather than the powdered form. Thus, the epidemic appeared to have been caused by the pushers themselves when they shifted their marketing strategy, which in turn was due to changes in supply and demand. Even if the pushers had continued to sell only cocaine powder, the number of addicts would have increased somewhat over the years, but it would not have had the sudden, devastating pattern it now shows.

CHANGES IN THE UNITED STATES

The pattern of rapid rise in the use of freebase cocaine, followed by an increase in the use of medical facilities, is now being seen in the United States, particularly on the east and west coasts. In the U.S., there has been a rapid development of "basehouses," fortified buildings or apartments where base users can come in safety and smoke cocaine base until their financial or physiologic resources are exhausted. (9,10) These appear to be very difficult for the legal authorities to stop, in part because of the danger and because there are enough delaying tactics to give the operators of the basehouse time to get rid of the evidence before arrest.

Low priced "crack" cocaine is rapidly increasing in its availability. It appears that the ability of the drug traffic system to get the drug into the Bahamas and North America far exceeds the ability of those countries to interdict the traffic, so that for the forseeable future, we must assume that there will be a relatively abundant supply of cheap cocaine. This is unfortunate, because it has been demonstrated that easy access to inexpensive drugs hastens their adoption, use, and consequences. (11) However, even though interdiction will not be completely adequate, it is still a cornerstone of national policy in order to keep the supply as small, and therefore the price as high, as possible. Under the assumption of freely available, inexpensive freebase cocaine, what are the alternatives for controlling the problem of drug abuse? We attempt to supply some suggestions by applying three public health models.

PUBLIC HEALTH APPROACHES

An Etiologic model

The first model to be considered is the epidemiologic model for thinking about possible causal factors: host - agent - environment. Because of the special role of the drug pusher, who acts as a vector of the drug, we will incorporate into this model a fourth component called the vector (usually this is used to mean such things as mosquitos in classical public health) (Figure 1).

110

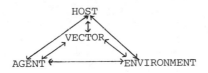

Figure 1. The epidemiologic triad, augmented

The HOST is a human being who is a potential user of the drug, cocaine. Most people, at least at some age, are potential cocaine users. However, males between the ages of 15 and 40 especially those who are not very successful in their society, are at highest risk. Nevertheless, it would be a mistake to assume that any sex or socioeconomic group is free from the risk of cocaine abuse. In the case of freebase cocaine, the risk of addiction is so high with successful use that we cannot consider many persons to have a "natural immunity" to the drug. Significant exposure must be equated with probable addiction.

Because there is little natural immunity to freebase cocaine, we must concentrate on preventing exposure to the drug. It is here that public education has the best opportunity for success, and the recent deaths of two respected athletes may help to strengthen the public message. Whether public health workers or the media will be believed when they warn of the dangers of cocaine abuse remains to be seen, but it is one of our best hopes.

In addition, Gawin and Kleber, as well as others, have shown that many cocaine users apparently have underlying mood disorders, such as depression, and these persons may be using cocaine as a kind of antidepressant. (12) They found that this type of user used less cocaine than the previously normal persons. This is interesting because of the similarity of physiologic actions between cocaine and some of the tricyclic antidepressants. (13)

An Educational Model to Alter Host Factors

Human beings are social and psychological creatures, who are often under tension, with feelings of powerlessness and defeat. One dangerous aspect of cocaine, along with some other illicit drugs, is that in addition to giving pleasure, it can provide at least for a short time, a sense of competence and power. Cocaine may help people to feel they have certain characteristics they do not possess in real life, such as power and success. Both of these effects are powerful attractants. How can the host be rendered more resistant to cocaine?

The "health belief model" may shed some light on how to make persons more resistant to starting cocaine use. (14) The developers of this model derived their theory from the study of why persons did or did not seek certain preventative medical interventions, such as immunizations. They described certain kinds of beliefs that apparently had to be present before people would take actions to prevent a disease or problem.

First, there must be the conviction that the disease or problem is serious. That is, people must believe that if they become addicted to cocaine, or even if they try it once, it could be serious to their survival, health, and happiness. Those who have experimented in the past with cocaine powder as a recreational drug and have not had serious adverse effects often have difficulty believing that the new form of cocaine is really as serious as is being asserted. Likewise, it was commonly thought, at least until recently, that cocaine was not very dangerous. The sudden, cocaine-related deaths of two popular U.S. athletes may change this perception somewhat, but a large amount of misinformation still remains.

A second message of the health belief model is that the person must believe that she or he is at risk for acquiring the serious problem in question. The "it will never happen to me" syndrome, described by Elkind as the personal fable," is commonly found in adolescents and young adults, particularly males who try to prove themselves grown up and powerful. (15) The stories of the two athletes may convince some that they are at risk, but it is likely the educational process must start early before the children enter adolescence.

Third, the health belief model states that the person must believe that the preventive measure will be effective. In this case, it would seem obvious that not using the drug would be a successful prevention. What is less obvious is whether those most at risk will see that merely "trying" freebase cocaine, i.e., experimenting with it, is a violation of the non-use principle.

Fourth, in the health belief model, the individual must see the preventive method as being of reasonable cost, with few, if any, negative side effects. It is not clear that avoiding cocaine, especially freebase cocaine, will be seen as free of costs and negative effects. Peer pressure and overblown descriptions of the pleasures of cocaine use may lead young people, or those feeling depressed and powerless, to conclude that "depriving" themselves of the drug is too great a cost. What is absolutely clear from the research on drugs is that social and psychological factors are critical in the choice whether or not to use drugs, and in the selection of which drugs to use. Also, because of the lack of convincing evidence that educational strategies alone will be sufficient, we must look beyond the host to the agent and the environment of the drug problem.

The AGENT here is cocaine, a topical anaesthetic that acts in the brain by blocking the reuptake of the neurotransmitters dopamine and noradrenalin at neural junctions and also, apparently, by sensitizing receptor sites to the effects of dopamine and noradrenalin. (16) However, the form of cocaine and the route of administration are critical in determining the central effects of the drug and its addictive potential. The freebase form of cocaine passes quickly through lipid membranes into the central nervous system. Also, the euphoric effect of cocaine is strongly dependent on how fast the blood level of the drug rises; in fact, the speed of rise in blood cocaine level appears to be even more important in producing euphoria and addiction than the ultimate blood level achieved. (17,18) Snorting cocaine powder gives a slower rise in blood cocaine level because of the limited absorption area in the nose and because the cocaine has a direct sympathomimetic effect that produces vasoconstriction and reduces the rate of absorption from the nose. By contrast, inhaled freebase cocaine is exposed to the tremendous surface area of the lung, whence it is rapidly absorbed and begins to reach the brain in as little as 8 to 12 seconds. A peak blood level of cocaine is quickly achieved, producing intense euphoria, which is ultimately followed by an equally intense period of dysphoria, including depression and anhedonia. Therefore, not only the basic structure of the agent is important, but also the form of the cocaine and its route of administration are critical to the ultimate effects it produces.

Even as the AIDS virus attacks the immune system in such a way as to reduce its capacity to resist infection, cocaine acts on brain physiology in a way that reduces the ability of the user to resist more cocaine. After a period of intense stimulation of the pleasure centres of the brain, dysphoria follows euphoria and instead of excess pleasure there comes a period of depressed ability to feel pleasure, referred to as anhedonia. This is a condition in which the usual sources of pleasure, food, sexual stimulation, music, friendships, sunsets, etc., no longer produce sensations of pleasure. In fact, the only thing which can produce pleasure and diminish the depression during this stage is more cocaine. The stronger the cocaine stimulation the stronger the psychophysiologic exhaustion and need afterward. By this self-reinforcing mechanism, as well as because of the memory of the "high", the cocaine addict finds it almost impossible to resist more cocaine if it is available.

The ENVIRONMENT is another critical element to consider in cocaine use and abuse. The most important single environmental factor is the availability of the drug. Use will be non-existent or limited if the drug is not available or is very expensive. For some the difference in cocaine availability may determine whether they use any psychoactive drugs at all; for others, if cocaine is not available or affordable, they may turn to other drugs.

Cost must be considered not so much in terms of dollars per gramme of cocaine as in terms of dollars per "hit" or dollars per given level of euphoria. There seems little doubt that the cost of cocaine has dropped considerably over the last few years in terms of dollars per hit or dollars per effective high. Part of this may be a drop in cost per unit weight of street drug. Also, the drug now being sold may be more pure than what was sold a few years ago. The biggest gain in effect per dollar, however, has been the switch in the marketed form, from hydrochloride powder to the freebase form. Thus, in an increasing number of areas of the U.S., the environment offers a cleverly marketed, potent product for as little as $5 to $10 for a rock of freebase cocaine, putting it within the reach of many children.

The cocaine "sales force" appears to be far more extensive and decentralized than for most other drugs, which means that modern marketing methods are going beyond the product to the sales organization also; this phenomenon needs more study. It takes a massive and dependable supply system to encourage this kind of decentralization, so that the "wholesalers" can make sufficient profit merely by making the freebase form and selling to the retailers.

The fear of acquiring AIDS by the use of infected needles is causing many to switch their allegiance away from injected drugs, and some of these are finding freebase cocaine a satisfactory alternative to injected heroin.

The VECTOR, the cocaine pushers and middlemen, have played a more active and important role in freebase cocaine than for most other drugs, because they totally changed the marketing strategy by revising the form in which the drug is sold. The anopheles mosquito became an effective vector of malaria because the malaria parasites can undergo a certain developmental change within the mosquito's body. Modern cocaine pushers have become effective vectors because they make a critical change in the agent while it is in their possession.

Pushers also have created another clever marketing scheme. The "rocks" of freebase cocaine are sold in small plastic vials. The pushers now give money back when these are returned, which has enlisted many people to return the vials for cash, or, more likely, for more of the drug. This tends to bring the user back to the pusher, just as a return on bottles brings customers back to the store for refunds.

Methods of Control

One of the most used public health models for control is called the "levels of prevention." (19) These are presented here in a modified form which corresponds most closely to their current usage.

PRIMARY PREVENTION consists of methods to interrupt the causal forces before the disease or problem has become manifest in an individual. Primary prevention includes two related efforts, the first of which is called "health promotion." This refers to general activities to improve the nutrition, the environment, and the health knowledge and behaviour of an individual (an a community) so as to promote good health. The other part of primary prevention is sometimes called "specific protection" and includes specific technologic efforts to attack a problem or disease, such as immunizations for certain diseases or dietary supplements for specific deficiencies.

Primary prevention could include measures to eliminate cocaine from the environment, or at least to reduce its prevalence and increase its cost. These would include legal and educational efforts, although it remains to be shown that these will work well against the current marketing system for freebase cocaine. One could try to eliminate the vector or to limit its range. Whether legal efforts against the pushers, such as tougher laws against the possession and selling of cocaine, would reduce their number and effectiveness is uncertain. One hope is that it will be possible, through education, to "immunize" most people against wanting to start using the drug. Educational programmes against cigarette use have had, at best, modest success, but the success has been greater in preventing the onset of smoking than in smoking cessation. It remains to be seen whether the same will be true for education against cocaine use. If education is to work, it will probably require starting with primary school children and addressing each of the elements in the health belief model. However, teaching children "how to say no" may help, as well as teaching them to clarify their own values. (20) The more spiritual issues, such as meaning and purpose in the user's life should not be avoided. (21)

SECONDARY PREVENTION implies the early detection of a problem before it has become symptomatic, and then providing the necessary treatment to prevent negative consequences. Good medical examples are the early detection and treatment of hypertension or cervical cancer. Secondary prevention has not been prominent in drug abuse, partly because the user seldom seeks help at an early stage, and by the time his problem is obvious to others, it is by definition past the time for secondary prevention.

Psychiatrists have claimed that the earlier in the addiction the patient is started in treatment, the better the outcome is likely to be. (22) Although self-selection may play a role here, i.e., the earlier a person comes for treatment the more concerned and motivated he may be, it also is easier for a user to return to his normal life if he has not pilfered goods or injured other persons while using cocaine.

Efforts toward implementing the concept of secondary prevention of cocaine should include publicity as to where a person with a cocaine problem can find the confidential help he or she needs. Drug hotlines may provide a very important referral source to get patients into treatment, although most of the callers have serious enough problems it may no longer be considered as secondary prevention. Also, every effort should be made to enable one seeking help to be seen as soon as possible, and to be treated with confidentiality rather than condemnation.

TERTIARY PREVENTION describes medical and public health efforts that occur after the disease/problem has already become evident in an individual. It includes both therapy and rehabilitation, the former having a more medical connotation and the latter implying efforts to enable a person to return to normal, or at least productive, social functioning.

The best therapy for cocaine abuse has not been clarified. One useful adjuvant has been the discovery that tricyclic antidepressants such as imapramine and desipramine remove the craving for cocaine. However, they do not remove the remembrance of past pleasures from cocaine use, so that unless a patient has other goals with which the cocaine is interfering, and hence has a strong desire to be free from the addiction, these medications are of limited value. If, however, the patient had a prior mood disorder or other psychiatric problem and was using the cocaine as a form of self-treatment, the tricyclic antidepressants may be very effectively used to treat the underlying disorder, as well as reducing the craving for cocaine.

Psychotherapy has a role, but it is limited if no clear psychopathology has developed. (23) Often, however, cocaine will leave a legacy of paranoia or depression, which may be made worse by the remembrance of things the patient did while using cocaine. Psychotherapy may be important under these circumstances as well.

The role of group therapy is not settled at the present time. Some believe that most, if not all, cocaine therapy must be done in the outpatient setting, because the patient must have help in adapting to and resisting the pressures and temptations to which he or she will have to return to using the drug. (24) Group therapy has been seen to be of benefit here, but it is not a panacea either. Suffice it to say that both treatment and rehabilitation will be very difficult and will be unsuccessful for many abusers unless the accessibility of the drug is markedly reduced. Until that time it would seem advisable to use every level of prevention and to apply careful evaluations to each modality, in order to determine how best to use scarce resources.

Other models could be used. For example, Perry and Jessor suggested a behavioural health promotion model emphasizing the difference between health enhancing behaviours, which are to be encouraged, and health-compromising behaviours, which are to be discouraged. (25) They also emphasized that action programmes should not be limited only to drugs but should focus on a broad range of factors in the behaviours, personality, and environment of adolescents that influence physical, psychological, and social health. To this list I would like to add the "spiritual," partly because religious young people have consistently been shown to have less drug use. (26) Also, however, I believe the best ultimate force against the drug threat is for our young people to believe that their lives have meaning, purpose, and hence, value. This does not necessarily imply a religious meaning, although it may, as emphasized by the Viennese psychiatrist Victor Frankl who founded logotherapy or, as it is sometimes known, "the third Viennese school of psychiatry." (21)

In conclusion, the forces promoting illicit drug use in the West, both external and internal, are so powerful (and well funded) that mere "programmes" are not likely to stop the problem. Ultimately, only a fundamental strengthening of the beliefs, values, and commitments of the societies involved can accomplish what is needed. However, the public health models discussed here can help us to analyze what needs to be done and to channel our efforts efficiently and effectively.

REFERENCES

1.   Newsweek Magazine, June 16, 1986, p. 20.
2.   R. K. Siegel, Cocaine Smoking, J. Psychoactive Drugs 14:271-355 (1982).
3.   L. N. Robins, The natural history of adolescent drug use, Am. J. Public Health 74:656-657 (1984).
4.   P.M. O'Malley, J. G. Bachman, L. D. Johnston, Period, age, and cohort effects on substance abuse among American youth, Am. J. Public Health 74:682-688 (1984)
5.   J. J. Forno, R. T. Young, C. Levitt, Cocaine abuse: the evolution from coca leaves to freebase, J. Drug Education 11:311-315 (1981).

6. D. F. Musto, Lessons of the first cocaine epidemic, Wall St. J., June 11, 1986.
7. J. F. Jekel, D. F. Allen, H. Podlewski, et al., Epidemic freebase cocaine abuse: Case study from the Bahamas, The Lancet 1:459-462, March 1, 1986.
8. R. Byck, personal communication.
9. Nassau Guardian, Cocaine freebase houses sweeping South Florida, Jan. 6, 1986.
10. Newsweek Magazine, Crack and crime, June 16, 1986, pp. 16-22.
11. There are no recent studies, but several references which state this include: R. Byck, Testimony before the Select Committee on Narcotics Abuse and Control, House of Representatives, Ninety-sixth Congress, July 24, 26, October 10, 1979. The studies reported by the National Institute of Drug Abuse and elsewhere show a curious lack of interest in reducing the availability of the drug.
12. F. H. Gawin, H. D. Kleber, Cocaine use in a treatment population: patterns and diagnostic distinctions, National Institute on Drug Abuse Research Monograph Series 61, National Institute on Drug Abuse, Washington, D. C. (1985).
13. R. Wilbur, A drug to fight cocaine, Science 86:42-46 March, 1986.
14. I. M. Rosenstock, What research in motivation suggests for public health, AJPH 50(2):295-302 March, 1960.
15. D. Elkind, "The Child and Society," Oxford University Press, New York (1979).
16. F. H. Gawin, H. D. Kleber, Cocaine abuse treatment, Arch. Gen. Psych. 41:903-909 (1984)
17. D. Paly, P. Jatlow, C. van Dyke, et al., Plasma cocaine concentrations during cocaine paste smoking, Life Sciences 30:731-738 (1982).
18. R. Zahler, P. Wachtel, P. Jatlow, R. Byck, Kinetics of drug effect by distributed lags analysis: an application to cocaine, Clin. Pharm. and Therap 31(6):775-782 (1982)
19. J. Mausner, S. Kramer, "Epidemiology: An Introductory Text, 2nd Ed." W. B. Saunders & Co., Philadelphia (1985).
20. J. V. Toohey, Activities for the Clarification of Values in Drug and Substance Abuse Education: A Manual for the Instructor, Arizona State University, Tempe (1985).
21. V. Frankl, "Man's Search for Meaning," Washington Square Press, New York (1963).
22. M. Neville, Treatment issues, medical and criminal perspectives in the realities of the provision of treatment to cocaine addicts in the Bahamas, presented at the First International Drug Symposium, Nassau, Bahamas, 20-22 November, 1985.
23. H. D. Kleber, F. H. Gawin, Cocaine abuse: a review of current and experimental treatments, in "Cocaine: Pharmacology, Effects, and treatment of Abuse, 1984, National Institute on Drug Abuse Research Monograph Series 50," National Institute on Drug Abuse, Washington, D.C. (1984).
24. H. D. Kleber, F. H. Gawin, The spectrum of cocaine abuse and its treatment, J. Clin. Psych. 45(12):18-23 (1984).
25. C. L. Perry, R. Jessor, Doing the cube: preventing drug abuse through adolescent health promotion, in "Preventing Adolescent Drug Abuse: Intervention Strategies, National Institute on Drug Abuse Research Monograph Series 47," DHHA, ADAMHA, Washington, D.C. (1983).
26. J. D. Swisher, H. Teh-wei, Alternatives to drug abuse: some are and some are not, in "Preventing Adolescent Drug Abuse" Intervention Strategies, National Institute on Drug Abuse Research Monograph Series 47" DHHS, ADAMHA, Washington, D. C. (1983).

THE TRANS-SHIPMENT SOCIETIES: THE CARIBBEAN EXPERIENCE

THE BAHAMAS AND DRUG ABUSE

Brian G. Humblestone

Sandilands Rehabilitation Centre
Nassau, Bahamas

David F. Allen

National Drug Council
Nassau, Bahamas

The Commonwealth of the Bahamas forms an archipelago of some seven hundred islands and cays covering a total land area of 5,353 square miles of the Atlantic Ocean. Of these, about twenty-nine islands are inhabited. The islands and cays which are distributed over a total area of 100,000 square miles, lie between latitude 20° to 27° North and longitude 72° to 79° West or stretching approximately 50 miles off the west coast of Florida southwards to some 90 miles from the northern coast of Haiti.

The present population is approximately 225,000, 65% of whom live on New Providence Island on which the capital, Nassau, is situated. The population of the Bahamas is predominantly a young one, with 43.6% of her people being 15 years of age and under (1980 census).

Tourism is by far the largest activity and accounts for about 75% of GNP. The Government Service, banking, insurance, fishing, and agriculture account for the bulk of the remainder of economic activities.

The Bahamas has not always had an illegal drug problem. Records of the Sandilands Rehabilitation Centre, the only psychiatric hospital in the Bahamas, show that in the 1960s the major substance of abuse was alcohol, with no evidence of illegal drugs involvement. Alcoholism was and still is a serious problem in the Bahamas. In 1967, 60% of male admissions to Sandilands were due to chronic alcoholism manifested by alcoholic psychosis. The clinical picture of alcoholic psychosis consisted of severe confusion with extreme paranoid delusions. For example, a patient described being in Hawaii with paratroopers dropping nets over him. To free himself, he had used a cutlass to cut the net away. Unfortunately, a policeman, unaware of the delusion interrupted the actions and was seriously injured.

In the late 1960s and early 1970s, a number of young American tourists were admitted to hospital with LSD psychosis. Following the trend, eventually a number of young Bahamians were admitted with the same diagnosis. However, the use of LSD did not become widespread.

In the early 1970s, marijuana use became fairly widely accepted in the This undoubtedly causes more drug abuse.

119

Bahamas, as in many other countries, and the number of admissions which were drug related increased. Admitted in acutely disturbed states, the patients were diagnosed as having cannabis psychosis. Around 1971-1972, the admission rate of cannabis psychosis for males under 30 years was about 20 cases per annum. The phenomenon of cannabis psychosis is controversial but the experience in the Bahamas was discussed by Spencer (1).

In the mid-seventies, other illegal drugs came on the scene and the pattern of drug abuse changed in that more patients were using a number of drugs at the same time. Methaqualone enjoyed a period of popularity and was known on the streets as Quaaludes, Mx, or Disco Bisquit. Methaqualone's ability to potentiate the action of alcohol made it both dangerous and alluring because it was possible to get 'stoned' almost immediately for the price of just one drink and one pill.

The other drug appearing at that time was phencyclidine (PCP or Angel Dust). Producing violent acting out, often against their own person, young men were admitted with bizarre symptoms. For example, a young man was admitted because he persistently beat his head against the wall. This dreadful drug had a powerful impact but because of inadequate drug screening it is still unknown whether a number of deaths of young men appearing in the media obituary columns had resulted from phencyclidine use.

It was also in the mid-seventies that cocaine hydrochloride arrived on the market in increasing quantities, rapidly gaining popularity. Usually snorted at parties, the drug was a symbol of being "cool" or "with it." Compounding the situation, cocaine was used in conjunction with all of the other drugs, producing psychoses, acute intoxifications, and periodic bizarre violent behaviour due to the disinhibition and release of aggression. According to one drug addict interviewed in the prison, a good night out consisted of smoking some grass, drinking a Guiness, taking a quaalude and snorting some cocaine.(2)

The next major change in drug abuse in the Bahamas was in 1979 when ready to smoke freebase was introduced. Use of this form of cocaine was still limited, although stories were heard around town. It was not until 1983 that the first cases were hospitalized, there being about 69 that year. The following year 523 cases were admitted. According to addicts and as described by Jekel et al., it seemed that the major increase in application for treatment was due to the change from snorting to freebasing. Rock or crack cocaine is more addicting with increased pathology.

This consequence of a serious freebase cocaine epidemic appeared in the Bahamas about two years before the crack cocaine upsurge in the U. S. The cocaine rock is made by heating the cocaine hydrochloride with baking soda which precipitates the freebase or crack crystals. These are then smoked in a water pipe or home-made apparatus called a camoke.

What are the factors leading to this change from an illegal drug abuse situation to an horrendous epidemic of cocaine abuse? Firstly, being on the doorstep of America, the Bahamas is subject to the trends and developments occurring in that country. The drug use of the 1960s and the emphasis on sexual freedom undoubtedly had its effect on the Bahamas. Often media and literature underplay the effect of the sexual revolution on drug abuse. But drugs, especially marijuana and cocaine, lead to disinhibition and more heightened sexual desire. Hence they have a definite aphrodisiac quality. at least at the beginning of the use of these drugs. With continued use, a corresponding disinterest in sexual pleasure occurs.

Secondly, the Bahamas is a major tourist resort area with a "party-time" atmosphere which encourages released superego restraints and "living it up."

Thirdly and perhaps most importantly, the Bahamas is located on the trans-shipment route for illicit drugs from South America to North America. Where-as previously these drugs passed through the area, now a percentage of them is left behind on the islands, feeding a powerful cocaine epidemic.

Finally, a low threshold for the tolerance of dysphoria leads to drug abuse. Psychological factors such as family conflict, disruption and changes in lifestyle, and loss and bereavement are associated with dysphoria or labile mood states. However, another factor contributing to feelings of dys-phoria in the West is poor nutrition and lack of exercise. Junk foods have excessive amounts of refined carbohydrates which deplete thiamine and other vitamin B supplies in the brain which in turn lead to anxiety.

PREVALENCE OF COCAINE

Prevalence studies of cocaine, crack, or rock addiction are difficult because:

1. The covert nature of the use. Initially used in cocaine camps, or base houses where there was secure geographical definition, addicts now use private homes or less suspicious locations.

2. Indicators of cocaine use are difficult to assess. By the time the major signs appear, the addiction is at an advanced state. For example, the alcohol drunken state is fairly obvious in most per-sons, whereas some cocaine addicts are able to hit all night and still look almost normal. Of course in the advanced stages of the addiction a person may be emaciated with signs of avitaminosis.

3. Using the rates of patients attending treatment centres for the study of prevalence may be misleading, because the high availabil-ity of cocaine in the community allows the addict to feed his/her habit, thus obviating the need to seek treatment, at least in the eyes of the addict.

This curse of the high availability of high quality, inexpensive cocaine is perhaps the most powerful contributing factor to our pervasive cocaine addiction rates.

POSSIBLE BRAIN MECHANISM OF CRACK/FREEBASE COCAINE

Freebase or crack cocaine is purer than cocaine hydrochloride and pro-duces more intense highs followed by more intense lows. Also it leads to addiction faster than hydrochloride. Our experience in the Bahamas shows that there is an 80% probability of addiction after the first intense high which may or may not occur on the first hit or use of the drug. A number of cocaine addicts claim that they had to try a number of times before reaching that first intense high. The reason for this is not clear, except to hypo-thesize that they may require a period of time to familiarize themselves with the smoking process.

There is much controversy concerning the possible brain mechanisms of crack cocaine. It is safe to say that it is now believed that the craving or hunger for cocaine so common among chronic crack or freebase cocaine users is biologically based through subtle changes in brain bio-chemistry. Accor-ding to Rosecan (3) this has undermined the traditional distinction between "psychologically" addicting substances like cocaine and "physiologically" addicting substances such as heroin. We now believe that the craving for cocaine has psysiological roots.

It is also believed that cocaine produces euphoria by disrupting certain chemical messengers like dopamine in the pleasure centres of the brain. These chemical messengers transfer signals from one nerve ending to the receptors in the next nerve cell. The tranmitter chemicals are then partially reabsorbed in the first nerve ending, replenishing its supply and preparing the cells for further message transmission. It is generally accepted that cocaine blocks the reuptake of the chemical transmitter and allows the continuous transmission of the message of pleasure until the chemical breaks down. Eventually, in the chronic freebase or crack user, the dopamine supplies in the nerve cells are depleted and the highs are less intense and the crashes more painful and severe. Thus cocaine, while promising euphoria, actually delivers dysphoria and almost immediate addiction.

As the nerve cells are depleted of dopamine, the receptor cells become supersensitive, establishing the biological basis for cocaine craving. Of course, when a heavy user stops using the stores of dopamine are replenished after a period of time.

TREATMENT ISSUES

Detailed discussion of treatment methodologies are found in the papers by Gawin, Washton, and Clarke as well as by McCartney and Neville, in this volume. There is general agreement that treatment must involve abstinence from cocaine and other drugs along with regular urine testing. Group and individual psychotherapy are necessary for the emotional growth and development of ego-strength to say "No" to drugs. Generally the preferred mode of treatment is an outpatient setting because of cost-effectiveness and the normalization concept of having the addicted person face his/her life in the community vis-a-vis the family, other relationships, and the workplace. Addicted persons, with poor support systems, severe psychopathology, or who have failed outpatient treatment should be treated as inpatients.

If this hypothesis that the craving in cocaine addiction is essentially biologically based, like hunger and thirst, is valid, then it seems feasible that pharmacotherapy could be therapeutic. Recognizing this, cocaine treatment programmes at Columbia, the Presbyterian Medical Centre, and Yale University are using anti-depressants. According to Kleber and Gawin, "Animal research on neurotransmitter and receptor changes following chronic cocaine suggest long term effects are possibly reversible by treatment with tricylic anti-depressants." (5) They also reported elevated growth hormone and decreased plasma prolactin in human cocaine abusers which is in keeping with adrenergic and dopaminergic receptor changes in animals. Using the anti-depressant desipramine in the treatment of cocaine addiction, they found patients using desipramine had a lower relapse rate than those receiving a placebo. It is also feasible that cocaine addicted persons with underlying bipolar disorder could be helped by lithium.

Conversely, a number of treatment experts do not use drugs in the treatment of cocaine addiction. In the Bahamas, we have found a mixture of the two views. Patients with serious craving or cocaine depression are given anti-depressants or other psychotropic medication while those with no clear psychopathology are treated without the use of drugs.

Cocaine is an illegal substance, and therefore a user is in a strict legal sense, breaking the laws regarding possession of illegal substances. Thus it is our personal opinion that mandatory treatment should be implemented with cocaine addicted persons. Using the Bahamas Mental Health Act as a basis for legal committal for treatment, we have successfully treated a number of persons with severe alcoholic problems. Our conviction is this same process could be applied to the users of freebase of crack cocaine to initially require them to undergo treatment for their addiction and thus preventing them from further destroying their lives and the lives of those around them.

REFERENCES

1. D. J. Spencer (1971) Cannabis induced psychosis, International Journal of the Addictions (6), pp. 322-326.
2. M. Neville, Drug abuse in the Bahamas, a report to the Commission of Inquiry, Nassau, Bahamas, 1985.
3. E. Eckholm, Cocaine Treatment, Experts Dialogue, New York Times, Monday, Sept. 8, 1986.
4. Ibid.
5. H. D. Kleber, F. H. Gawin, The Spectrum of cocaine abuse and its treatment, J. Clin. Psych. 45(12):18-23.

EPIDEMIC FREEBASE COCAINE ABUSE: A CASE STUDY FROM THE BAHAMAS

James F. Jekel

Yale University School of Medicine
New Haven, Conn.

David F. Allen

National Drug Council
Nassau, Bahamas

Henry Podlewski, Nelson Clarke, and Sandra Dean-Patterson

Sandilands Rehabilitation Centre
Nassau, Bahamas

Paul Cartwright

Community Psychiatry Clinic
Nassau, Bahamas

ABSTRACT

    Beginning in 1983, a sharp increase was noted in the number of new ad-
missions for cocaine abuse to the only psychiatric hospital and to the primary outpatient clinic in the Bahamas. For the two facilities combined,
the new admissions for cocaine abuse increased from none in 1982 to 69 in
1983, then to 523 in 1984. Although there was some evidence for a rise in
cocaine use during this time, due to increased availability and a drop in
price, a primary cause of this medical epidemic appeared to be a switch by
pushers from selling cocaine hydrochloride, which has a relatively low addicting potential, to almost exclusive vending of the freebase form, which
has a very high addicting potential and produces severe problems quickly.
Although the rate of freebase cocaine use is rising around the world, this
is the first reported medical epidemic due almost exclusively to freebase
cocaine use.

INTRODUCTION

    Cocaine has long been a part of the Western illicit drug scene, although
until recently its high cost restricted its use primarily to persons with
considerable financial resources. Within the past decade, there has been an
increase in the use of cocaine in the United States and the United Kingdom,
perhaps due in part to the fact that it was not generally perceived as

being as dangerous or as addicting as heroin.(1-4) However, data are ac-
cumulating to suggest that cocaine is a very dangerous drug and produces
more psychological dependence than does heroin. (5) Such data come from
both clinical studies (6,7) and from animal studies. (8)

Howard, et al., reported a case of acute myocardial infarction following
cocaine abuse in a young woman with normal coronary arteries. (6) Pollin
reports "...data from the National Institute on Drug Abuse's DAWN system
indicate that there was a 91% increase in cocaine-related deaths from 1980
through 1983." (9) Kleber has suggested that certain drugs have a low pro-
clivity for producing compulsive-addictive behaviour (e.g., less than 15%
of people using that drug become addicted); examples would include alcohol
and marijuana. At the other extreme are drugs that cause a high proportion
of users (over 90%) to develop a compulsive-addictive pattern; examples
include nicotine and heroin. Significant to this report is his suggestion
that cocaine may tend toward either extreme, depending on the method of use.
For example, if the use pattern is either by nasal inhalation (snorting) or
by mouth, including chewing coca leaves as is often done in South and Central America, a relatively small proportion of users may develop the compulsive-addictive pattern, whereas if it is smoked (freebased) or injected
(shooting), a high proportion of users develop the compulsive-addictive
pattern. Thus, a switch in pattern of cocaine use from snorting to freebasing could produce a major increase in the number of addicted persons, without a change in the overall prevalence of users.

The epidemiology of cocaine abuse is not well understood, in part due
to the difficulty in obtaining cooperation in studies of illegal drug use,
and partly because the pattern of cocaine use appears to be changing.(1,2,4)
Some have called the current American drug situation an "epidemic", because
cocaine use appears to be rising steadily. (2) However, it is more accurate
to consider the recent trend a "long term secular trend," due to gradual
societal changes, rather than an "epidemic" which reflects a rather sudden
imbalance between the forces promoting a disease or problem and those re-
tarding its change. In this paper, we report a change in cocaine use in
the Bahamas that meets the epidemiologic criteria for an epidemic of cocaine
abuse.

The study was initiated by physicians in the Bahamas who were concerned
about an apparent rapid increase in cocaine abuse in clinical settings.
They were assisted by a Fulbright Faculty Research Fellow (JFJ). Various
data sources were examined retrospectively to determine the size and character of the drug problem to decide the accuracy of the clinical perception
of a recent large increase in cocaine-related admissions to psychiatric
facilities.

METHODS

There is only one psychiatric hospital in the Bahamas, the government-
run Sandilands Rehabilitation Centre (SRC) on the island of New Providence.
Patients are referred to that hospital from the rest of the Bahamas (the
Family Islands). In 1980, almost two-thirds of the Bahamian population
lived on the small (about 7 by 21 miles) island of New Providence, where the
majority of people live in the capital city of Nassau. The other two hospitals on New Providence, and the one acute hospital on Grand Bahama (the
only other hospital in the nation) seldom accept drug abuse patients, and
they have low censuses for psychiatric patients in general. Therefore,
most of the drug abuse patients admitted to hospital go to SRC.

Likewise, the primary government-sponsored community mental health
(psychiatric) clinic in the Bahamas is the Community Psychiatry Clinic (CPC)
in Nassau. Most patients who do not go to private psychiatrists or other

private physicians use the outpatient services of either the SRC or the CPC. The primary exceptions are two smaller government clinics in Freeport and Eight Mile Rock, Grand Bahama, which saw a total of 47 cocaine addicts in 1984, only 14% of the cocaine addicts seen by Bahamian mental health clinics that year, and only 9% of the cocaine addicts seen at government facilities. Drug abuse patients seen in emergency rooms are routinely referred to SRC. Therefore, the data from the CPC and the SRC regarding the incidence of psychiatric problems are more complete than could be obtained in most areas of the world. Unfortunately, age and sex specific population data from the 1980 Bahamian census of the population were not yet available, precluding our converting the data on incident cases into rates. However, because the population was essentially stable over the relatively short period in this study, the data on trends of new cases are almost as interpretable as rates. For the purpose of this study, an incident case of cocaine abuse was defined as the first admission to either the CPC or the SRC for cocaine abuse, even if other diseases were present. If cocaine were the predominant drug of use for a polydrug user, the person was considered to be a cocaine abuse admission.

Data Sources

The Community Psychiatry Clinic publishes a monthly summary of the number of new and repeat patients by diagnosis and sex. This study focused on new patients in an attempt to estimate the incidence of treatment of new cases. These monthly data were added to study the number of new patients by sex and diagnosis for each quarter of a year. Alcoholism, non-cocaine related drug abuse, and cocaine related drug abuse were studied from the beginning of adequate records in 1982 through the first half of 1985.

Likewise, the admission logs of the Sandilands Rehabilitation Centre (SRC) were studied to obtain: 1) the total number of admissions each month from 1980 through 1984 by sex; 2) the number of these patients admitted for alcoholism and/or drug dependence; 3) if drug dependence, whether or not cocaine was the primary drug; and 4) whether or not this was the first or a repeat admission to SRC. These data were analyzed by quarter of the year during the study period. Admissions to both facilities for alcoholism showed only a slow, steady increase over the study period and therefore will not be discussed further.

The total number of new cases does not reflect the incidence of new cases of drug abuse among the relatively small wealthy population of the Bahamas, who usually seek health care from other sources, including the nearby United States. Moreover, some drug abuse cases from the Family Islands were undoubtedly treated in situ by local physicians. However, because there was no evidence for a change in the accessibility of care or the referral patterns, the marked change in the pattern of new admissions undoubtedly reflects real changes in the size of the community problem.

Duplicate measurement was another potential research problem. It was possible that patients could have been admitted to both the CPC and the SRC. There was no central data system that could examine these data for duplications. However, some physicians work at both facilities, and on the basis of their observations, the problem of overlap in these data is considered minimal. We found confirming evidence in that, during the study period, only four drug patients admitted to the SRC were referred from the CPC. Likewise, in discussion with the majority of the few private psychiatrists on New Providence, it was clear that only a small number of Bahamian drug abusers they see are not referred to either the CPC or the SRC. Therefore, because of the lack of alternative medical sources for most of the Bahamian population, we concluded that the combined incidence data on new drug abusers from the CPC and the SRC represent the large majority of the new abusers seeking medical treatment in the Bahamas. Here the term "drug abusers" was used to mean drug users whose use had led them into severe enough problems to require their seeking medical assistance.

RESULTS

Community Psychiatry Clinic (CPC)

The CPC was started in 1980, and it started keeping monthly statistics in January, 1981. For the first year, no distinction was made between new and returning patients, so 1981 statistics could not be used for the purposes of monitoring incidence. Therefore, the analysis of data from this clinic begins in 1982. In 1982, only four patients with a drug-based problem, none of which were cocaine related, were admitted to the CPC. In 1983, however, there was a total of 37 cocaine-based CPC admissions, in addition to 18 drug admissions where cocaine was not considered to be the primary cause for admission. In 1984, the numbers rose still higher to 299 cocaine-based CPC admissions and 43 due to other drugs. During 1984, the average number of cocaine admissions per quarter was 75, compared with 9 the previous year, and none the year before that. For the first half of 1985, there was a quarterly average of 55 cocaine admissions and 13 due to other drugs, down slightly from 1984. During the early phases of cocaine's appearance at the CPC, some of the cocaine use may have been recorded only as "drug abuse" or "drug dependence," but the number of such cases would have been small. If, as was usual, the patient was a polydrug user, the predominant drug, i.e., the one that appeared to precipitate the problems for which the user sought help, was recorded.

Drug abuse increased from 1% of the clinic's patients in 1982 to 9% in 1983, and to 39% in 1984. For the first six months of 1985, it continued to be 31%. The big increase in new patients in 1984 was due almost entirely to cocaine dependence. In fact, for all practical purposes, the cocaine related new admissions began in the third quarter of 1983. The number of new cases of depression and/or schizophrenia by quarter was fairly stable over time, suggesting that the increase in drug patients was not primarily due to increased clinic awareness.

Sandilands Rehabilitation Centre (SRC)

As the only government psychiatric hospital in the Bahamas, the SRC receives patients from all over the nation. The SRC has a long tradition of treating acutely ill alcoholics as well as acutely ill drug dependent individuals. The majority of the Bahamian population (65%) live on the island of New Providence, and because of proximity, 86% of the 1984 drug admissions to SRC were from that island.

Although there were a few cocaine related admissions during the first three quarters of 1983, a marked increase began in the last quarter in 1983, which was only slightly later than the increase seen in the Community Psychiatry Clinic. The number of first drug admissions for which cocaine use was the primary cause increased from one in 1980 to two in 1981, none in 1982, 32 in 1983, and 224 in 1984. The number of first admissions primarily due to drugs other than cocaine was more stable, but also increased generally over time: 64 in 1980, 40 in 1981, 59 in 1982, 47 in 1983, and 86 in 1984. Although the admission logs for early 1985 were available, the tremendous increase in patient admissions, largely due to cocaine abuse, resulted in less complete recording of the admission number after November, 1984. Therefore, data were estimated for the last quarter of 1984, based on the data from the first two months of that quarter.

Often a patient would be admitted with cocaine or other drug abuse
along with symptoms suggestive of an underlying psychiatric disease. Usually
the paranoia, hallucinations, etc., were due to the drug use. Therefore, in
all cases where cocaine or other drug abuse was indicated as being important,
the patient was considered a drug admission for the purposes of this study,
even though sometimes other diagnoses were listed also.

Mode of Cocaine Use

Most of the cocaine use in the Bahamas in 1984 was "freebasing", smoking
the ready to smoke freebase rocks, which accounted for about 98% of the cocaine related referrals that year. In the eastern U.S., ready to smoke freebase cocaine is called "crack." Freebase cocaine is produced when the powder form, cocaine hydrochloride, is treated with an alkali, usually sodium
bicarbonate. It is volatile with modest heating and is easily absorbed
through the lungs and rapidly transmitted to the brain.

The somewhat complex method of making freebase cocaine was used by some
experienced addicts in the early 1980s, but many users either did not know
how to make the freebase cocaine or did not wish to go to the trouble of
making it, so the predominant form of use at that time was snorting the hydrochloride powder. By 1984, however, the pushers were selling only the
freebase form, which was smoked using a home-made pipe called a "camoke."
Using this method, up to 80% of the cocaine reaches the brain, and the rush
can begin in 8-12 seconds, producing the "hit" or "rush", which is a short
period of intense ecstatic pleasure. This fleeting sensation of delight is
most powerful on the first use of cocaine, and although the addict always
seeks to repeat this feeling, it becomes elusive, and the same high is not
reached again.

The most common pattern of usage varied from a few hours, when the user
may consume 4 to 5 grammes of cocaine, to a few days of intermittant use,
usually over the weekend, when up to 10 grammes of cocaine may be consumed.
Most patients reported using cocaine while congregating with friends at
freebasing parties or "base houses." Solitary use occasionally occurred.

Clinical Spectrum

Cocaine dependent individuals usually presented for help during or
after some crisis. These crises were usually financial, social, medical,
psychological, or a mixture of these reasons. The characteristics of each
were:

Financial. After spending their total financial resources, the addict
often sought help, either voluntarily or under pressure from family members,
friends, or employers.

Social. Because of the need to support their habit, freebase addicts
often encountered problems with the law, through stealing, violent behaviour,
etc.,and were referred by the courts for treatment. Usually the victims of
the stealing, trickery, and blackmail were family members, friends, or employers who subsequently initiated police involvement. If the case reached
formal court proceedings, frequently the addict would be referred for treatment.

Medical. Cocaine addicts presented with a number of medical phenomena
such as epileptic fits, severe itching (the cocaine bug), loss of consciousness (tripping out), cardiac arrythmias, vertigo, pneumonias, gastrointestinal symptoms, and avitaminosis associated with severe malnutrition. A
number of pregnant addicts were referred from maternity wards.

Psychological. The cocaine addicts often presented with severe depression, manifested by unkempt appearance, insomnia, anorexia, withdrawal, and suicidal ideation. In the period 1984 to mid-1985, there were at least ten cocaine-associated deaths, five of which were suicides. Cocaine psychosis was a frequent presentation at the time of referral. This was usually manifested by severe agitation, impaired judgment, extreme paranoid ideation, intense denial, violent behaviour, threats of a suicidal or homicidal nature, and bizarre hallucinations, usually of the auditory and visual type. The patients were contentious, and in periods of lucidity tried to mislead the physician or manipulate the evaluation. Usually a family member or friend was useful to provide consensual validation of the psychotic state.

Demographic Characteristics

For 1984, 19% of the cocaine admissions to SRC were women and the rest were men. About 86% of all first drug admissions were from the island of New Providence (Nassau). In 1984, the age of drug abusers on first admission to SRC varied from 11 to 56 years for males, with a mean of 24.8 years. The cocaine users were slightly older than other drug users, 25.8 years compared to 22.5 years. In 1984, the comparable figures for women were from 15 to 39 years with a mean of 24.4 years. The overwhelming majority of the patients, including the drug abuse patients, were Bahamian.

Other Forms of Surveillance

Data were obtained from the first full year of operation of the Sandilands Rehabilitation Centre's Drug Clinic (1984). Because most of these drug patients were referred from the SRC following discharge there, including them in the data reported here would be duplicative and therefore was not done. However this clinic kept records of all drug users and the type of drugs they were on. Approximately 60% of the patients were using both cocaine and cannabis, although it was usually the cocaine use which precipitated the hospital admission.

An attempt was made to see if the Bahamian vital statistics would be useful for surveillance purposes, particularly whether or not it would be useful to monitor suicides and drug related deaths. Neither of these diagnoses is usually recorded on the death certificate in the Bahamas, apparently for legal as well as cultural reasons. Therefore, in consultation with persons from the Department of Statistics, Government of the Bahamas, it was decided not to use the vital statistics as a surveillance method.

Police statistics showed there was some increase in street drug arrests in 1984, but not the magnitude suggested by the clinic and hospital admission data. For example, for the four years prior to 1984, the annual average of drug arrests was equal to 1094, with no clear trend over time. There were 1501 arrests for 1984, an increase of 37%. It is possible that the police statistics underestimated the increase in volume of drug traffic, due to limitations of police resources. However, the medical impact seen in 1984 was far greater than changes in the volume of drugs arrests would indicate.

DISCUSSION

In the Bahamas, surveillance of the public psychiatric facilities provided documentation of an epidemic of drug abuse, especially cocaine abuse, requiring medical care. Often these drug abusers also used marijuana and alcohol to control adverse symptoms from cocaine use. There was either a major change in the incidence of new drug users, especially cocaine users, or in the method of use. Our data cannot directly decide between these two options, but they show that in early 1983, something upset the previous drug use equilibrium, suddenly forcing hundreds of individuals into medical facilities for treatment of complications of drug abuse that were not appearing previously. The most obvious hypothesis was that cocaine was suddenly introduced for the first time, or was introduced in a much greater volume than previously, probably with a significant drop in unit price, due to market forces. The testimony of former addicts who were using cocaine as early as the 1970s confirms that cocaine powder was available, if costly, for many years but that some time in late 1982 and/or early 1983, cocaine rather suddenly became much more plentiful, due to greater production in South America. As a result, the street price of cocaine in Nassau dropped to approximately one-fifth of its prior price.

Further discussion with ex-addicts who were familiar with the street scene since the late 1970s revealed one other critical bit of information. At approximately the same time that the cocaine became more plentiful and less costly, the pushers switched their vending practices from the powdered form of cocaine (called "snow"), which was used mostly for snorting (nasal inhalation) or shooting (injection) to the pure alkaloid form (called "rocks" or "freebase" or "crack") which is used exclusively for smoking (freebasing). In fact, it quickly became more difficult to obtain the powder at all in Nassau. By making this change, the pushers were essentially forcing all cocaine users to become addicts, and therefore they would buy far more of the drug from the pushers. Many of the pushers were addicts themselves, and dealt cocaine to feed their own habits. Selling freebase cocaine guaranteed an eager, even desperate market for the increasingly available cocaine. It is not known to what extent the motives for this marketing change came from the desire to create a captive market of addicts, or to what extent it was a response to user demand. However, the basic outlines of this scenario were confirmed by everyone we talked with.

A related hypothesis concerns the potency of the drug. Smith reported that an important reason for more cocaine deaths in the San Francisco area was "higher potency cocaine." (10) Siegel reported that the recovery of cocaine freebase from pure cocaine hydrochloride, using various street kits ranged from 41% to 72%, and although the kits removed some of the adulterants, lidocaine and ephedrine were partially left with the cocaine. (1) Ex-addicts indicated that the street cocaine powder in Nassau had usually been cut about 50% before sale. Although we had no direct data about the purity of the cocaine, the sources for the cocaine remained similar, and as indicated above, the process of extracting the freebase was not perfect. Therefore, changes in levels of purity seemed to be an inadequate explanation for the observed findings.

CONCLUSIONS

Thus we conclude that the "medical epidemic" of cocaine-related physical and psychiatric problems in the Bahamas was related to the interactive effect of greater availability of cocaine at lower cost and the switch in the method of vending from powder to almost exclusively the treated, freebase form. This does not imply that there was no freebasing in the Bahamas before 1983, or that it would not have increased without intervention by the pushers. However, because of the ease of use of powder as well as the expense factors, the clinical abuse in all probability would have been an upward secular trend, rather than an epidemic, and mixed forms of cocaine use would have continued rather than changing almost exclusively to freebase smoking.

Not only is greater availability of cocaine a threat to a society, but the form in which it is sold is critical to the epidemiological pattern of medical demand. Availability at low cost combined with vending of freebase cocaine appears to be an especially dangerous combination. Monitoring the method of vending may be critical for "target" Western nations, as well as developing nations (particularly producer or trans-shipment countries such as the Bahamas and some South American countries).

Surveillance of medical facilities, especially psychiatric facilities, may be a relatively quick and effective form of monitoring some aspects of the drug situation in a country. Surveillance of the drug situation in producer and/or trans-shipment countries may be an important early warning method for target nations, such as the U.S. and the U.K. Therefore, it may be in the self-interest of target nations to assist the struggling producer and trans-shipment countries not only to control the problem but also to maintain an intensive surveillance system.

As freebasing becomes more popular in the U.S. and the U.K., and particularly if the vending pattern switches to the freebase form (and there is evidence that this is happening in certain U.S. cities), emergency rooms, mental health clinics, and psychiatric hospitals should prepare for an unprecedented load of drug addicts.

REFERENCES

1. R. K. Siegel, (1982), Cocaine smoking, J. Psychoactive Drugs, 14(4): 271-355.
2. L. N. Robins, (1984), The natural history of adolescent drug use (editorial) AJPH 74(7):656-657.
3. P. M. O'Malley, J. G. Bachman, L. D. Johnston, (1984) Period, age, and cohort effects on substance abuse among American youth, Am J Pub Health, 74:682-688.
4. J. J. Forno, R. T. Young, C. Levitt, (1981) Cocaine abuse - the evolution from coca leaves to freebase, J. Drug Education 11(4):311-315.
5. J. V. Spotts, F. C. Shontz, (1984) Drug-induced ego states. I. Cocaine: phenomenology and implications, Int J. Addictions 19(2):119-151.
6. R. E. Howard, D. C. Hueter, G. J. Davis, (1985) Acute myocardial infarction following cocaine abuse in a young woman with normal coronary arteries, JAMA 254(1):95-96.
7. H. D. Kleber, F. H. Gawin, (1984) The spectrum of cocaine abuse and its treatment, J Clin Psychiat, 45(12):18-23.
8. M. A. Bozarth, R. A. Wise (1985) Toxicity associated with long-term intravenous heroin and cocaine self-administration in the rat, JAMA 254(1):81-83.
9. W. Pollin (1985) The danger of cocaine (editorial) JAMA 254(1):98.
10. D. E. Smith (1984), As quoted in Medical News, Dec. 10, 1984, p. 6.

ACKNOWLEDGEMENTS

The authors acknowledge the support and assistance of the following persons: Hon. Norman Gay, M. D., Minister of Health, Commonwealth of the Bahamas; Ms. Barbara McKinley-Coleman, National Drug Council; and Mr. Mozes Deveaux, Sandilands Rehabilitation Centre. This report is dedicated to Mr. Chrysostom Finlayson who, in addition to providing critical insight for this paper, gave his life in the war against drugs.

This study was supported, in part, by a grant from the Council for International Exchange of Scholars, U. S. Information Agency.

A shorter, edited version of this paper appeared in the March 1, 1986, issue of The Lancet; this version is printed with their kind permission in this volume.

DRUG ABUSE 1975-1985: CLINICAL PERSPECTIVES OF THE BAHAMIAN EXPERIENCE OF

ILLEGAL SUBSTANCES

Nelson A. Clarke

Sandilands Rehabilitation Centre
Nassau, Bahamas

INTRODUCTION

In 1969, Sandilands Hospital was the only in-patient psychiatric facility in the Bahamas. Hospital records show that for that year, there were twenty-two persons admitted to hospital who were treated for drug related conditions. In 1970, twenty-seven drug related cases were admitted to hospital for treatment and in 1971, there were forty-five, and in 1972, thirty-four.

Throughout the period 1969-1972, the hospital records indicated that within the group of persons being treated for drug related conditions, males outnumbered females by a wide margin. The substances concerned were mainly marijuana (cannabis) and Quaaludes (Methaqualone). LSD (diethyl lysergic acid) and cocaine were mentioned in a few cases but were difficult to substantiate.

Spencer (1971) reported in his paper on cannabis-induced psychosis that a dozen young men treated at the Sandilands Hospital appeared to be associated with marijuana smoking. He described the symptoms as aggressive behaviour, gross psychomotor overactivity, bizarre grandiose delusions, passivity and amnesia for the onset for the illness, and discussed the possible role that marijuana may have had in the causation of the illness. He concluded that the type of illness described, psychosis, and its course were not typical of schizophrenia although the two illnesses appeared similar, clinically.

Podlewski noted in his paper "Drug Abuse in the Bahamas: A Position Statement, February, 1974," that alcohol was by far the most popular and most frequently abused substance amongst all age groups. Although there were indications that cocaine was available in the Bahamas, only a small but increasing number of individuals were using the drug. The method of use at that time was snorting (inhalation of the finely ground powder). He stated that this situation appeared to be quite new.

THE COCAINE FREEBASE PROBLEM

It can be stated with some certainty and a degree of caution that prior to 1978 individuals using cocaine as a recreational drug were chiefly snorting the hydrochloride powder.

There is some evidence that cocaine was in no way a popular drug on the Bahamian scene at that time, and chronic use seems to have been limited to a small number of individuals who were affluent enough to afford the drug. The market price of a gramme of cocaine at that time was approximately B$100-120.

It is thought that the phenomena of using cocaine in its freebase form began to grow popular during the early part of 1979. No one is quite sure exactly when this took place, but according to individuals in treatment, 1979 seems to have been the year when it happened.

It was noted that the number of admissions to psychiatric hospital began to increase for cocaine related problems, that is, psychotic states, induced by cocaine, as well as those persons admitted for treatment for cocaine dependence. During the latter part of 1979 and early 1980, the increase was gradual; but by the end of 1980, this increase had become much more dramatic.

The evidence suggests that several factors were responsible for the increased level of cocaine consumption and the resultant increase, suggesting an increase in cocaine trafficking through the islands. An increase in the number of persons arrested locally for possession of cocaine indicated an increase in the popularity of cocaine among Bahamians. Information from persons in treatment suggested a marked decrease in the price of cocaine on the local market.

The use of freebase cocaine continued to gain popularity throughout 1981, 1982, and 1983. In 1983, it was reported by patients in treatment that the price of cocaine on the local scene had decreased further: a gramme of cocaine could be bought for $30 to $40 and even obtained free in some instances.

In 1982 and 1983, most of the patients presenting themselves for rehabilitation for a cocaine dependence problem were males between the ages of 18 to 30 years old. Although all social classes were represented more than fifty of these patients were from the lower socio-economic groups.

Over 1983 and 1984 a number of important changes were noted. Firstly, the number of patients seeking help for cocaine dependence had increased from 69 patients in 1983 to 220 patients in 1984. This dramatic change strained the existing facilities, the Sandilands Rehabilitation Centre and the Community Psychiatry Clinic, to the point where they found themselves unable to cope with the demand for treatment. Secondly, more females were presenting themselves for treatment for cocaine dependence: 9 in 1983 and 51 in 1984. Thirdly, the age range of persons of both sexes presenting for treatment had widened. Persons as old as 50 years were among the group and persons under the age of 15 were becoming increasingly common.

A final observation noted during this period was that the increased number of individuals who were suffering from mental illnesses as a result of cocaine use reported that their method of use was freebasing. It is incidental perhaps, but important to note that most of these individuals reported concomitant use of either alcohol or marijuana and more commonly, both alcohol and marijuana.

COMMON PATTERNS OF FREEBASE COCAINE USE

The most common patterns of use of freebase appeared to be daily use, with the period of time per day actually using ranging from 1 to 2 hours to 4 to 5 hours of intermittent use. During this time it is often reported that one user may consume as much as 3 to 5 grammes of cocaine. Less commonly, users may spend 2 to 3 days, usually over the weekend, in constant, intermittent use.

Most patients report that they use their cocaine in the company of others, however, the heavier users often prefer to freebase in solitude. A variety of social phenomena have occurred which facilitates the use of cocaine freebase: the basehouses and base camps are two such entities where cocaine may be easily obtained and used in relative safety.

THE VARIETY OF CLINICAL PRESENTATIONS

The clinical presentations of individuals with cocaine addiction/dependence problems varies from the occasional user, who appears to enjoy his habit, but who still claims to have some degree of control over his use to the severely addicted individual who has lost all control and whose existence is governed by powerful urges to use cocaine incessantly.

A number of individuals present themselves for help because of problems arising from their use of cocaine; they consider that the cocaine problem itself is temporary, secondary, and incidental. These problems are often related to job performance, marital difficulties, legal problems and other domestic problems.

Frequently individuals are admitted to psychiatric wards with illnesses which are thought to be cocaine related. The illnesses commonly seen are severe depressive states, schizophreniform psychoses and manic episodes.

SYMPTOMS COMMONLY REPORTED BY COCAINE DEPENDENT PERSONS (FREEBASERS)

The symptoms may be divided into two main categories: (1) physical symptoms, and (2) psychological symptoms. A further division of these two major groups may be made into (a) acute symptoms, which occur during and immediately after a bout of freebasing up to approximately 72 hours after use; and (b) late symptoms, which appear after one has been a regular user approximately one to three months, that is, long-term effects.

Commonly reported acute physical symptoms include: headaches, dizziness, coughing, increased muscle tension, abdominal cramps, diarrhoea, blurred vision, palpitations, increased body temperature with sweating, insomnia.

Long-term physical effects commonly reported include: weight loss, loss of muscular power, skin rashes, menstrual irregularities in females, impotence in men.

Commonly reported acute psychological symptoms include: feelings of power and increased abilities, increased feelings of self-esteem and importance, hypersensitivity to auditory and visual stimuli, misinterpretation of auditory and visual stimuli, paranoid ideas, visual and auditory hallucinations, usually of a paranoid flavour. Powerful urges to re-engage in using cocaine are very common.

Long-term psychological effects and late psychological symptoms include: loss of drive and motivation, increased irritability, depressed mood, loss of self-esteem, and feelings of worthlessness, suicidal ideas.

PROFILE

The profile of an individual with a cocaine problem presenting for help and rehabilitation is as follows:

He is usually male and between the ages of 18 to 30 years; he is unem-

ployed and has a history of previous drug abuse including marijuana or alcohol or both in most cases. He may have sold drugs on a small scale to
support his cocaine habit. He may be in a stable relationship. He may have
been arrested for drug related offences. Relationships within his nuclear
family (i.e., mother, father, sisters, brothers) are poor. He usually has
the attitude that his cocaine problem is his one and only drug problem and
that his use of other substances is inconsequential.

CASE HISTORIES

     Case histories may be helpful in understanding the problems, and appre-
ciating the clinical aspects of the cocaine problem and presentations.
Following are three such examples.

Case (A)

     K is a young man 14 years old who lived with his mother and two siblings
prior to his commital to hospital. He admitted to having used marijuana
amost daily since he was nine years old and said he was introduced to cocaine freebase when he was 12 years old, by an older friend. Admission to
hospital had to be compulsory because he refused to get help voluntarily.
When confronted by his mother about his drug problem he became angry and
abusive.

     His mother stated that he stopped attending school three months prior
to his admission to hospital. He said that he had lost interest in school
long before that time. His performance at school had been very poor over
the period of the last four years. After leaving school, his cocaine habit
grew from occasional use on weekends to a daily habit.

     He often experienced severe cravings for cocaine, resorting to stealing
valuables and money from his mother in order to buy the drugs. When this
was not possible, he smoked marijuana which he felt helped him to endure
the cravings. He had experienced several episodes of disordered thinking,
with paranoia, and persecutory auditory hallucinations, during which he had
displayed violent behaviour. He stated that he had long ago stopped enjoying the effects of cocaine and could not really understand why he seemed
to desire it constantly. He was irritable and constantly in a depressed
mood. After three days in hospital, he absconded by jumping over the hospital fence.

     This case exemplifies the devastating effect of the addiction of
cocaine on a teenager and his family.

Case (B)

     M, a 26 year old male, was admitted to hospital after he had stolen
$300 from his place of work, the second time he had stolen such a large
amount of money in order to support his cocaine habit. He was admitted on
a voluntary basis. He stated that he felt that his cocaine problem needed
to be solved or he might get himself into further trouble. He was single,
lived at home, and was a high school graduate.

     He had tried to stop previously but had only managed to stay away from
cocaine for a period of two weeks. He had a responsible job in a government
ministry, and earned a reasonable salary which was almost entirely spent on
cocaine. He had been using cocaine from the time he was 20 years old, first
at parties and only by snorting, usually obtaining it free from friends.
He was introduced to freebasing by an older brother at age 23 and for a
time, they used together. As he developed a liking for the drug, and did

not wish to share with his brother, he began to use it alone and avoided sharing his cocaine.

At the time of admission, he was using cocaine daily, and often stayed at home one or two days per week, either because of his use or in order to use. He claimed that although he found the crash after using cocaine for a number of hours uncomfortable, and unpleasant, he solved this problem by using alcohol in large quantities in order to get to sleep, and to help to alleviate his depression and guilt feelings. Since his introduction to cocaine he had stopped using marijuana and stated that marijuana had never been a favourite drug of his and that since discovering freebase cocaine, he felt no desire to use marijuana at all.

M was successful in completing a ten-week period in the rehabilitation programme and is now in the follow-up programme; he has been drug free for almost one year. This case history illuminates the fact that some persons with cocaine dependence may be successfully rehabilitated, even though their problems seem severe.

Case (C)

J is a 29 year old female, married, but separated. The mother of three children aged 13 years, 9 years, and 1 year, she was admitted to residential drug rehabilitation programme in September, 1985; when she claimed that she sincerely wanted help for her freebase cocaine problem, she said that she had been using cocaine continuously during the previous week. When she had been introduced to cocaine in early 1983 by a friend, she tried snorting the hydrochloride powder. Shortly after her introduction to snorting, she tried freebase cocaine and says of that experience, "I liked it so much that I never wanted to stop; it was such a wonderful feeling."

She said that she had managed to stay away from cocaine for two months once, but drifted back to freebase cocaine. She regularly spent $150 per day on cocaine, and when she had no money, she would perform sexual favours for money or drugs. She was unemployed but was able to get money from friends and family members.

At the time of admission to the rehabilitation programme, she was romantically involved with a drug dealer, whom she claimed she intended to marry. Although she was able to complete the three month inpatient period, her insight into her drug problem remained poor, and her desire to stop using drugs never seemed genuine. She did not appear to understand the necessity for follow-up therapy. J left the inpatient programme a drug free individual and went to California to reunite with her husband.

Once in California, she grew increasingly unhappy with the marriage situation and eventually discovered places where cocaine freebase could be obtained. The first occasion of use resulted in a binge of several days' duration when she stayed away from home for two days. On her return home, her husband forgave her but warned her that if it happened again, she would be sent back to the Bahamas. It happened a week later. The third time it happened, she was sent back to the Bahamas.

Since her arrival in early 1986, in her own words, "I have freebased every day." She was readmitted for inpatient care in late August, 1986, and is now awaiting formal admission to the Drug Rehabilitation Programme.

When asked about her reasons for her relapse, she claims that whenever she argued with her husband, she felt tempted to return to cocaine freebase. She admitted that whenever things did not go her way, she felt intense cravings for cocaine. During her first admission for drug rehabilitation, she had refused to learn methods for coping with cravings for cocaine.

J now understands that in order for her to give up cocaine, she will need to carefully explore her past relationships with cocaine and other drugs and find ways of coping successfully with cravings, and without drugs.

These clinical vignettes are samples of the plethora of clinical presentations within the spectrum of cocaine addiction/dependence.

COCAINE PSYCHOSIS

Theo Manschrek

Harvard University
Cambridge, Mass.

Psychosis has been a relatively uncommon reaction to cocaine in this century: with freebase, however, this reaction is no longer unusual. The purpose of this chapter is to highlight the rising abuse of freebase cocaine (also known as crack) from the standpoint of the clinical psychopathology of psychosis, arguably one of the most serious of cocaine manifestations. I would first like to define cocaine psychosis; second, to summarize some of what we know about this disorder; third, to relate that knowledge of the Bahamian experience with freebase psychosis; and finally, to suggest some ways in which our knowledge should develop.

What is cocaine psychosis? Let me define it as severe psychopathology induced by cocaine that can result in disturbances affecting awareness, perception, thinking, and behaviour such that the individual affected may be described as out of touch with reality. In terms of critical psychotic symptoms, the patient may be hallucinated, confused, or disoriented, deluded, incoherent, aggressive, and/or prone to repetitive behaviours such as staring or other stereotypic movements. Associated symptoms could include irritability, suspiciousness, euphoria, depression, lability of emotional response, reduced attentiveness, increased sensitivity to visual and auditory stimulation, impulsiveness, and disturbed judgment.

Several features suggest that cocaine psychosis is indeed a most severe form of psychopathology. First, psychosis reflects intense brain dysfunction. We are aware, for example, that profound changes in a variety of neurotransmitters occur in response to cocaine, presumably in concert with the clinical psychopathology (Mule, 1984). Some have theorized that kindling mechanisms may be involved in the development of cocaine psychosis, possibly involving electro-chemical or even structural changes in the brain (Post and Kopanda, 1976). Second, the psychosis is also severe because of associated features such as potential suicide, death by accident, social withdrawal and isolation, and violent behaviour (National Drug Task Force, 1984). Third, there are possible long term complications of cocaine psychosis. The psychiatric ones, in addition to residual symptom patterns, include the high risk of relapse associated with the dependence producing nature of the drug. Social consequences such as child neglect, crime, disrupted careers, broken homes, and job loss create difficulties that make recovery and rehabilitation complicated. Injury from accidents and/or overdoses also take their toll. Also, the ominous possibility of residual defects of memory, attention, and perception cannot be overlooked.

I would like to propose that psychosis represents an important epidemiologic barometer of the extent and seriousness of the cocaine problem. Further, I would propose that the occurrence of psychosis in individual cases may be an important predictor of chronic residual symptomatology.

Ambivalence Toward and Reappraisal of Cocaine's Danger

Even the possibility that cocaine psychosis occurs, makes it surprising that the public image of cocaine has been ambivalent with respect to its dangerousness as a Lancet editorial observed in 1983. However, the recent surge of cocaine cases in the Bahamas underlines the dangerousness of the drug and more importantly, the danger of our ignorance. In 1984, the Bahamian National Drug Task Force documented the widespread nature and consequences of cocaine abuse. Recently, Professor James Jekel of Yale and colleagues from the Bahamas have concluded that a nationwide epidemic of freebase abuse exists related to the switch from the hydrochloride to the freebase form of cocaine (Jekel, et al., 1986). Freebase in short, is forcing a reappraisal of the ambivalent public stance toward cocaine. Why?

There are several facts that support reappraisal. First, the bioavailability of smoked freebase cocaine and its potent effects on the brain are at least equivalent to intravenous cocaine and possibly superior. Second, the purity of freebase is as high or higher than most other forms of available street cocaine. These two factors influence the dangerous cycle of dependence-producing repetitive administration of cocaine and enhance the likelihood of greater acute and chronic brain toxicity. While snorting cocaine may have low addictive potential, all indicators suggest that smoking freebase or crack has high addictive potential. Third, certain consumer attractions are noteworthy. Freebase cocaine is easy to use, does not require needles, and is readily available. It is low in cost, and is often sold already prepared (like crack in the United States) on Bahamian streets. In sum, these facts suggest that whatever can result from snorting cocaine hydrochloride can probably happen more readily and with greater severity with freebase cocaine abuse. The basic principle is that as drug potency and availability increase, more widespread consequences can become evident in the normal population, and especially among those who may be at risk for psychotic disorders (Kleber and Gawin, 1984). In short, the risk of serious consequences has grown.

Prior Observations on Cocaine Psychosis

What do we know about freebase cocaine psychosis? Unfortunately, we know little. Freebase psychosis is a relatively new phenomenon. We must, therefore, depend on a limited literature largely consisting of case reports from the earlier part of this century concerning psychosis associated with cocaine hydrochloride to gain even indirect insight about it. Let me discuss a few selected points from that literature to illustrate some of the problems and concerns, then turn attention to current lessons from the Bahamian medical experience with freebase psychosis.

A spectrum of phenomenologic presentations occurs with cocaine intoxication and with cocaine psychosis. The range that has been described runs the gamut from presentations resembling acute schizophreniform conditions and mania (which are thought to be more likely with chronic heavy use) to delirious, toxic psychoses (which appear to occur after large acute doses) (Cohen, 1984). With certain exceptions, then, the nature of the psychosis appears to be at least somewhat related to the dose and the duration of use. This variability of responses suggests that different factors besides the drug itself play critical roles in the development and sustaining of the psychosis. We are aware, for example, that prior disorder, including psychosis, may influence the occurrence of these disturbances. Robert Post (1975) of the Nation-

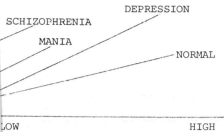

DEPRESSION
SCHIZOPHRENIA
MANIA
NORMAL

LOW   HIGH

LEVEL OF COCAINE DOSAGE

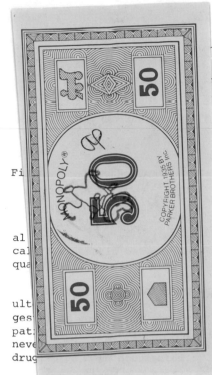

Fi... ...tion of dose and psychiatric state in cocaine
...ogy.

al... ...n the United States has proposed a hypotheti-
cal... ...isting psychiatric disorder and duration and
qua... ...re 1).

...ine is rarely used alone. In addition to adult... ..., marijuana, alcohol, and other drugs are in-
ges... ...ria of the crash effects. And in cases where
pat... ...question of etiology becomes more complex;
neve... ...e to cocaine use alone, involving no other
drug...

Cocaine psychosis is generally short-lived, running from hours to days,
occasionally longer. Often psychotic symptoms occur in association with the
acute intoxication but much less commonly they remain after the intoxication
ends. What we are learning about freebase is changing this view.

There is a great variability in treatment response to cocaine disorders;
and consistent and clear guidelines for treatment of cocaine psychosis are
not available. Lithium, neuroleptic medication, and a host of other treatments, including no treatment at all, have been attempted.

Another important factor in understanding cocaine psychosis is that,
sadly, there is a tendency to relapse, paralleling the pattern of dependence.
Also relevant is the similarity of cocaine effects to those of other central
nervous system stimulants, such as amphetamines (Beherot, 1970; Schiorring,
1981). The literature on amphetamine psychosis indicates remarkable similarities in psychopathologic consequences, pharmacologic responses, and behavioural effects of amphetamines and cocaine. Two characteristics of amphetamine abuse are especially noteworthy. One is the tendency to violence that
accompanies amphetamine intoxication; and the other, the frequency of unusual social isolation and withdrawal (Ellinwood, 1971). Both characteristics
have been observed with cocaine.

Finally, there are residual states of symptom and cognitive impairment
about which we know little. What are the chronic impairments associated
with cocaine psychosis? What consequences arise from the tendency to relapse? It is too early to judge as yet. However, there is the possibility
of a dementia-like problem, or at the very least, diminished cognitive ability associated with cocaine psychosis (Maier, 1926). In 1908 Gordon, in re-
viewing cases of cocaine addiction noted the following: "...alongside of
hallucinations and delusions, there is a gradual and progressive decrease in
intellectual force and moral sense. Dementia is the final outcome."

Clearly, freebase has greatly altered our perception of the cocaine pro-

blem. The experience in the Bahamas with respect to cocaine psychosis indicates that the Bahamas is on the cutting edge of understanding and grappling
with the seriousness and widespread nature of freebase or crack. Let us turn
our attention to the urgent lessons from the Bahamian medical experience with
freebase.

The Bahamas and Freebase Psychosis: Initial Observations

As noted, cocaine related disorders have been on the increase in the Bahamas for some time (Jekel, et al., 1986) and that among patients admitted
with cocaine disorders to the Sandilands Rehabilitation Centre, the proportion experiencing significant psychosis is large, perhaps as high as 25-30%
(Manschrek, et al., in preparation). Moreover, freebase is associated with
98% of cocaine admissions.

The Bahamian clinical experience suggests that certain groups are vulnerable to psychosis. The preponderance of young adult patients, some even in
their teens and most in their prime, who are abusing freebase raises concern
not only for the disabilities associated with dependence, but also for the
special susceptibilities and complications associated with cocaine psychosis.

Another vulnerable group consists of patients with pre-existing psychotic
or mental disorders. Such patients are frequently admitted for disruptive
and/or violent behaviour of a severe nature, markedly unlike prior admissions.
Such patients vex the clinical staff with their unusual refractoriness to
standard treatments, such as anti-psychotic medication and reduction of environmental stress.

Other vulnerable groups include those with pre-existing neurologic conditions, such as epilepsy, head injury, and mental retardation. For example,
a Bahamian patient suffering from frontal lobe syndrome with marginal adjustment at home developed severe cocaine abuse and psychosis and has been so unmanageable that intitutionalization became necessary. Finally, there does
not seem to be any immunity provided by social class. In the United States,
the perception has been that cocaine is for the well-to-do. With lowered
costs, and as the figures from the Sandilands Centre suggest, social class
certainly provides no protection against addiction.

Virtually every feature reported in the literature on psychosis has been
reported to occur in cocaine psychosis. Because of space, let me mention a
few relevant examples from the current freebase experience. In the United
States, cocaine psychosis frequently has an affective component, with even
manic forms of presentation reported perhaps more commonly than schizophreniform conditions. In the Bahamas paranoid disturbances are common reactions
to freebase. These have a range of associated features including disorganized behaviour, heightened mood, and violent or aggressive behaviour. Often
paranoid patients report experiences of being spied on by police or other authorities through holes in the walls, windows, ceilings, and so on. This reaction is called "peeping" in the Bahamas, and has been observed in other
settings.

Another important observation noted by Dr. David Allen, is what he describes as ethical fragmentation. Ethical fragmentation refers to profound
changes in lifestyle, such as the selling of one's body, car, or furniture;
lying, stealing, and becoming a con-artist. Such changes are frequent among
cocaine addicts and usually mark a distinct departure from previous behaviour. A breakdown in physical health related to reduced attention to instinctual needs often accompanies ethical fragmentation, and I would call
this disturbing condition instinctual fragmentation. Many patients seem to
forego the normal instinctual pleasures such as sex (contrary to popular
lore) and even ingestion of water and food because of preoccupation with

cocaine intoxication. In the extreme, patients lose considerable weight, and their general self-care and condition deteriorate to dangerous levels.

While no systematic investigation is yet available to document the issue, Bahamian patients who have experienced cocaine psychosis frequently report diminished short and long term memory long after abstaining from cocaine. Because of earlier reports of possible dementia such as in Gordon (1908), these observations should be taken seriously, and a collaborative study is underway to examine them.

Cocaine psychosis is an example of a serious public health challenge and it demands a range of effective responses. The Sandilands Centre and other facilities are experiencing vast increases in service demands because of co-caine, largely because of freebase. The clinical experience at these centres is unique in that large numbers of patients have been treated for freebase disorders. To increase our understanding of cocaine psychosis and other cocaine problems, I recommend that efforts be made to create a confidential record of all new cases on these services so that we can systematically col-lect clinical, laboratory, neuropsychiatric, and cognitive data and follow-up information on individuals who present with cocaine problems. This data base would be useful for education about the cocaine problem, would permit systematic controlled investigations of various treatments, and would be an important basis for research on specific groups of patients. For example, this approach could allow us to examine those individuals whose primary pro-blem is active psychosis, those with evidence of residual problems, perhaps memory and attentional difficulties, and those with special vulnerabilities such as schizophrenic disorders whose use of cocaine may drastically compli-cate an already serious condition. Those with multiple drug problems con-stitute another group about which useful observations could be made.

Conclusion

The benign image of cocaine as a recreational drug needs to be reap-praised. We need to increase our knowledge of freebase cocaine, cocaine psychosis, and other cocaine-related psychopathology. We need to develop research for systematic assessment of cocaine psychosis and psychopathology. Freebase cocaine has challenged our ignorance.

REFERENCES

Bejerot, N., 1970, A comparison of the effects of cocaine and sympathetic central stimulants, British Journal of Addictions, 65:35-37.
Cohen, S., 1984, Cocaine: acute medical and psychiatric complications, Psy-chiatric Annals, 14:747-749.
Editorial, 1983, Images of cocaine, Lancet II:1231-1232.
Ellinwood, E. H., 1971, Assault and homicide associated with amphetamine abuse, American Journal of Psychiatry, 127:1170-1175.
Gordon, A., 1908, Insanities caused by acute and chronic intoxications with opium and cocaine, JAMA, 51:97-101.
Jekel, J., Allen, D., Podlewski, H., Dean-Patterson,S. , Clarke, N., Cart-wright, P., Finlayson, C., 1986, Epidemic cocaine abuse: A case study from the Bahamas, Lancet, I:459-462.
Kleber, H., Gawin, F., 1984, The spectrum of cocaine abuse and its treat-ment, Journal of Clinical Psychiatry, 45(12, Section 2):18-23.
Maier, H. W., 1926, "Der Kokainismus," G. Thieme, Leipzig.
Manschreck, T. C., Laughery-Flesche, J. A., Weisstein, C. C., Allen, D. F., Mitra, N., Characteristics of cocaine psychosis, (in preparation.)
Mule, S. J., 1984, The pharmacodynamics of cocaine abuse, Psychiatric An-nals, 14:724-727.

National Drug Task Force, 1984, "Report of the National Task Force on Drug Abuse in the Bahamas (David F. Allen, Chairman)," Bahamas Government Publication, Nassau, Bahamas.

Post, R. M., Kopanda, R. T., 1976, Cocaine, kindling, and psychosis, <u>American Journal of Psychiatry</u>. 133(6):627-633.

Schiorring, E., 1981, Psychopathology induced by "speed drugs", <u>Pharmacology, Biochemistry, and Behaviour</u>, 14(Suppl 1):109-122.

COCAINE AND THE BAHAMIAN WOMAN: TREATMENT ISSUES

Sandra Dean-Patterson

Sandilands Rehabilitation Centre
Nassau, Bahamas

"How can there be a single cause of anything so complex as an arbitrarily defined collection of behaviour characteristics? We are not so naive as
to believe any longer that any historical act, a war or a revolution, for
example, has a single cause...When the Zen master says the 'self is like
the sound of one hand clapping,' he means that it is as impossible to conceive of the self without the environment as it is to conceive of the sound
of one hand clapping without the other hand." (Dumont, 1968:27,39)

Substance abuse in the form of alcohol abuse has long been with the
Bahamas. Indeed in the 1970s alcohol abuse was identified by the newly independent Bahamas as its foremost social and health problem and a major con-
straint to social development. In 1974, the then Minister of Health
referred to the extent to which Bahamian society was being eroded by alcoholism in the following way:

> "In our everyday life one sees only too often many instances
> in which the family unit is destroyed as a result of either
> the father or mother or in some cases both parents falling
> prey to alcoholism. The unfortunate children left on their
> own without material or emotional security in most cases be-
> come the juvenile delinquents and social dropouts that later
> inhabit our prisons and our mental insitutions." (Roker, 1974:11)

Twelve years later, the very same comments could as easily be made about
a new substance that is being abused at epidemic proportions. (Jekel, et al.,
1986) and indeed the Prime Minister of the Bahamas commented in 1985:

> "Sadly, the drug scourge has taken a turn for the worst over
> the past decade. Not only has it taken a toll on the social
> fabric of society, but it has also undermined the physical
> well-being of many citizens and has successfully sapped the
> economic resources of all affected countries, be they devel-
> oped or developing."

Celetano et al. (1980) tell us that alcoholism and drug dependence present very similar methodological and interpretive problems in research,
have similar groups at risk, and similar psycho-social characteristics of
users. There has, however, been little cross-fertilization in this area
and Celentano and his associates call for a new epidemiology of substance

abuse to address the many existing substantive and methodological problems and thereby provide a more realistic assessment of the extent of the problem of substance abuse among women. The Bahamas offers a particularly unique situation to enable this cross-fertilization.

The crucial significance of the socio-cultural aspects of drug abuse has been stressed in several publications. In 1980, W.H.O. pointed out :

> "Drug dependence is in its every aspect embedded in culture...
> it is absolutely necessary to maintain a consistent relation-
> ship between policies and programmes on the one hand and on
> the other the socio-cultural setting in which any response is
> being planned...drug dependence is a cultural phenomenon.

This paper seeks to uncover the socio-cultural aspects of drug dependence in the Bahamas with a particular emphasis on women and their critical treatment issues.

As Shalloo (1963:199-320) tells us, "To isolate the individual for study is to do violence to the elementary canons of scientific methodology." To paraphrase Szaz (1961), "Our adversaries are not demons, witches, fate, alcoholism or drugs. We have no enemy whom we can fight, exorcise, or dispel by 'cure.' What we do have are 'problems in living, whether these be bio-logic, economic, political, or socio-psychological.'"

The perspective used in this paper is societal and "universalistic" highlighting those societal conditions in the Bahamas that gave such rise to the phenomenon of cocaine abuse and identifying some of the forces that mil-itate against women succeeding in treatment -- identifying structural twists and turns in the path that has provided the stimulus to. cocaine abuse and the success or failure in treatment.

Clearly as the main issue here is treatment, the role of physiological and psychodynamic factors is not denied; it is felt however that socio-cul-tural factors are crucial and all too frequently ignored. Human phenomena are always multi-dimensional and clearly individual issues must not be ex-cluded. There are however basic socio-cultural forces that impinge on and contribute to the problem; these are all too often ingnored or minimized. The way in which alcohol has been used historically (Patterson, 1977) and abused over the last thirty years laid very fertile soil for the entrance of cocaine to the Bahamian society in the 1980s.

Like most of the developing world, the Bahamas is experiencing a period of rapid societal change and transformation. No discussion of a social problem in the developing world can omit recognition of the effects of this process. In fact a great deal of the literature on the developing world points to growth of social problems as a response to social change (Clinard, 1973). Societies undergoing this kind of change provide the best seed bed for alcoholism (Madsen, 1973) and in this author's opinion, other substance abuse.

Parsons (1957) suggests that frustration is endemic in today's industri-alized western society and accompanies the modernization process. He points in particular to the family and occupational systems as manifesting this vulnerability. In the Bahamas, we can observe this process at work -- the extended family is being undermined. A telling and tragic illustration of the decline of the extended family is the long waiting list for the Geria-tric Hospital -- the way in which the announcement of a death in the Geria-tric Hospital elicits phone calls to the hospital by frantic relatives hoping to utilize that space for their aging mother and father. Such an occurrence and mind set would have been unheard of in previous years.

Young people migrate to urban areas or, living in the urban areas, decide that they want their own home. The absentee father, a feature in many families (McCartney, 1971; Patterson, 1986) was not a "problem" when the extended family could be called upon. Today the responsibility falls more and more on the woman who is at the same time increasingly drawn into the labour force. The rising figures of female alcoholism and now cocaine abuse are symbolic of this new status and the resulting pressures on today's women.

Whereas in previous periods, status and prestige had to come from such characteristics as age, fishing exploits, singing, story-telling, one's education, one's family -- today status seems to inhere in one's possessions, one's profession, and location in the social structure and/or in the case of women, one's relationship to individuals with certain professions or location in the social structure. One can argue that the stresses that accompany societal change result in the need for mechanisms that enable one to cope with new changes. Factors such as the weakening of kinship bonds, the tendency to see people in terms of what they possess, the illusory nature of "success," unemployment, etc., all operate to make available mood-altering substances desirable ways of altering one's consciousness and dealing with these new and sometimes threatening realities. Andrew Weil (1972) gives a thought-provoking discussion of man's need for changing his consciousness and the role mood-altering substances play in this.

Historical analysis readily demonstrates how alcohol came to be defined as the Bahamian mood-changer. Historically, the Africans brought with them to the Bahamas drinking practices that were ceremonial and ritualistic while the colonizers brought with them functions that were predominantly utilitarian but with hedonistic and ceremonial dimensions. We can hypothesize that this and other historical processes contributed to changing the drinking patterns of the majority black population by adding an instrumental dimension to the celebrative ones. In psychological terms, the oppressive and dehumanizing experiences, first of slavery and later of colonialism, had obvious consequences psychologically in terms of feelings of self-esteem. Social inducing factors were however equally strong. The economic role of alcohol during the slavery period and more particularly during the days of colonialism when it took on an even more influential role in the islands' revenue, was a persuasive one. While at a societal level, alcohol played a central role in the revenue via customs duties, at an individual level, it is very likely that it was used as partial payment for services via the truck system (Patterson, 1977). The process of diffusion (Lauer, 1973) is also manifest in the exposure of the Bahamians to cultures manifesting heavy drinking -- English plantation owners of the 18th and 19th Centuries, and more recently in the 20th Century, the fun-loving and perhaps equally heavy drinking tourists on whom the Bahamas depends for most of its revenue.

Societal arrangements that institutionalized inequality of opportunities for many of the population resulted in the growing migration from the more sharing life of the Family Islands to the more isolated, competitive life of the capital city; a move from the Gemeinschaft to the Gesellschaft, from the unstructured, easy-going life in the Family Islands to the clock economy of wage-earning Nassau. One can readily hypothesize that today's Bahamians are experiencing tensions as a result of the societal disruptions created by the social changes that seem to be inevitable accompaniments of modernization and urbanization. (Bacon,1971; Clinard, 1973) During periods such as these traditional coping mechanisms may not be available or may not be very effective and alternative mechanisms may be sought. If the cultural definition of alcohol is one of that of a mood-changer, one can immediately see how alcohol can be looked upon to provide a very accessible and immediate way to change one's consciousness, to insulate one from the stresses and strains of the new pace of one's life and the new relationships involved,

(Weil, 1972) and in this way cope with rapid change.

Shared definitions of alcohol and its use are part of the cultural heritage passed on from generation to generation (Patterson, 1976). For Bahamians today, alcohol is associated with having a good time, with parties and celebrations, undoubtedly a legacy of its associations with harvest times and festivals. Weddings, christenings, Junkanoo, constitute only some of the occasions synonymous with alcohol use. In Bahamian men, alcohol use is associated with virility and machismo, the development of which is seen to include drinking and the ability to hold one's liquor a source of prestige in a society that had, historically, offered its people few opportunities apart from the parenting of children and drinking, to demonstrate adulthood. The role of alcohol as a social lubricant is made manifest in a visit to a bar where one can find groups of men sitting or standing "catching one" with their friends.

Alcohol use is therefore very much integrated into Bahamian culture, performing a variety of functions for the individual and for society and helping to provide a very fertile soil for the growth of other mood-changers such as marijuana and cocaine.

It is interesting to note that a survey of admissions to the only psychiatric hospital in the Bahamas in 1961 and comparing admissions to psychiatric hospitals in the Caribbean revealed that only 2.8% of the admissions that year were alcohol related admissions. The largest number of admissions in that year were schizophrenics who accounted for 56% of the admissions. By the late 1960s, and early 1970s, however, the picture had changed dramatically. In 1969, 41% of the admissions to the psychiatric hospital were for alcohol or alcohol related problems. When one compares this with the U.K., in 1969 alcohol related admissions there accounted for 2-3% of admissions and in the U.S. where alcohol related admissions accounted for 20% of the admissions, one readily recognizes the impact of alcoholism for the Bahamas.

If the 1970s have seen tremendous growth and prosperity for the Bahamas, they have also certainly seen the flowering of Bahamian womanhood, educationally, professionally, and vocationally. Indeed, the drive for movement and self-actualization has caused many local politicians to worry and comment on the larger representation of women in night classes offered to allow working people an opportunity for higher education. These politicians express concern as to the plight of the Bahamian male who seems to be being left behind (Nassau Guardian, 1977).

Fifty years ago, women were not expected to go on for higher education. The classic example is the writer's own mother who, despite being consistently top of the class throughout high school and being offered a place in a local law firm was refused the opportunity by her father who saw no point to girls having higher education.

Today the roles have become redefined and women have moved on to open more and more doors educationally and in other spheres. Bahamian women today have become the mainstay of families, of organizations, of businesses. There are women in top Civil Service positions, indeed, the highest post in the Civil Service, that of Secretary to the Cabinet, was until recently held by a female. There are women with positions of Permanent Secretary of various ministries, Directors of Departments, Magistrates of Courts, and the current president of the Chamber of Commerce is a female.

Sociologists frequently speak of women operating more on the expressive plane while men operate more on the instrumental. Bahamian women are to a great extent successfully combining these two planes. This does not seem however, to be happening without cost (Patterson, 1986).

With the blurring of sex differences in the vocational area, it is not surprising to see a blurring of the sex differences in the area of substance abuse. The experience of the developed world mirrors this. A longitudinal study of alcohol use and abuse in the Bahamas has documented this. While drinking was found to be normative behaviour for many Bahamian women in 1969, there was still a significant difference between the sexes, with more men than women drinking and more men than women drinking to excess. In 1969, 43% of females were found to be abstinent as compared with 17% of the males, while in the higher intake groups, males outnumbered females substantially. Confirmation of this pattern historically was the report by the respondents of very similar abstinent roles for their parents: 19% of the fathers and 47% of their mothers had been abstinent (Patterson, 1978).

Interestingly, in 1977 when the same women were re-surveyed, more women said that they drank and female drinking patterns can be better described as light drinking rather than non-drinking. Only 28% of the women said they did not drink.

Neither were the drinking patterns of the heavy drinking females in the survey in 1977 very different from those of their heavy drinking male counterparts except that for women the heavier drinking is concentrated in the 30-39 age group whereas for men it was in the 50-59 age group. Very similar proportions of men and women drank on weekends (19% vs 20%) and even on a daily basis (8% vs 7%).

Robins and Smith (1980) found support for these differences in their review of the literature on sex differences found in longitudinal studies of alcohol and drug use. They point out that while men do have more drinking and drug problems than women, that these differences are narrowing and may disappear in time. They note, however, that similarities in frequency and cause of drug use are greater than the differences viz substance abuse in both sexes seems to be associated with early deviance, origins in deviant families and associated with the user's peers.

The Bahamas was in the 1970s a country in rapid transition from the Gemeinschaft to the Gesellschaft, a country whose psychosociological heritage was one of the normative use of mood-altering substances, a country of 700 islands whose geography made the trans-shipment of cocaine from Latin America to the U.S. a very simple matter. So to the Bahamas of the 1980s came an epidemic of cocaine abuse.

Jekel et al. (1986) have documented an epidemic of cocaine abuse that hit the Bahamas in the Eighties, peaking in 1984 with almost as many female drug abusers being admitted to the psychiatric hospital as female alcoholics (48% alcoholic females vs 46% drug abusing females). Prior to 1984, there had been a steady but small number of drug related first or repeat admissions. In 1979, 9 drug dependent women were admitted and in 1980, 4; in 1981 and 1982, 6 such women were admitted. In 1983, there was a more than 300% increase to 21 and by 1984, the numbers had increased to 46. In 1984 there were twice as many female first admissions for drug dependence as repeat admissions.

During 1985, drug abuse was the most common cause of admission for women: 58 women (a 600% increase in 6 years) entering the only inpatient treatment facility in the Bahamas for drug related problems. Up to July of 1986, 32 women had been admitted to the inpatient unit suggesting that at least as many women are likely to be admitted in 1986 as in 1985.

While it is true that many more men are cocaine abusers than women, it is hypothesized by treatment personnel that women are sustained longer in the community and can delay coming for treatment longer than men can, and this has certainly been the experience with female alcoholics, so admission figures therefore may not necessarily mirror the correct ratio of male/female users. Men frequently end up resorting to crime to sustain their habit and end up facing jail sentences and/or seeking treatment rather than go to jail, or being made to seek treatment by the Court system.

The testimonials of female drug abusers, however, tell a tragic tale. Many of them sustain their habit by selling their one commodity, their body, and involve themselves in different sexual deviations in order to support their habit.

What Allen (1986) terms "ethical fragmentation" is classically manifested in the female cocaine user in her utilization of her body. This utilization is the key factor in the current poor prognosis that female cocaine users seem to exhibit. The double standard which enables men to boast of their sexual deviations and exploits and sexual "war stories" and to be able to leave these behind, does not allow the woman to do so. As she comes through the treatment process, recalling the level to which she was forced to sink in order to support her habit, she seems unable to integrate this behaviour and ends up returning to self-destructive behaviour.

The study of data is derived from 58 women admitted in 1985 and 32 women admitted by July, 1986, who were interviewed at the time of their intake to the psychiatric facility by social service staff and a review of patient notes. The average age of the 1985 sample was 27 with the youngest female 17 years old and the oldest 65. In 1985, 54% of the women admitted were between 25 and 34 with 35% between 18 and 24 and 9% were over 35. In 1986, more younger women came forward for treatment; the average age was 25 years with about half the women in the 18 to 24 age group and the other half in the 24 and over age group. The oldest woman admitted in 1986 was 38 years old and had been admitted in 1985 also.

Most of the women admitted using cocaine first between the ages of 14 and 17. One woman in 1985 admitted to beginning to use cocaine when she was 11. All were freebasing cocaine at the time of admission although several admitted to having snorted. The number of years of use ranged from 1 to 2 years to 6 to 8 years. In 1985, a third of the women admitted using cocaine for 1 to 2 years, while a third admitted to using cocaine for 3 to 5 years. Three of the women admitted in 1985 said they had been using cocaine for 6 to 8 years.

In keeping with other research findings that women are more likely to be unemployed or dependent on others for their support (Eldred and Washington, 1985) the employment picture of Bahamian women is very similar to that found by other researchers. Female drug dependent patients have also been found to have more restricted job opportunities than men and to fall into lower educational categories (Burt et al, 1979). Again the profile of Bahamian women bears this out. Of the 58 women admitted in 1985, only 2 had jobs and of the 32 women admitted in 1986, only 4 came into hospital still in employment.

The majority of these women had held service oriented jobs, having been maids, janitresses, waitresses, or clerical staff. A few had been self-employed, e.g., straw vendors, beauticians, etc.

Unlike U.S. findings that drug dependent women are more likely to be separated or divorced, the majority of the Bahamian women admitted to hospital were single. Only 10% of the women admitted in 1985 and 25% of those admitted in 1986 were or had been married. Those that were in relationships tended to be in relationships with men who were themselves involved with drugs.

Most of the women however, did have children. In 1985, 40% of the women had one or two children; 28% had 3 or 4 children; and 5% had 5 or more children. In 1986, the picture is similar. Of the 32 women admitted, only 4 had no children. Twelve women had one or two children; 6 women had 3 or 4 children; and 5 women had 5 or more children.

In fact it was the fear of losing their children or accusations of child neglect that precipitated several of the women into treatment. Indeed, the Department of Social Services reports a high percentage of the children taken into care as being the children of women who are dependent on cocaine (Green, 1986). The case of Julie (name changed) is characteristic of this. She is a 25 year old woman admitted first in 1984 for 1 month. After leaving hospital, she returned to drugs, was readmitted for a few days in 1985 and again for a week in 1986, after which she absconded. She is currently living with a drug dealer with free access to cocaine. In 1984, she had been accused of child neglect and came into hospital in order to keep her children. She has now however been declared an unfit mother and her children placed in care. She has attempted suicide several times. She claims to have begun using cocaine because of her unhappiness in relationship with her boyfriend.

The religious breakdown of the cocaine dependent females in 1985 reflects the religious makeup of the country with Baptists, Anglicans, and Roman Catholics accounting for the majority of the admissions with fairly even representation: 27% Anglican, 25% Baptist, and 24% Roman Catholic with the remainder being Seventh Day Adventist, Pentecostal, Methodist, Jehovah Witness and in 1986, the proportions were similar.

The women reported using cocaine daily or as often as they could get it. One patient reported she would use it 24 hours if she could get it. The majority of the women admitted being polydrug users, with combinations of cocaine and marijuana, cocaine and alcohol, and cocaine, marijuana and alcohol. They seemed to have used these other drugs to help modify the effects of the cocaine. Not surprisingly, given the cultural integration and acceptance of alcohol, the most commonly used other drug by the cocaine dependent women was alcohol. Interestingly, those women who had been through the treatment programme and were continuing to be cocaine-free, admitted to still using and in many cases feeling the need to use alcohol.

The majority of the women reported being first introduced to cocaine by friends. The remainder had been introduced by boyfriends or spouse or other family members. Four of the women in 1985 had been introduced to cocaine by a drug pusher. In 1986 the picture was similar except that more of the women reported being introduced to cocaine by a drug pusher.

Interestingly only 2 of the women reported freebasing alone. The majority report freebasing cocaine with friends; 20% of the women admitted in 1985 admitted to freebasing with their boyfriend or spouse.

About half of the women admitted in 1985 came into the treatment facility for the first time, about a third were being admitted for the second time, and the remainder were admitted into hospital for their third or more time. Indeed, three women were being admitted for their 8th or 9th time. The 1986 picture is similar, with three women being admitted for their 4th or 5th time. Many of the early U.S. studies showed that drug dependent women do not remain in treatment as long as men, nor are women as well prepared for a drug free life after treatment as men (Giona and Byre, 1975; Levy and Doyle, 1974). That women have different treatment needs than men is recognized by several researchers (Beschner and Thompson, 1981; Burt et al. 1979; Reed and Morse, 1979). Minkoff (1973) has identified what he sees as "learned helplessness" in the female sex who is socialized to be dependent and passive and nurturing rather than competitive and self-sufficient. This model is not applicable to the Bahamian woman who is on the contrary, raised to be independent, to be able to survive by a mother who has herself had to work in most instances and wants her daughter to be "independent."

The writer recalls the case of a mother who brought her 13 year old daughter, the eldest of six children, to the Adolescent Clinic for behaviour problems. The child had begun to act out against the pressure by her mother who was herself unmarried and had her daughter taking care of all the younger siblings as well as attending typing and music lessons. When questioned as to why there were no complaints about the 11 year old son who could be seen running about the streets, the mother replied, "I don't have to worry about my son, he will find a woman to take care of him. I don't want my daughter to suffer the way I did. She is going to be able to take care of herself." The pressure is on today's Bahamian woman to be independent and to take care of herself.

At the same time however, most Bahamian girls are still raised to hold onto the belief that they must have a man in their life, a legacy of romanticism which encourages unrealistic expectations in women especially. The romantic ideal is one of ecstatic love, the prince on the white horse, centralizing sexual longing yet placing its satisfaction outside the realm of the possible (Wilson, 1983). Bahamian girls are socialized and encouraged to be nurturing and to care for younger siblings but at the same time are expected to get whatever skills they can to become self-sufficient.

Women tend to put greater emphasis on interpersonal relationships and interactions than men. They are shown to be more responsive than men to reactions to important people in their lives (Chodorow, 1978). The major reason women gave for entering treatment was to maintain relationships with drug free boyfriends or family. Most of them entered inpatient drug treatment for short periods of time, ranging from a few days to a few weeks. Several women absconded after a few days or discharged themselves against medical advice.

The story of Jane (name changed) paints a tragic picture of the life of a female cocaine abuser. She is 23 years old. The produce of a one-parent family with a visiting father, she is the fourth child of eight. Two of her sisters are freebasing. She has a brother who is an alcoholic. She has vivid memories of parental violence. Her father drank heavily and would frequently beat her mother who would frequently beat Jane and her siblings. She left school in the eighth grade and consequently has few job skills. She has had a few jobs as a domestic. She began using marijuana at 14 as a result of peer pressure and admits to having used a $2 bag of marijuana daily whenever she could get the money. She began using cocaine at 16 and snorted for 3 months before beginning to freebase, having been introduced to it by her boyfriend. She described being willing to hit 24 hours a day as long as she has the money. She reports using 3 to 4 grammes of cocaine a time. The most she recalls spending on cocaine is a day is $600. She admits having obtained money by selling her body walking the streets at night, directing her attention to men in their forties, fifties, and sixties who she feels most confortable seducing. She was suicidal at the time of admission with very low self-esteem, feeling she is a disgrace to herself and to her family. Admission to hospital was precipitated by pressure from her mother from whom she had stolen $100 and who was threatening to report her to the police if she did not seek help. She had returned home to her mother's house after a fight with her boyfriend who had beaten her badly. She claims to take cocaine because of problems in her relationship with her boyfriend.

The typical drug dependent female is likely to be young, single, unem-

ployed, but with children who may or may not live with her. She is lacking
both vocational skills and education. Her inability to get or to keep a job
parallels her inability to form and sustain love relationships.

The pattern seems to be one of repetitive admissions for short periods
of time with either no change or short-lived change in drug behaviour. Of
58 admissions in 1985, 32 are known to have returned to drugs.

Many of these women who were being readmitted for drug dependence attributed their return to drugs to involvement with male friends or problems in
interpersonal relationships, substantiating what has been found in studies
of drug dependent females elsewhere (Chodorow, 1978).

As pointed out earlier, most of the women had children who were cared
for by grandmothers or other relatives while they were in treatment. Some
women blamed the responsibilities and pressures of single parenthood for
making them susceptible to drug use. In some cases children had been removed from their mother's care and placed in foster care.

The women were usually from broken homes or reconstituted families,
though some were from two-parent homes. There was a mixture of positive
and negative family relationships. Many of the young women were themselves
victims of sexual assault as a child or had become involved in sexual activity at an early age, usually with a much older man. Many of them during
periods of drug usage were battered by their boyfriends. Research with
some women at Odyssey House has revealed a similar picture.

Characteristic of these female drug abusers was isolation, loneliness,
low self-esteem, feelings of having been taken advantage of by friends and
partners.

TREATMENT ISSUES

These women point to neglect of children, lowered self-esteem, dependency
and disruption of family life as resulting from drug use. Most have had to
use their body as their one commodity to obtain drugs, to survive and acknowledge this as the only alternative they felt they had.

Many of them have very poor feelings about themselves, blaming themselves
for their circumstances and generally communicating hopelessness and a sense
of powerlessness in their situation. Interestingly, a poor self concept and
and low self-esteem has been identified in other drug dependent female populations (Cotten, 1979; Reed and Moise, 1979). Beckman (1976) found female
alcoholics to have lower levels of self-esteem and higher levels of dependency and anxiety than drug dependent men. Many of the women admitted in
1985 and in 1986 demonstrated this low self-esteem, manifesting this in
their speedy return to using drugs even in instances where their environment
seemed fairly supportive and circumstances seemed much in their favour.

For many women, drug dependency can be an escape from the awful realities
of their drug dependent life -- creating a vicious cycle. Their self-esteem
constantly battered, their bodies frequently abused, many turn to drugs to
make them feel good about themselves. Cocaine gives them, albeit fleetingly
and temporarily, a sense of power, a sense of being in control, of being
able to do anything, be anything.

Complicating treatment for women is the fact that during their drug using
periods they have violated the societal sexual taboos, they have violated
the social stereotypes of expected feminine behaviour. They recognize that
society defines women who behave as they behaved as "promiscuous," "cheap,"
"babbits." Several researchers have pointed out that it is these "shoulds"

about socially acceptable feminine behaviour that drug dependent women use
to harshly judge themselves and other women (Colten, 1979; Gomberg, 1979;
Reed and Moise, 1979; Wilsnock, 1973).

Cocaine is readily recognized as a very sexual drug. Women describe
the highly sexual sensations they are able to achieve especially with
cocaine. Smith et al. (1979) report that occasionally after intravenous use
of central nervous system stimulants such as cocaine, both men and women
report engaging in sexual acts that in retrospect in the non-drug state are
viewed as dystonic, humiliating, and guilt-producing. In the Bahamas this
occurs with greater frequency with freebasing. Women in particular are made
by pushers and other men to perform a wide range of degrading sexual acts in
order to obtain cocaine. It is their use of their bodies that enables women
drug users to keep out of the criminal system. Where men have to steal and
perform other criminal acts (local magistrates report that three-fourths of
the crime seen in their courts are drug related)* in order to sustain their
habits, women sell their bodies and make themselves available for whatever
is required of them sexually. Especially difficult to cope with are requests
from husbands or boyfriends who are themselves on drugs and use their mate
to obtain drugs. Mandonaro et al. (1982) put the problem for drug dependent
women well:

> "They feel they are bad and failures because they have not
> lived up to the ideals set for them. Drug use further re-
> inforces this negative self-image and makes it even more
> impossible to achieve the ideals that are unrealistic in
> the first place. The women gradually repeat a vicious
> cycle of blame, self-hate, and self-destruction, and
> lose sight of any personal strengths or qualities that
> would support positive changes."

He goes on to recommend that staff members working with drug dependent women
need to be able to take a balanced, informed approach to offset this negative judgmental cycle. Rigid attitudes and values of counsellors concerning
female sexuality and feminine behaviour both reinforce and collude with the
dependent woman's immobilizing blame and low self-esteem.

The women who freebase have an even more difficult time since cocaine
and sexuality are so intertwined. The confrontational self-disclosure
methodology of group therapy developed to help male patients to stop denying
problems is therefore likely to cause more harm than good for the drug dependent woman as she is confessing behaviour that violates social and sexual
taboos. Research by DeLeon and Baschner (1977) confirms that this kind of
group therapy is embarrassing for the drug dependent female. It increases
their depression and low self-esteem and accounts in this writer's mind for
the frequent dropping out of treatment and leaving hospital against medical
advice. Many begin to accept their status, accept their negative image of
themselves and return to their self-destructive behaviour. For those women
who complete their treatment, the test comes for them in the outside world
where the significant other or others in their environment frequently remind
them of where they have been. Remarks are made in their workplace, remarks
are made by friends which remind them of the level which they reached during
their drug using period.

The traditional concepts of masculinity and femininity while under
attack in the modern Bahamas in terms of educational and vocational opportunities, still remain firmly entrenched in the sexual arena. The pressure
remains on boys to be aggressive, competitive, active -- "Lock up your cows

---

* Personal communication from the magistrates of local Magistrate's Courts.

'cause my bull is going to be loose tonight," is a joke with underlying seriousness for the proud mother of a teenage boy.

The pressure on the girl on the other hand, especially after puberty, is to be feminine, pretty, passive in courtship, to "keep her dress down." Bahamian society still buys the sexual stereotypes of masculinity and femininity. "Bad" girls do, "good" girls do not. The cultural attitude towards the sexual life of women is one of denial. In earlier years there was denial of its very existence. Today there is still a tendency to deny that it is as important or as urgent as the sexual life of men. The differential handling of the sexuality of men and women has resulted in a double standard, a "men will be men" attitude. A women, however, who has all the sexual experiences a man has, is suspect.

Gender role and identity are shaped by cultural forces. Culturally defined cues pass on to girls how they are to behave. Little girls are encouraged to be babies, to be dependent, to flirt; little boys are told not to be a baby but to act like a man. Little girls are praised for how they look while little boys are praised for how they act, whether they are brave, whether they are strong. According to Julian Wood (1982), "Boys ... are encouraged to measure their masculinity via woman-hating, rapacious sexuality. This pressure to be a sort of Tarzan-cum-Ripper is sedimented into the history of how to be male."

Unlike, however, the more developed world where women tend to be brought up to be dependent, history dictated different messages to the Bahamian women. In slavery, she had to take care of the family; her man was sold or made to act as a stud elsewhere. James Boggs writes of the U.S. but what he says has applicability to Bahamian society, "Black men from the time they were captured back in Africa with the help of the tribal chiefs, lost their dominance over the black woman. For a long time in the U.S. in black slave society there was no dominance of the female by the male. Both were entirely dominated by whites." In the Bahamas this dominance continued in colonialism when it was difficult for her man to find a job, so the Bahamian woman had to work to feed the family. Indeed, as late as the 1940s men left their families to work abroad on the projects in Florida or in Latin America. On the Family Islands, men were away for long periods fishing, and the women had to manage. Consequently the Bahamian women have a cultural heritage of work and they have passed this on to their daughters. She has not however, passed on to her daughter such a positive and strong heritage for good male/female relationships. The high cultural acceptability of wife-beating is a strong indication of this (Patterson, 1985), and it is in this same area of male/female relationships, of sexuality, that we see drug dependent women having the greatest difficulty in treatment. Research done in the 1970s on alcohol use among Bahmaians endorses the power and significance of male/female relationships for the Bahamian woman who is alcohol or drug dependent. The case of Julie is classic in this regard. In the longitudinal study of alcohol use and abuse in the Bahamas referred to earlier, 32 year old Julie was one of the respondents interviewed. She had been a light, weekend drinker in 1969, but in 1977, reported drinking daily. She was drinking rum daily and beer once or twice a month. Though she was drinking more in 1977 than she had in 1969, she felt in 1977 she was drinking less than previously. "It gives me too many problems -- makes me fight all the time."

Though single, she had had three children by 1977, had changed her residence four times during the eight year period and had not been able to sustain a stable relationship. She admitted drinking to help her relax, when things got her down, to help her forget her worries and to celebrate. Unstable, injurious romances had taken their toll and she was reclassified from the light drinker she had been in 1969 to a heavy drinker in 1977 and clearly at risk for alcoholism.

A review of the American literature of women and drugs reveals that women are traditionally underserved in most drug treatment programmes and relegated to subservient roles in therapeutic settings (Beschner, et al., 1981). They point to the need for special attention to the intake and assessment of female clients, counselling procedures, coordination of programmes, and community services, special medical and health care to female clients, the utility of family therapy, special vocational training of unskilled female addicts and the development of parenting programmes to assist children of addicts.

The freebasing epidemic in the Bahamas calls for an additional component to the treatment services for women. The persistence of the drug dependent female's drug habit and her poor prognosis for treatment is related to the role assigned by society to women. Low self-esteem has been identified as the one reliable correlate of both drug and alcohol abuse. Women who feel worthless cannot resist the temptation of chemical relief to relieve this feeling. That relief however is temporary and only ends up increasing the feelings of worthlessness. Drug dependent women have accepted and internalized societal stereotypes of a woman's role and cling more rigidly to this picture than do non-drug dependent women (Baldinger et al., 1972; Colten, 1980; Gomberg, 1979; Wilsnack, 1973). Treatment programmes for women must reflect and respond to the special needs of women.

SUMMARY

The Bahamas is in the Eighties a country whose psychosociological heritage is one of the normative use of mood-altering substances, a country whose geography has attracted the trans-shipment of cocaine through its 700 islands and cays. It is a country in which development and social change has been such that Bahamian women have taken on a new identity in the world of work and are as much at risk for abusing mood-altering drugs as are Bahamian men.

Increasingly in the Eighties have women fallen prey to the cocaine epidemic documented by Jekel et al. (1986). Unfortunately, the prognosis for the treatment of the female who is dependent on cocaine will not be good if the special needs of the female addict are not taken into consideration.

The power that sex stereotypes and sexual taboos have in the culture must be recognized and strategies built in to enable women to work through these and find ways to feel good about themselves. The whore/madonna complex is a classic example of the double standard. Jean Charles Letaurneau puts this double standard well:

> Almost at the origin of human society, woman was subjugated by her companion. We have seen her become in succession beast-of-burden, slave, miner, subject -- held aloof from free and active life, often maltreated, oppressed, punished with fury for acts that her male owner would commit with impunity before her eyes.

It is not enough however that the drug dependent female be helped to break through her low self-esteem and gain or regain a positive self-image. Self-esteem is also influenced by feedback and reinforcement from one's environment. Consciousness-raising and sensitization to this problem is necessary also for the people in that individual's network or relationships as well as for the community at large so that neither will perceive and/or treat her as a "babbit," a woman of loose morals. Negative responses by

her peers and associates batter her newly-acquired sense of being, refusing
to allow her to be the new person she has become, and facilitate the fulfil-
ment of the prophecy, forcing her back into drug dependency.

This self-fulfilling prophecy is a script that women stick to and maintain by continually perceiving all that happens or is said to them in the
same negative way. It is our challenge then as therapists to help the drug
dependent female to rewrite her script and confront her drug free life,
replacing the old prophecy with her own, built on a better understanding of
herself and a non-judgmental acceptance of her past, even in the face of
negative judgments from her peers.

Willard Gaylin tells us that it is part of the wonder of the human being
that even the state of hopelessness can be used to generate hope. The challenge for us as therapists is to enable that hope to be generated in a
community willing to allow healing.

REFERENCES

Allen, D. F. (1986) Modes of Use, Precursors, and Indicators of Cocaine
    Abuse, this volume.
Bacon, S. (1962), Alcohol and complex society in "Society, Culture, and
    Drinking Patterns," D. J. Pittman and C. R. Snyder, eds., John Wiley
    and Sons, New York.
Baldinger, R., Goldsmith, B. M., Capel, W. C., and Stewart, G. T. (1972)
    Potsmokers, junkies and squares: A comparative study of female values,
    International Journal of Addictions, 7:153-166.
Barton, A. H. (1963) Social organization under stress, in "Disaster Research
    Study No. 17," National Academy of Sciences, National Research Council
    Disaster Research Corp., Washington, D. C.
Bechman, L. J. (1975), Women Alcoholics: A review of social and psychological
    studies, Journal of Studies on Alcoholism, 36:797-824.
Bechman, L. J. (1976) Alcoholism problems and women: An overview in "Alco-
    hol Problems in Women and Children," M. Greenblatt and M. A. Schukit,
    Grune and Stratton, New York.
Berger, P. L. and Luckman, T. (1967) "Social Construction of Reality,"
    Anchor Books, New York.
Beschner, G. M., Reed, B. C. and Mondanaro, J. (eds.) (1981) "Treatment
    Services for Drug Dependent Women, Vol. 1," U. S. Printing Office,
    Washington, D. C.
Beschner, G. and Thompson, P. (1981) "Women and Drug Abuse Treatment Needs
    and Services: NIDA (Adm) 81-1057 Services Research Monographical Series"
    U. S. Government Printing Office, Washington, D. C.
Boggs, James (1970) unpublished manuscript quoted in Stambler, S., "Women's
    Liberation," Ace Books, New York.
Boothroyd, W. E. (1963) The female drug addict: a profile, Addictions 10:36-
    40.
Brandon, N. (1969) "Psychology of Self-Esteem," Bantam, New York.
Burt, M. R., Glynn, T. J., Sowder, B. (1979) "Psychosocial Characteristics
    of Drug Abusing Women," National Institute on Drug Abuse, Rockville, MD.
Celetano, D. D., McQueen, D. V., and Chee, E. (1980), Substance abuse by
    women: a review of the epidemiological literature, Journal of Chronic
    Diseases 33(6):383-394.
Chambers, C. D., Griffey, M. S. (1975) Use of legal substances within the
    general population, the sex and age variables, Addictive Diseases: An
    International Journal 2(1):7-19.
Chodorow, N. (1978) "The Reproduction of Mothering: Psychoanalysis and the
    Socializing of Gender," University of California Press, Burbank.
Clinard, M. B. (1973) "Crime in the Developing World," Wiley and Sons,
    New York.
Colten, M. E. (1979) Description and comparative analysis of self-perceptions

and attitudes of heroin-addicted women, in "Addicted Women: Family Dynamics, Self-Perceptions, and Support Systems," National Institute on Drug Abuse, Rockville, MD.

Cuskey, W., Berger, L., and Densen-Gerber, J. (1977) Issues in the treatment of female addiction: a review and critique of the literature, Contemporary Drug Problems 6(3):307-371.

DeLeon, G. and Baschner, G. M. (eds.) (1977) "The Therapeutic Community: Proceedings of Therapeutic Communities of America Planning Conference," National Institute on Drug Abuse, Rockville, MD.

Dumont, M. (1968) "The Absurd Healer," Science House, New York.

Eldred, C. A. and Washington, M. N. (1976) Interpersonal Relationships in heroin use by men and women and their role in treatment outcome, International Journal of Addictions 11(1):117-130.

Gaylin, W. (1979) "Caring" Avon, New York.

Goina, C. and Byrne, R. (1975) Distinctive problems of the female drug addicts: experiences at IDAP, in "Developments in the Field of Drug Abuse," E. Senay, ed., Schenkman, Cambridge, Mass.

Gomberg, E. S. (1976) Alcoholism in Women, in "The Biology of Alcoholism, Vol 4" B. Kissin and H. Begleiter, eds., Plenum Press, New York.

Gomberg, E. Franks, W., eds., "Gender and Disorderly Behaviour: Sex Differences in Psychopathology," Brunner Megel, New York, 1979.

Green, Leila in a speech to the Ministry of Education (Bahamas) Conference as quoted in the Nassau Guardian, April, 1986.

Greenblatt, M. and Schuckit, M. A., eds., (1976), "Alcoholism Problems in Women and Children," Grune and Stratton, New York.

Jekel, J, Allen, D. F., Podlewski, H., Dean-Patterson, S., Clarke, N., Cartwright, P., Finlayson, C. (1986) Epidemic Freebase Cocaine Abuse: Case Study from the Bahamas, The Lancet, March 1, 1986, pp 459-462.

Ladner, J. (1971) "Tomorrow's Tomorrow," Anchor Doubleday, New York.

Lauer, R. H. (1973) "Perspectives in Social Change," Allwyn and Bacon, Boston.

Levy, S. J., Doyle, K. M. (1974) Attitudes towards women in a drug abuse treatment programme, Journal of Drug Issues 4:428-435.

Maddox, George (1962) Teen-age drinking in the United States in "Society, Culture, and Drinking Patterns," D. J. Pittman and C. R. Snyder, eds., John Wiley and Sons, New York.

Madsen, W. (1973)"The American Alcoholic: Nature-Nurture Controversy in Alcohol Research and Treatments," Charles C. Thomas, Springfield, Ill.

McCartney, T. (1971) "Neurosis in the Sun," Executive Printers, Nassau, Bahamas.

Minkoff, K., Bergman, E., Beck, A., (1973), Hopelessness, depression, and attempted suicide, American Journal of Psychiatry, 130:455-460.

Mandonaro, J. (1977) Women: Pregnancy and children and addiction, Journal of Psychedelic Drugs 9(1).

Moise, R., Reed, B. G., Ryan, V. S. (1982) Issues in the treatment of heroin addicted women: a comparison of men and women entering two types of drug abuse programmes, International Journal of Addictions 17(1):109-139.

Moise, R., Reed, B. C., Connell, C. (1981) Women in drug abuse treatment programmes: factors that influence retention at very early and later stages in two treatment modalities, International Journal of Addictions 16(6):1295-1300.

Parsons, T. (1957) "The Social System" Free Press, Glencoe, Ill.

Patterson, S. D. (1976) "Alcohol Use and Abuse in the Bahamas: A Socio-Cultural Study," unpublished substantive paper, Brandeis University, Waltham, Mass.

Patterson, S. D. (1978) "A Longitudinal Study of Changes in Bahamian Drinking Habitd, 1969-1977," Dissertation, Brandeis University, Waltham, Mass.

Patterson, S. D. (1985) "Violence and the Bahamian Woman," paper presented to the Caribbean Federation Mental Health Conference, Nassau, Bahamas.

Patterson, S. D. (1986) "The Mental Health of the Bahamian Woman" in press.

Pittman, D. J. and Snyder, C. R. (eds.) "Society, Culture, and Drinking

Patterns," John Wiley and Sons, New York, (1962).

Reed, B. G, Kovach, J., Bellows, N., Moise, R. (1981) The many faces of addicted women: implications for treatment and future research, in "Drug Dependence and Alcoholism, Vol I: Biomedical Issues," A. J. Schecter, ed., Plenum Press, New York.

Robins, L. N., Smith, E. M. (1980) Longitudinal studies of alcohol and drug problems and sex differences, in "Alcohol and Drug Problems in Women, Vol. V, Research Advances in Alcohol and Drug Problems," O. J. Kalant, Plenum Press, New York.

Shalloo, J. P. (1941) Some cultural factors in aetiology of alcoholism, Quarterly Journal of Studies in Alcoholism, 2:464-478.

Smith, D. E., Buxton, M., Daminonm, G. (1979) Amphetamine abuse and sexual dysfunction in "Amphetamine Use, Misuse, and Abuse," D. E. Smith and D. R. Wesson and M. Buxton, eds., G. K. Hall, Boston.

Spencer, J. (1970) Alcoholism: How serious is the problem? Bahamas Mental Health Association Magazine, November, 1970.

Szaz, T. (1961) "The Myth of Mental Illness," Delta, New York.

Weill, A. (1972) "The Natural Mind," Houghton Mifflin, Boston.

W. H. O. (1980) "Drug Problems in the Socio-Cultural Context: A Basis for
P   Policies and Programme Planning," G. Edwards and A. Aris, eds., W.H.O. Paris, France.

Wilsnack, S. C. (1973) Sex role, identity, in female alcoholism Journal of Abnormal Psychology, 82:253-261.

Wilson, E. (1983) "What is to Be Done About Violence Against Women," Penguin, London.

Wood, J. (1982) Boys will be boys, New Socialist, No. 5, May/June, 1982.

TREATMENT APPROACHES TO COCAINE ABUSE AND DEPENDENCY IN THE BAHAMAS

Timothy McCartney

Sandilands Rehabilitation Centre
Nassau, Bahamas

Michael Neville

Sandilands Rehabilitation Centre
Nassau, Bahamas

As far back as 1975, many of us in the Bahamas were becoming more aware that there were chemicals being abused other than alcohol, marijuana, or prescription drugs. Slowly in our society, heroin, P.C.P., but more significantly cocaine, were rearing their ugly heads. In 1982, the cocaine avalanche began to make its presence known in the Bahamas and by 1983, a full-blown epidemic had arrived.

Mental health professionals were disappointed with the results of their intervention and rehabilitative techniques. It was then that we started looking around elsewhere and found, unfortunately, that the scientific literature gave an alarming rate of rehabilitative rate of failure of cocaine addicts.

From our investigations of most hospitals, centres, institutes, and clinics, treatment approaches are just variations on the same theme. Meetings, seminars and correspondence from hospitals and institutes with many professionals in private practice, showed frustration, confusion, and the lack of any real innovation.

There were some glimmers of hope however, we had a visit from one of the world's foremost experts in chemical dependency, Dr. Sidney Cohen who shared his experience and knowledge. We have also been in contact with the Palm Beach Institute and the Familization Therapy of Co-Dependency Programme. We had also received positive information from the Lincoln Hospital Acupuncture Programme.

In essence, most programmes follow what we call the "Sidney Cohen Model" for treating acute toxicity and chronic use, viz:

1. "Immediate stoppage of the drug as significant withdrawal symptoms can be treated with sedatives. Long-term removal from the cocaine environment and cocaine-using friends is necessary. Relapse is a possibility after months, perhaps years, of abstinence.

2. "Usually individuals come for treatment after a catastrophe: a psychotic break, serious impairment of health, loss of job, or family, or a serious suicide attempt. They may be highly motivated to stop initially, but over time their motivation tends to wane. An important aspect of therapy is to maintain the strong desire to remain abstinent.

3. "Sometimes family or concerned friends see the destructive nature of the cocaine usage before the user acknowledges it. In such instances, group confrontation may be persuasive. This is not an accusatory confrontation, rather, it is a friendly, loving meeting which takes the place of an impending catastrophe. It should be led and monitored by the therapist. The goal is to change the drug-using behaviour and to instill a positive attitude toward treatment.

4. "Lifestyle changes have to be made. The person must switch from cocaine to gratifying, non-drug activities. These will not match cocaine in providing peaks of pleasure, but they will be more sustaining.

5. "The client's physical condition may have deteriorated as a result of extended cocaine use. Any nutritional deficiency ought to be corrected and an optimal physical state achieved.

6. "Pharmacotherapy. Treatment of depressions may be required.

7. "Group therapy with other cocaine or poly-drug abusers has been helpful and where groups of recovered and recovering people are available, they can be a helpful therapeutic device."

Sidney Cohen believes that the "goal of emergency management is to keep the patient alive for a short period of time while permitting the liver and blood enzymes to metabolize the cocaine." (1) He suggests the use of Valium or phenothiazines "like Thorazine" or perhaps Inderal (propanolol) which has not proven to be that helpful.

It would appear, then, that there are four basic steps that are used in most programmes that cater to the needs of the addict:

1. Detoxification and medication when indicated.

2. Confrontation and abstinence.

3. Support of the family and "significant others."

4. Group psychotherapy.

Most cocaine rehabilitation programmes in the Bahamas follow this model. The authors initially utilized the same approach and found alarming failure rates. It was then decided to set up a small carefully screened "research-cum-therapeutic" approach, with significant flexibility, adaptability, and risk-taking, in order to find a methodology that at least provided much higher hope of rehabilitation for the cocaine user. We believed that the method should be simple in its application, cost effective, long-range, humanitarian, and hopeful.

The fundamental question that must be answered before we begin a discussion of treatment approaches is whether there is anything to treat. Many persons see addictions as moral issues rather than medical illnesses. It was certainly that way in the past when alcoholism and drug addiction were far more likely to be dealt with by a prison sentence than a trip to hospital. The statistics in 1891 (2) for example, demonstrate that of 1392 inmates of San Quentin Jail, 44% were said to be addicted to opium and 49%

to alcohol. It is easy to view this as a medieval, even barbaric way to deal with addictions, but we need to examine whether we have changed and whether we have improved the care and prognosis for these groups.

The major strides were probably made first by Alcoholics Anonymous in 1935 who accepted that addictions were some sort of illness. The definition then adoped by W.H.O. in 1952 (3) further underscored the need for treatment and with Jellinek's famous book "The Disease Concept of Alcoholism" (4), the idea was pretty much accepted.

Therapists blossomed in many parts of the world and the Bahamas was no different. Alcoholics Anonymous has been active for many years and Narcotics Anonymous has recently grown in numbers and stature. A special unit for the treatment of alcoholism was opened in 1967 and a drug treatment unit in 1985.

The major concern for therapists, however, is despite the increasing acceptance that addictions are diseases, the ability to treat addictions has not moved forward at the same rate. The ability to treat has at times been so disappointing that one is forced to re-examine the initial question: "Are addictions disease processes?"

The characteristics which satisfy the demand for a theoretical disease model have been proposed by Hershan (5). The first would be that the disease should seem to be related to a relevant underlying pathological process, and secondly, that the person so afflicted cannot decide by willpower alone, not to have the pathological process and consequent illness. He cannot or will not will it away.

These concepts when viewed in the knowledge of cocaine addiction and therapy models provide something of a paradox. It would seem that the more severely the individual is addicted with all the physical and mental sequelae, the easier it is to fit them into the disease model. The early user, with good insight and motivation is much harder to fit into the disease model. It is here that we find the paradox for it is the patients who do not readily fit the disease concept that are accepted into the treatment programmes and those that are more severely afflicted, often rejected.

There are countries where the disease concept is being rejected with harsh penalties being imposed on traffickers and users alike. The clamour for the mandatory death sentence and long prison terms for addicts is a far cry from the disease concept and yet seems to echo the statistics from San Quentin (2) in the past.

In the Bahamas with a 30-bed unit and a twelve week rehabilitation programme it is possible to treat 150 addicts a year. Unofficial guess estimates at the prison have suggested that there may be as many as five hundred inmates with a drug problem. If this is so, then due to our inability to effectively treat and obtain funding for larger units, more addicts are likely to go to prison than are likely to be admitted to hospital. If this ratio is correct and found to be so in other countries, then we are left with some awkward questions: Can we improve our ability to treat addictions and hence argue that more beds should be made available or should we discard the disease concept and allow addictions to be regarded and dealt with as a moral issue as in the past?

There are those who already feel that by accepting addictions as diseases we have moved the responsibility for getting well to the medical profession. We all know that the addict must to some extent heal himself and perhaps this very ambivalence is counterproductive. It was to this question that we developed "The Therapeutic Learning Process " (TM) as a viable methodology in the treatment of substance abuse.

163

The Therapeutic Learning Process (TM) is a non-culturally biased, holistic methodology that extracts from existing theories ideas about the development of personality and behaviour of human beings but more specifically in this context, drug abusers. The Therapeutic Learning Process (TM) means simply a "process of self-healing" or "learning how to heal one's self." It operates under four distinct areas:

1. The origin and development of the person.
2. The socialization process as it relates to:
    a. interpersonal relationships
    b. professional client interaction
    c. self-actualization
3. physio-bio-chemical concerns exploring the mind-body link
4. conflict resolution (problem-solving) and growth enrichment (wellness) of human beings

We utilized these insights and developed this methodology after evaluating the specific needs of our clients. As regards the cocaine issue in the Bahamas, we realized the following:

1. Cocaine is a highly addictive drug and is ideally treated in an inpatient, highly structured environment.
2. Cocaine itself creates an intensive craving for the drug and gives severe psychological and physical dependency. Replacing the intensity of one drug with another drug was counterproductive to total rehabilitation. We believed, however, that if the addict had become psychotic, or there was strong medical urgency, appropriate drugs should be used to remedy this and then no drugs given at all. We found that electro-acupuncture was an ideal vehicle to minimize the cravings for cocaine and restore the body's resources to its natural function. We now use both methods of acupuncture.
3. Our experience indicated that although group psycho-therapy has definite merit, individuals get "lost" in groups. Even when many of the defenses have been broken down, there are still many areas of sensitivity and privacy that the person will not share in a group setting, especially in a tightly-knit small community such as the Bahamas. We opted for intense individual psychotherapy (using the TLP methodology) and then prepared the addict for eventual group psychotherapy and/or meeting with their families.
4. After two weeks intensive therapy as inpatients, clients were given weekends at home or in their usual environment so as to practice their adjustment techniques in the real world. These weekend excursions were carefully reviewed and analyzed. More effective coping mechanisms and problem-solving techniques were taught if required.
5. Random urinalysis during inpatient stay and six-month's follow-up was used as an indicator or either relapse into drug use of a drug-free existence.

At present we are experiencing a fifty-two percent (52%) recovery/rehabilitation rate. We believe that it is still too soon to become too dogmatic about these results. Our sample after three years of using this method is just 165 and the time span is too short. A larger sample of clients, continuing rigid follow-up, and a longer time period will provide the necessary data for more valid and reliable appraisal.

We believe that we have been creative in our endeavours. We are constantly revising and improving our methods and/or adding to our programme. The primary site of St. Augustine's Monastery provides the ideal atmosphere for this total approach. The future will judge our interventions by the return to society of fully-functioning, drug-free human beings.

REFERENCES

1. S. Cohen, Health Hazards of Cocaine, reprinted from "Cocaine Today".
        Drug Enforcement, Fall, 1982.
2. San Quentin Prison Report, 1981.
3. W.H.O. 1952 Tech. Rep. Ser. World Health Organization No. 48, Expert
        Committee on Mental Health, A/C Sub. Committee, 2nd report.
4. E. M. Jellinek, "The Disease Concept of Alcoholism", College and Univer-
        sity Press, New Haven, Conn., 1960.
5. H. Hershan, The Disease Concept of Alcoholism-a Reappraisal, in "Pro-
        ceedings, 30th International Congress on Alcoholism and Drug Depen-
        dence," Lausanne International Council on Alcoholism and Addictions,
        Amsterdam, 1972.

COCAINE UPDATE

Michael H. Beaubrun

University of the West Indies
Port of Spain, Trinidad

INTRODUCTION

Like alcoholism, cocaine dependence can be arrested but relapse is only
one snort away and indeed treatment objectives must include avoidance of all
mind-altering chemicals from then on. This usually requires a complete
value reorientation or spiritual rebirth.

Cocaine, the drug fashion of the past decade, has been called "The
Great Deceiver." I prefer to think of it as the new Simon Legree, bringing
back a new slavery to our islands, the like of which has not been seen since
the days of the middle passage of two centuries ago.

SHORT HISTORY AND DESCRIPTION

Cocaine is found in the leaves of the shrub erythroxylon coca which
grows abundantly on the eastern slopes of the Andes in Bolivia and Peru. It
has been chewed by the Indians of South America for centuries to give them
energy and for religious ceremonies. The leaves contain from 1% to 2% cocaine which is only slowly released when chewed mixed with lime, producing
moderate stimulation and few problems. For many years the Incas of Peru
allowed only priests and nobles to use it but after the Spanish conquest, it
spread to all classes.

About 100 years ago, a German chemist, Albert Niemann, perfected a
technique for extracting cocaine hydrochloride, a white crystalline powder
readily absorbed from mucous membranes anywhere in the body. Nasal ingestion of this powder, nicknamed "snow", began to cause a serious problem of
addiction in Europe and North America but its high price, limited availability and the growing knowledge of its dangers helped to contain the first
epidemics. The vasoconstrictor action limited its rate of absorption nasally
but even when snorted it is highly addictive.

Cocaine is a member of the tropane family of alkaloids, a central nervous system stimulant like the amphetamines but shorter acting and also a
local anaesthetic. It is rapidly absorbed from all mucous membranes and
the rate of its absorption is faster than it can be detoxified or excreted
so that toxic blood levels may be readily attained.

Ten to twenty percent of the cocaine is excreted unchanged depending on
acidity. The rest is rapidly metabolized. The main urine metabolites are

ecgonine and benzoylecconine. They are excreted in amounts equal to one-quarter to one-half of the original dose in 24 to 36 hours.

PROCESSING AND USE

The leaves of the coca bush are gathered in the mountains of Peru, Bolivia, Equador, and Colombia and processed in primitive laboratories to a crude paste containing 50% cocaine sulphate. This is a bulky smelly product heavily contaminated with the sulfuric acid, kerosene, and other materials used in the process. This is the coca paste known as basuco in Colombia and pallido in Bolivia. It is smoked by the youth of those countries with disastrous results. It is said to be even more addictive than the pure cocaine hydrochloride or even its freebase.

The coca paste is usually further refined in illegal laboratories to cocaine hydrochloride which is more easily concealed and transported. It used to be extremely expensive, selling at about US $150 per gramme, but is now being sold in purer form at cheaper and cheaper prices. Currently it is being sold for TT $10 for about 200 mgms on the streets of Trinidad's Port of Spain.

The cocaine powder may be snorted or injected intravenously, but it cannot be smoked with any significant effect. It is therefore further processed by heating with a strong alkali and ether to the volatile "free base" which can be easily vapourized, and inhaled or smoked. The method of freebasing rapidly produces high blood levels of cocaine with greater toxicity and more rapid dependence than nasal ingestion. Its effect is equal to or greater than intravenous injection.

Freebasing has become the main method of use in Trinidad and is usually done with "Zooch" pipes made from tiny brandy bottles, which have suddenly become very scarce.

Trinidad and Tobago Cocaine Epidemic of the 1980s

During the 1980s Trinidad and Tobago has seen an explosive increase of cocaine use and in the past three years this has reached epidemic proportions. Before 1980 it was not a police problem and before 1984, not a significant feature of hospital admissions. This is true not only of the General Hospital but also of St. Ann's Hospital, the only psychiatric hospital.

The reasons for the epidemic may be summarized under factors affecting availability and factors affecting demand. The availability factors would be:

a. Location along main trafficking routes, e.g., Peru, Bolivia, Colombia to the U. S.
b. Growth of organized crime.
c. A relatively high per capita income in Trinidad and Tobago making it a marketing target.
d. The low price of available cocaine.
e. The purity of available cocaine.
f. The high prevalence of freebasing.
g. An apparent marketing decision which seems to have been made by the traffickers in 1982/83 to provide cocaine pre-processed as ready to smoke rocks for freebasing.
h. And now we are threatened with an invasion of coca paste which is not only dirt cheap but even more addictive than cocaine hydrochloride or freebase.

Demand factors include:

a.   The cocaine mystique born of its history, its appeal as a status symbol, its use by pop stars, sports heroes, and others socially accepted and admired. Our own cultural lifestyle with its accent on hedonism is also a powerful demand factor.
b.   The "non-addictiveness" myth.

The Non-Addictiveness Myth

Perhaps one of the first things to update about cocaine is the myth of its being non-addictive. It is one of the most enslaving substances known to man. Only heroin and the new synthetic "designer drugs" can rival or outstrip it. Yet the Bible of American psychiatry, DSM-III, the diagnostic statistical manual, third edition, of the American Psychiatric Association, deliberately omits cocaine dependence from its drug dependence classification. For most drugs there are separate categories for abuse and dependence. There is a category for cocaine abuse, but none for cocaine dependence!

The reasons for the myth may be traced to traditional concepts of addiction. Drug addiction, or dependence, was formerly conceived of as being neatly divided into psychological dependence and physical or physiological dependence. It was thought the true addiction required that the presence of physical dependence be demonstrated. The criteria for this dependence were tolerance and withdrawal symptoms. It was alleged that there was no tendency to develop tolerance to cocaine and that there were no withdrawal symptoms.

Psychological dependence was considered to be a learned behaviour pattern and biological scientists were inclined to under-rate its significance, failing to see that not only could such learning be powerfully enslaving, but that the learning was mediated by biological factors such as the reward of stimulating the median forebrain bundle of the hypothalamus, the so-called "pleasure centre."

Jeri and Noya

Latin American cocaine experts have for years warned of the tremendous addictive potential of cocaine. In particular, Dr. Nils Noya of Bolivia and Dr. Raul Jeri of Peru have pointed out that withdrawal symptoms did in fact exist. Their findings were treated with scepticism because of what has been called the N. I. H. (not invented here) syndrome.

Studies carried out in under-developed countries are often not considered serious or trustworthy due to the mistaken prejudice that such studies "do not meet the standards of scientific research." Therefore, what Noya and Jeri have said for years is only now being accepted because it has only recently been confirmed by North American scientists.

New research indicates that tolerance does indeed develop to some effects, e.g., loss of appetite, but not to others. On the other hand, increased sensitivity may develop to other effects, e.g., seizures. This may be the reason for some of the cocaine deaths in status epilepticus after apparently non-toxic doses.

A recent study of outpatients by Frank Gawin and Herbert Kleber of Yale showed that some of these patients had ingested more than one gramme of cocaine, an amount usually considered lethal. This suggests that some tolerance had been developed.

Withdrawal Symptoms

Withdrawal symptoms frequently seen include:

- irritability
- anxiety, depression, sometimes crashing depression
- hunger
- boredom, fatigue, lassitude
- hypersomnia with long disturbed sleep
- intense drug craving which tends to be intermittent
- inability to concentrate
- impotence

But it would be a mistake to measure the craving in terms of the presence or absence of measurable physiological withdrawal or tolerance which often seem negligible. It is the extent to which behaviour is controlled by these compelling urges to take the drug over and over to the neglect of everything else.

Last year Dr. Roy Wise and associates at Concordia demonstrated the direct action of cocaine on the pleasure centre of rat brains produces more intense need than heroin. Given free access laboratory rats will select cocaine over food or water to the point of death.

It is important for us to understand that even after the pleasure has turned to pain and dysphoria that the learned compulsive behaviour persists.

NATURE OF DEPENDENCE

How then can we best conceptualize the nature of dependence? In the past it has been called: 1) a moral problem; 2) a symptom of neurosis; 3) a. learned maladaptive behaviour; and 4) a disease or illness.

The latter two concepts have replaced the first two and in practice, the disease concept of drug dependence has proven the most serviceable operationally in treatment settings.

There has, however, been some semantic confusion about the term "disease." Du Pont has warned that this concept while useful for treatment, may carry undesirable messages for prevention and public education. To some, it may seem to imply that, if you are not genetically susceptible to the "disease", then you may go ahead and sniff cocaine with impunity -- and that is certainly not a message we wish to convey. The degree of risk may vary from person to person, but everyone is at risk.

It is therefore most important that people understand what we mean when we conceptualize drug dependence as disease. The word disease is used here to mean "a deviation from health" in its broadest bio-psycho-social sense as defined by W.H.O. It is a dimensional not a categorical model. In other words it is something you can have in small degrees like hypertension, obesity, or diabetes. It represents a continuum from the normal to the pathological. It is not an all or none phenomenon like measles, chicken pox or pregnancy that one either has or one doesn't have. However, once control of intake becomes impaired it can never be regained.

DRUG EFFECTS

I will deal only briefly with specific drug effects. All drug effects are dependent on a variety of factors, eight of which are listed below:

1. The quantity or strength.
2. The manner of use (snorted, smoked, or injected).
3. The reasons for use.
4. The mental set and setting.
5. The susceptibility of the user, which may be genetically determined.
6. The personality of the user and his/her "reaction type."
7. Drug experience and tolerance or sensitivity.
8. Concurrent use of other drugs.

The short term effects of a small dose of cocaine when the drug is snorted last from 20 to 40 minutes and include:

1. Euphoria.
2. Increased energy.
3. Enhanced mental alertness.
4. More sociability.
5. Decrease in hunger and fatigue.
6. Indifference to pain.
7. Feelings of capability and power.
8. Heightened sex interest.

The effects of a larger dose are:

1. Intensified "high."
2. Erratic, bizarre, violent behaviour.
3. Increased blood pressure and heart beat.
4. Rapid breathing.
5. Dilated pupils, sweating, pallor.
6. Rise in body temperature.

The effects of a toxic dose are:

1. Restlessness, agitation.
2. Overactive reflexes, inco-ordination.
3. Tremor, twitching, muscle spasms.
4. Delirium.
5. Chest pain.
6. Nausea.
7. Blurred vision.
8. Death.The average lethal dose is 1.2 gm.

Death may result from:

1. Respiratory paralysis.
2. Suffocation.
3. Hypertension - C.V.A. (including subarachloid haemorrhages).
4. Cardiac arrythmias or infarction.
5. Hyperpyrexia.
6. Convulsions leading to status epilepticus.
7. Accidents or suicide.

The psychiatric effects of chronic use include:

1. Anxiety, dysphoria, suspiciousness.
2. Restlessness, insomnia.
3. Hallucinations (cocaine bugs, snow lights, etc.).
4. Paranoid delusions.
5. Bizarre, violent behaviour.
6. Homicide or suicide.

The medical effects of chronic use include:

1. Malnutrition.
2. Dental caries, periodontal abscesses.
3. Vitamin deficiencies: B_1; B_6; C.
4. Rhinorrhoea, septal necrosis.
5. Frontal siusitis.
6. Hoarseness, aspiration pneumonia.
7. Impaired pulmonary function.
8. Hepatitis B.
9. AIDS.

Rhinitis, septal necrosis, frontal sinusitis tend to be associated with snorting. Hoarseness, aspiration pneumonia and impaired pulmonary function with freebasing, and hepatitis B and AIDS are associated with the injection route.

ACTION OF COCAINE

Cocaine interferes with catecholamine neurotransmission. It is a dopamine agonist, blocking the re-uptake of dopamine and noradrenaline so that there is a build up of these catecholamines at the synaptic cleft. Dopamine, a precursor of noradrenaline, is found in the corpus striatum and in that part of the hypothalamus regulating hunger and thirst, also temperature, mood, sleep and sexual arousal. It activates the arousal system through the ascending reticular network and also activates the median forebrain bundle of the hypothalamus believed to be the "pleasure centre."

Dopamine Depletion Hypothesis

Early last year, Mark Gold and associates at the Fair Oaks Hospital, Summit, New Jersey, hypothesized that the debilitated chronic abuser syndrome was a dopamine depletion syndrome and demonstrated high prolactin levels in chronic abusers, indicating low dopamine levels. They alleged that the recurrent urges to take the drug were due to dopamine depletion and not to psychological dependence.

They advocated the use of Bromocryptine, a dopamine receptor agonist, for treating the chronic abuser. We have attempted to replicate this study here in Trinidad. A medical student, Ms. Joy Parris, as an elective project, undertook a pilot study of Bromocryptine with promising results, but we were unable to demonstrate significant hyper-prolactinaemia except in one female patient. Despite this, there was a noticeable improvement in the patient's craving scores and we think the method holds promise for the management of chronic cocaine abusers.

BASIC TREATMENT PLAN

Most "addictionologists" agree that the basic Alcoholism Treatment Model is the one best suited to cocaine addicts. It is conventional to divide treatment into four phases: 1) pre-treatment; 2) detoxification; 3) rehabilitation; 4) follow-up. Each phase has its particular skills and the methods of one stage would be inappropriate at another stage. The most important phases are probably pre-treatment and follow-up, although we are inclined to spend all our resources on detoxification and rehabilitation.

I shall not describe the details of these phases of treatment here. I do want to point out that the goals of the rehabilitation phase are attitude change, a new self image, a new lifestyle, and a new value system. We are talking about thought reform or the phenomena of religious conversion. These are essentially techniques for "washing out" or abolishing learned behaviour and substituting new behaviours and values. Our behaviour therapy skills will need refinement but we would do well to study again the techniques of religious conversion. Two landmark works of the past should be looked at again.

Battle for the Mind

Over twenty years ago, William Sargant wrote "Battle for the Mind," a book which sold over 200,000 copies. He studied the mass conversion techniques of Wesley and other charismatic religious healers of the past who frightened their congregation with hellfire and damnation and worked them into a state of great excitement and near collapse and then offered them salvation. He compared this with the experience of Pavlov's caged dogs in Leningrad. Pavlov's dogs had been conditioned by months of patient work, but when the River Neva was in flood, the dogs were trapped in their cages by the rising waters. They were found swimming about at the tops of their cages with the water close to the roofs, a terrifying ordeal. Many of the dogs went into a state of collapse or stupor and all of their newly conditioned behaviour seemed to be washed out. Sargant's observations led him to develop the technique of excitatory abreaction for battle neurosis in World War II. This is similar to the behavioural technique of flooding or implosion used by some behaviour therapists today.

Persuasion and Healing

Jerome Frank, Professor of Psychiatry at Johns Hopkins in 1961 wrote another landmark book, "Persuasion and Healing." He studied the techniques of shamans and primitive religion all over the world and compared them with the techniques of modern psychotherapy. He came up with the following prescription for attitude change:

1. Confession of sins to a high status healer.
2. Indoctrination and repetition.
3. Removal of ambiguity.
4. The opportunity for identification.
5. Active participation and the opportunity for exercising initiative.

It is not without significance that all these five points are to be found in the treatment method used by both Alcoholics Anonymous and Narcotics Anonymous.

Roles and Responsibilities of Health Care Team

It is important that continuous training be undertaken of health care personnel, self-help workers and all other community agencies, both voluntary and professional. The Caribbean Institute on Alcoholism and other Drug Problems has undertaken this function for the Caribbean region over the past twelve years.

THE PRODUCER SOCIETIES: THE SOUTH AMERICAN EXPERIENCE

SOMATIC DISORDERS ASSOCIATED WITH THE ABUSE OF COCA PASTE AND COCAINE

HYDROCHLORIDE

F. Raul Jeri

National University of San Marcos
Lima, Peru

SUMMARY

Three hundred and eighty-nine users of coca paste or cocaine hydrochloride were studied. The coca paste was smoked, while usually the cocaine powder was inhaled. Use was intensive or compulsive in 89.2% of the subjects. These patients presented toxic manifestations of dysphoria (67%), hallucinosis (18.7%), psychoses (11.5%), or atypical syndromes (6.6%).

The neuropsychological disturbances which existed previously to cocaine use were personality disorders, cyclothymias, anxiety, schizophrenia, depression, childhood encephalopathies, mental retardation, and convulsive disorders. The vast majority of the patients used other drugs prior to the onset of cocaine abuse. Similarly, the great majority employed other chemical substances before, during, or after a cocaine binge.

The associations or complications seen in coca paste or cocaine abuse were malnutrition (82.4%), infections (35.6%), cardiovascular disorders (30.1%), respiratory disorders (24.4%), and epilepsy (5.1%). Eighteen illustrative cases are described, including epilepsy, heart arrest, cerebral abscesses, neurosyphilis, anemia, nephritis, toxoplasmosis, meningoencephalitis, prolonged coma, corneal ulcers, anorexia nervosa, porencephaly, and intracranial traumatic haemorrhages.

Less than half of the patients (46.2%) were followed up for two years, after having been treated by various techniques. Only 25.9% were abstinent and 24.3% had died by that time. Deaths were caused by infections, accidents, homicides, and suicides.

Excessive use of coca paste or cocaine hydrochloride is frequently associated with severe organic disorders, causing serious limitations and high mortality in these subjects.

INTRODUCTION

The considerable demand for cocaine from the western world has produced an enormous increase in production of the coca leaves in Bolivia and Peru and a marked increase in coca paste and cocaine hydrochloride consumption in Argentina, Bolivia, Brazil, Colombia, Ecuador, Peru, and Venezuela. In some of these countries, epidemics of coca paste and cocaine hydrochloride abuse have been described, including serious and persistent addictions to these compounds in many subjects. When the user is not treated, a characteristic natural history of the addiction develops, starting from experimental to compulsive use, accompanied by progressive personal and social deterioration.

Initially, few somatic complications or associations were observed in the intensive users; but morbidity increased as patients were followed up for several years. In this paper, I shall present some of the organic associations and complications seen in Peru.

METHOD

This report is based on observations made on 389 patients examined in several hospitals: Larco Herrera, Dos de Mayo, Central de Policia, San Isidro, San Juan de Dios, and San Antonio. A few patients came from the infirmary of Lurigancho jail and others from my private practice.

All patients were studied through a procedure which consisted of demographic data, biographical information, family history, present state, drug questionnaire, physical examination including neuropsychiatric evaluation, laboratory tests, and in some instances, psychometric and projective tests as well as symptom questionnaires. The clinical evaluation ended with a diagnostic formulation according to the criteria of the Diagnostic and Statistical Manual of Mental Disorders of the American Psychiatric Association.

There is no consensus about definitions of use, abuse, and dependency regarding coca paste and cocaine hydrochloride. Therefore, it is necessary to establish the characteristics of use in each individual case. In this investigation, users have been classified as experimental, recreational, circumstantial, intensive, or compulsive. Recreational consumers smoked paste or inhaled cocaine powder at parties or in small groups with the purpose of having fun. Circumstantial users have ingested cocaine for definite purposes while experimentalists have tried it just to find out what the immediate effects were. Intensive users abused the drug. Compulsive users demonstrated the syndromes of tolerance, craving, and withdrawal.

RESULTS

Symptoms on Admission

The symptoms on admission were related to the natural history of the disease, the antecedent pathology and the psychological and somatic complications associated with excessive use of coca paste or cocaine hydrochloride. Patients were classified on admission according to the category of use. Analysis of these admissions showed that most of the subjects were intensive and compulsive users, according to the Resnick and Schuyton-Resnick classification. (22) Therefore, although some of these patients had come to the hospitals because they had suffered from complications of circumstantial and recreational use (4.6%), the majority of them consulted or were referred for problems derived from intensive or compulsive consumption. Eightynine percent of the patients had become daily or high frequency users within six months to two years, confirming observations of several authors that paste smoking, cocaine sniffing, and cocaine injection rapidly escalates use to dependency if the environmental conditions allow considerable quantities of the drug to be easily obtained. (3-8, 14-17)

In recreational users, the symptoms on admission were: anxiety, anorexia, insomnia, loquacity, and psychomotor excitement. These disturbances motivated patients or relatives to come to the hospital. Patients who con-

178

sumed paste or cocaine hydrochloride for circumstantial purposes also presented with manifestations of the pre-existing disorders, such as depression,
cyclothymia, or adult attention deficit. Recreational and circumstantial
users included students, writers, artists or athletes who used the drugs to
improve physical or psychological performance. A high dose caused alarm in
both groups due to the intensity of the effects, such as tremors, palpitations, hyperhydrosis, insomnia, muscular rigidity, myoclonic jerks, and paranoid ideation.

Most psychophysiological disturbances were observed in the compulsive
and intensive users. These included: anxiety (81.8%), craving (80.1%),
anorexia (75.2%), insomnia (71.2%), and excitement (61.7%). It must be considered that these were symptoms on admission, some corresponding to residual
intoxication, others to hallucinatory and paranoid syndromes, with a smaller
group showing marked depressive manifestations associated to abstinence or
depression previous to intoxication. Anxiety and craving were observed
during the first contact with the patients, when most of them had stopped
administration of the drug for one or two days. Later the majority continued
to experience anxiety and craving when they thought or talked about paste or
cocaine. The main somatic disturbances found in excessive users were regis-
tered during the first physical examination, when many patients were still
under the influence of the drugs or had suffered complications. There was
a predominance of sympathomimetic manifestations including tachycardia,
mydriasis, hyperhydrosis, spasms, myoclonic jerks, and high blood pressure.
Later, many complications and associations were discovered such as thinness,
malnutrition, anemia, infections, respiratory disturbances, arterial hypertension, and generalized convulsions.

In this series, the diagnostic procedure was done according to the Diagnostic and Statistical Manual of Mental Disorders of the American Psychiatric
Association (DSM III), whenever possible using the five axes, permitting a
thorough evaluation of the patient.

Coca paste or cocaine hydrochloride users were assessed for clinical
conditions and associated psychological disorders. Abuse predominated considerably in this group (89.2% were intensive or compulsive users). The
predominant intoxication seen was the dysphoric reaction (67%) which
required repeated use of the drug to neutralize the unpleasant effects of
the crash.

Hallucinations were observed or reported by 18.7% of the patients and
11.5% developed acute psychotic reactions, some of them lasting several
weeks. Psychoses were characterized by multiple illusions and hallucinations, delusional ideation (predominantly paranoid), and bizarre behaviour
which could be aggressive, fugitive, or hyperactive and which was conditioned by the hallucinations or the abnormal ideation.

Regarding associated disorders, eighty patients (i.e., 20.6%) had presented with psychological disturbances prior to drug use. These included
predominantly cyclothymic disorders (reactive depression) (6.4%), chronic
anxious disorders (4.6%), schizophrenic disorders (3.9%), major depression
(2.8%), paranoid disorders (1.5%) and bipolar disorders (1.3%). If previous
personality disturbances are considered, 36.4% of the total group had manifestations of character disorders, predominantly of the anti-social, passive
aggressive, and schizoid varieties. Eight percent of intensive users were
mentally retarded. Summing up it can be said that 64.9% of this group of
patients had important psychological disturbances before beginning their
use of coca paste or cocaine hydrochloride. On the other hand, a considerable proportion of these subjects had used other drugs before employing cocaine. Only a few individuals began to smoke coca paste before using any
other compound. (See Table 1).

Table 1. Antecedent and associated use of other drugs in 348 cocaine
          dependent persons.

PREVIOUS USE	CASES	PERCENT
Cannabis	248	71.2
Alcohol	223	64.0
Tobacco	141	40.5
Amphetamines	125	35.9
Solvents	32	9.1
San Pedro (mescaline)	29	8.3
LSD	18	5.1

ASSOCIATED USE	CASES	PERCENT
Alcohol	275	79.0
Benzodiazepines	183	52.5
Cannabis	52	14.9
Barbiturates	24	6.8
Heroin	5	1.4
Pentazocine	3	0.8
Meperidine	2	0.5

Most recreational or circumstantial users employed cocaine to dispel
the intoxicating effects of alcohol or only used cocaine hydrochloride to
obtain euphoria or energy. Many subjects passed rapidly from the recreational to the intensive and compulsive use, but a modest proportion (4.6%)
remained in recreational use for several years, a few up to ten years. However, the majority (79%) of coca paste or cocaine hydrochloride users combined use of the alkaloid with alcohol, some to diminish the dysphoric effects of the paste, others to reduce the inebriating effects of alcohol when
they had drunk too much or when they wanted amelioration of the dysphoria
produced by the cocaine. Naturally these procedures were inadequate because a double intoxication is more damaging than one, and the half-life
of the substances is quite different. As alcohol is metabolized more slowly,
it is usual to see marked sedation when the effects of the cocaine wear off.
This explains why many automobile and motocycle accidents were seen after
combined alcohol-cocaine binges. Frequently the patients took benzodiazepines or hypnotic pills when they arrived at home so that they could get to
sleep. Very few subjects combined intravenous cocaine with heroin, pentazocine, or meperidine.

ILLUSTRATIVE CASES

I shall now present very briefly several illustrative cases of coca
paste or cocaine hydrochloride abuse associated with somatic complications
and associations.

Patient 1: Coca Paste and Epilepsy

A 37 year old male was admitted due to repeated major convulsions. On
examination, he was unconscious and had an extensive fronto-malar haematoma.
He had had other seizures, beginning a few months previously. There was no
personal or family history of epilepsy. While in hospital, he had nine
tonic-clonic seizures. The physical examination, including EEG, was normal.
He was moderately retarded and only completed six grades at school. The
seizures tended to appear when he smoked too much coca paste. He began to
use the substance 20 years previously. He had also experienced respiratory
difficulties. Diagnosis: Mental retardation, intensive coca paste use, cocaine epilepsy.

Patient 2: Alcoholism, Coca Paste, Hallucinosis, Epilepsy

A 44 year old male was admitted because of repeated convulsions. His
father had been an alcoholic and had committed suicide; an uncle was also an
alcoholic. The patient began to drink alcohol when he was 18 years old, and
within a few years he drank heavily every day. He has smoked coca paste for
the last five years at the rate of 10 to 15 cigarettes a day. He has developed tonic-clonic seizures after drinking or after using paste and alcohol.
He has had several hallucinogenic epidodes, mainly of visual hallucinations.
On examination, he had a moderate increase of blood pressure (150/100 mmHg),
and many scars due to burns. Fourteen years ago he had been run over by a
car while inebriated and had sustained a skull fracture, spending several
hours in a coma. Diagnosis: Alcoholism, intensive use of paste, probable
brain scar, alcoholic and cocaine epilepsy.

Patient 3: Alcoholism, Coca Paste, Cerebral Abscess

A 57 year old male was admitted because of aphasia and hemiplegia.
About 20 days previously, during a fight with a neighbour, he had been hit
on the forehead by the opponent's teeth. Seven days later, the wound was
infected and painful. Later, he experienced constant headache and sustained
a moderate fever (38°C). Gradually he became confused, was unable to talk,
and was unable to move his right limbs. On examination he had neck rigidity
in addition to the motor signs already mentioned. He had an infiltrate in
the right lung and an area of concentration of the radioactive material in
the left frontal region. Blood examination showed an increase in white cells
with a predominance of polymorphonuclear leukocytes. Treated with penicillin
he improved considerably. He has smoked coca paste for the last 10 years.
He also smoked marijuana and used alcohol and sedatives to excess. His wife
died from tuberculosis and he had the same disease while younger. Diagnosis:
Alcoholism and polydrug abuse, pulmonary tuberculosis, brain abscess, personality disorder: antisocial.

Patient 4: Pulmonary TBC, Coca Paste Abuse

A 26 year old male was admitted because he was spitting blood. Five
years before, he was hospitalized for coca paste abuse in a psychiatric in-
stitution, where the doctors found that he also had tuberculosis. He was
treated at that time, but after he was discharged, he continued to abuse
alcohol, coca paste, and tobacco. Three years prior to admittance, he
began to notice bloody sputum, which gradually increased. At the same time
he smoked coca paste every day and had no other interest in life. He had
dropped out of school failing in mathematics and science courses. When he
was 18, he began to smoke coca paste. His father was an alcoholic and had
died under mysterious circumstances. On examination, the patient was thin,
and had bilateral lung tuberculosis and multiple scars due to knife cuts
on his upper limbs. He had been detained several times for assault and
battery. Diagnosis: Alcoholism, coca paste abuse, bilateral lung tuberculosis, personality disorder: passive-aggressive.

Patient 5: Interstitial Nephritis, Cocaine, Pentazocine, and Heroin Abuse

A 29 year old male was admitted with generalized edema. On examination
he was a tall, thin man with marked anemia, thrombophebitis in his upper
limbs, urinary tract infection, and laboratory signs of liver damage. He
had difficulties with mathematics and science, but did finish high school.
At 13, he began to smoke marijuana. Soon after, he tried amphetamines and
cocaine. He inherited some money and travelled to several countries. In
Holland he was heavily involved with cocaine and heroin. He had several
accidents while injecting drugs: two with cocaine and two with heroin.
While using cocaine, he had a heart arrest and was resuscitated. When he

came back to Peru, he resorted to pentazocine because he could not obtain heroin. A kidney biopsy revealed interstitial nephritis. Diagnosis: Heroin, cocaine and pentazocine addiction, interstitial nephritis, personality disorder: passive-aggressive.

Patient 6: Suicide Attempt, Coca Paste Abuse

A 28 year old male was admitted to the emergency room with nausea, vomiting, diarrhea, generalized aches, paresthesia, and loss of hair. Six days previously he had felt lonely and tried to commit suicide by strangling himself, but the pain produced by the rope dissuaded him. Then he took rat poison containing thallium and an insecticide. Soon he felt nauseous, sore throat, and began vomiting and had diarrhea. He was taken to the hospital. In the following days, he had intense aches in the muscles of his lower limbs and difficulty in walking. Later, he lost all of his head and body hair. He had been born prematurely, in a dystocic labour, cyanotic and required resuscitation. During childhood, he was very withdrawn. At age 15 he began to drink alcohol frequently and to smoke tobacco. One year later, he began to smoke cannabis, and at age 20, coca paste alternating with alcohol. On examination he had hypoesthesia and hypopalesthesia in his upper and lower limbs. He had ideas of reference, and the urea and creatinine levels were elevated in his blood. Diagnosis: Obstetric encephalopathy, paranoid reaction with depressive episodes, alcohol, cannabis, and coca paste abuse, thallium poisoning.

Patient 7: Toxoplasmosis, Multiple Abscesses, Heroin, Pentazocine, Coca Paste and Cocaine Abuse

A 25 year old male was admitted for perineal aches. For eight months he had suffered pain in the genital areas and lower extremities, along with haematuria and diarrhea as well as dizziness and diminution of vision. Laboratory examination revealed toxoplasmosis. He was treated but stopped the medication because of nausea and gastrointestinal disturbances. Finally he was readmitted for lymph node enlargement and nodules in the limbs. He smoked 60 cigarettes a day, inhaled cocaine once a week, and cannabis two or three times a week, and used pentazocine and heroin in binges. He had sustained strong blows to his head during student protests. Several times he had purulent urethral discharge. On examination he had supurative infections of the nose, pharynx and skin, the lymph glands in the neck and limbs were very enlarged. He had inflammation of the epydidimus. The toxoplasmic haemaglutination was 1/256; a skin biopsy showed dermal microabscesses. He escaped from the hospital. Diagnosis: Polydrug abuse: heroin, pentazocine, coca paste, cocaine, toxoplasmosis, subacute epydimitis, multiple microabscesses, personality disorder: antisocial.

Patient 8: Coca Paste Addiction, Meningeal Tuberculosis, Prolonged Vegetative State

A 19 year old male was admitted with headaches. He had been an excellent student and athlete; at 18 he had been called to the military. There he learned to smoke cannabis and coca paste. He quickly became an intensive smoker. After six months, he needed to smoke paste every day. He was expelled from the Army and later no one could stop him from climbing through the windows and ceilings to get the drug. He sold all his belongings, robbed his own house, then those of the neighbours; later he became associated with a gang of bank robbers. He was detained, and while in prison, he exchanged his clothes, food, and money for paste. He contracted pulmonary TBC and tuberculous meningitis. In spite of that disease, he would exchange his medicines for the drug. Gradually he became worse and was taken to the hospital, where he deteriorated steadily. The infestation was controlled, but he developed obstructive hydrocephalus and a state of prolonged coma with

persistent myoclonic and generalized convulsions. At present, he is unconscious, cachectic, and rigid. Diagnosis: Pulmonary and meningeal tuberculosis, akinatie mutism, coca paste addiction, personality disorder: passive-
aggressive.

Patient 9: Coca Paste Abuse, Respiratory Distress

A 22 year old male was admitted for respiratory difficulty. For the
previous month, he had dyspnea while standing, lying, or sitting, which increased when he ate or made any effort. The distress was very considerable.
He began to smoke coca paste 18 months prior to admission. He also drank
alcohol. He would smoke 40-50 paste cigarettes in one night, until he felt
anxious, wanted more drug, felt that he was suffocating, shook all over, and
had paranoid ideation. He also experienced intense epigastric aches. A bad
student, he was absent from school many times and finally dropped out altogether. He began to drink alcohol at 13, and smoked cigarettes at 14. Several maternal relatives drank alcohol to excess. On examination, there were
no clinical, laboratory or radiological signs of peptic ulcer or lung infection. His cardiovascular system was normal. He had many scars on his limbs
and thorax due to knife fights and suicidal gestures. Indirectly, the police
informed the doctors that he had been a well-known thief since his adolescence. Some of his siblings were also failures in school and drank exces-
sively. Diagnosis: Moderate mental retardation, coca paste abuse, ventilatory difficulties produced by coca paste smoking.

Patient 10: Coca Paste Hallucinosis, Anemia due to Ancylostomiasis

A 26 year old male was admitted with dizziness and blurred vision. Two
months previously he began to experience rotatory vertigo, dimness of vision,
chills, fever, and excessive sweating, as well as diarrhea alternating with
constipation. The symptoms developed after smoking coca paste. He began
to smoke marijuana at 15 and soon afterwards, coca paste. Four years prior
he began to work for an oil company in the jungle. There he would smoke
paste every 28 days, for 2 or 3 days on end when he would smoke 40 paste
cigarettes per day. While intoxicated he had had many visual illusions and
hallucinations and paranoid ideas. After a binge, he would become depressed
and ashamed. He would spend one month's earnings in two days smoking coca
paste.(23) Many of his relatives were heavy drinkers, especially on the maternal side. On examination, he was thin and pale; his blood haemoglobin
was 7gms/dl with 3,000,000 red cells per cubic milimetre. His faeces showed
eggs of uncinaria, trichuris trichiura, and ancylostoma. Diagnosis: Coca
paste abuse with hallucinations, anemia due to ancylostomiasis.

Patient 11: Polydrug Abuse, Hallucinosis, Attention Deficit Disorder

A 23 year old male was referred because of bizarre behaviour. He was
found lying on the floor of the bathroom, a pistol at his side, with the
bathtub faucet running, and he was yelling at the top of his lungs. He saw
insects everywhere: spiders, snakes, and all kinds of poisonous animals.
Several hours previously, he had gone to an emergency room, saying that he
had been stung by an insect, and he thought that he had a red mark on his
skin. For four weeks he remained under observation in the hospital where he
kept peering at and pricking at his skin, seeing and feeling the insects.
Other people could see nothing. He had been born prematurely, weighing
1600 grms, and was in an incubator for one month. He was extremely restless
and destructive in childhood. When he was eight, he would smoke a whole
cigar. In school, he had difficulties with science and mathematics. He
was aggressive and impulsive during adolescence. He suffered five traumatic
accidents, some involving periods of unconsciousness. He went abroad to
study, but in 18 months, he only finished one term. Most of the time he
drank alcohol and used drugs. He was expelled from that country and not

permitted to return. He spent all the money his mother had in a joint account. At present, he goes out only with wealthy friends who drink alcohol
and use drugs. His father was a heavy drinker. Several maternal cousins
are considered extremely nervous. He inhales cocaine frequently. On examination he was thin, arrogant, and extroverted. He admitted that he had attacked several employers and destroyed part of their businesses. He is moderately retarded and had psychopathic and paranoid deviations on the MMPI
scales. Diagnosis: Cocaine hallucinosis, attention deficit disorder, residual type, moderate mental retardation.

Patient 12: Congenital Toxoplasmosis, Generalized Convulsions, Intensive Use of Alcohol, Coca Paste, and Cannabis

A 19 year old male university student was referred for excessive cough.
For several days previously he could not sleep because he coughed constantly.
He was born prematurely, weighing 1500 grms. He showed multiple behaviour
disturbances at school, resulting in his expulsion from several institutions.
From the time he was 15, he had suffered from repeated grand mal seizures.
At age 17 he began to smoke tobacco, cannabis, and coca paste, usually from
Friday through Sunday. While intoxicated, he was irritable and garrulous.
At 11, he was examined by a neurologist who found multiple intracranial cal-
cifications on X-rays of his skull. A Sabin-Feldman test showed a titre of
1/512 and indirect haemaglutination of 1/32 against toxoplasmosis. An EEG
demonstrated marked dysrhythmia with theta and delta waves, accentuated on
hyperventilation. The paternal grandfather smoked 40 cigarettes a day and
died of myocardial infarction. The mother had high blood pressure. A paternal uncle has a personality disturbance and at 46 lives off his mother's
earnings. A sister had generalized grand mal seizures when she drank alcohol.
On examination the patient was thin, reticent, and irritable; he had old
chorioretinitic patches on both fundi, and a persistent cough with rales
and wheezes. Diagnosis: Congenital toxoplasmosis, generalized convulsions,
coca paste, alcohol, and cannabis abuse, respiratory asthmatic reaction;
personality disorder: passive-aggressive.

Patient 13: Cocaine Abuse, Nasal and Ocular Lesions

A 31 year old male was referred for excessive cocaine sniffing. One
and a half years previously, he had begun to inhale considerable quantities
of cocaine hydrochloride. Gradually he consumed more and more of the drug.
During the last year, he would use 10 grammes in 15 hours, or about 400 mgs.
per hour. These excesses produced intense muscular spasms, visual, auditory
and cutaneous hallucinations, considerable sweating, dehydration, nasal
haemorrhages, and eye congestion. While intoxicated, he masturbated in front
of his parents. He had required emergency treatment for these complications.
He had tried to stop sniffing cocaine at least a dozen times, but he always
relapsed after a few days. He was treated by a psychoanalyst for six months,
but did not stop his drug use. He smoked marijuana when he was 16 as well
as taking LSD. Nevertheless, he finished business administration studies at
a local university. A brother was a high school drop-out and an uncle was
a chronic alcoholic. On examination he was a thin, assertive young man,
dressed in a very conceited way, presenting considerable congestion in the
nose and pharynx. There were also points of nasal haemorrhage and marked
conjunctival inflammation, especially around both corneas, and a small corneal ulcer on the right side. Diagnosis: Hystrionic personality, cocaine
abuse, nasal haemorrhages, corneal ulceration.

Patient 14: Cocaine and Alcohol Abuse, Post Traumatic Brain Disorder

A 24 year old male university student was referred because he consumed
excessive amounts of cocaine. According to his girl friend, from the time
they met she observed that he would sneak away to inhale cocaine when he was

at parties. Gradually he became indifferent to her, being only interested in friends who had the drug. He said that he had to consume cocaine to be able to stay awake at parties. He had become too sensitive to the sedative effects of alcohol. He had lost considerable amounts of weight in recent months. He had developed normally and had had a healthy life. For example, he had been a champion motorcyclist. When he was 19, he was sent abroad to study. There he consumed alcohol and drugs. One day, while intoxicated, he fell from a tree and suffered serious craneocerebral injuries. He was found deeply unconscious with decerebrate posturing. Immediately operated on, he had acute epidural and intracranial haemorrhages. He was in a coma for several weeks, but recovered slowly. When he came back to his home country, he was slow, rigid, ataxic, amnesiac, confused, and had bilateral pyramidal signs. Eventually he improved and returned to University to study elementary courses. He worked hard only at the last hour. On examination, two years after the accident, he was thin, iterative, very slow to move and to talk, and had bilateral frontal signs of brain involvement. Diagnosis: Cocaine and alcohol abuse in a previously healthy personality; Post-traumatic brain disorder.

Patient 15: Porencephaly, Epilepsy, Intensive Use of Cocaine and Cannabis

A boy, aged 16, was brought in by his father for seizures and drug abuse. Shortly prior to admission, he had lost considerable weight, was very irritable, and only wanted to be outside on the streets. He was treated at home with neuroleptics, anticonvulsants, and benzodiazepines. When he became extremely rigid, the parents decided to discontinue the treatment. He smoked tobacco at 12, and marijuana at 14. He tried coca paste at 15, but as he did not like it, he changed to snorting cocaine hydrochloride. At 15 he also began to drink alcohol. He was born by Caesarean section and left in an incubator. As a child, he cried constantly and had terrible temper tantrums. He was aggressive, irritable, violent, and destructive. Neuroradiological investigation disclosed porencephaly. He was hyperactive during school and failed many courses, especially mathematics. At 13 he stole his father's car and had several crashes. He was expelled from school for drug possession, and he has also been detained by the police several times for carrying and consuming drugs. On examination, he was thin, unstable, and an ataxic young man whose tendon reflexes were hyperactive and asymetrical. Diagnosis: Dystocic labour, porencephaly, attention deficit disorder, excessive use of cocaine, cannabis and alcohol.

Patient 16: Excessive Alcohol and Cocaine Use, Anorexia Nervosa

A 20 year old woman, an unemployed secretary, was referred because of weight loss and excessive drinking. She failed courses at school, repeating one complete year and had behaviour problems such as teasing the teachers and smoking in the bathroom or on the bus. She began to masturbate at six, and had sexual relations at 16. She has irregular menses, and has been amnerroheic for several months. She began to smoke tobacco at 12, and to use cannabis at 16, and cocaine at 18. Lately, she drinks considerably and, induced by her boyfriend, snorts cocaine to facilitate love-making. Several times she has been unable to walk because of drunkenness. During the day, she would skip lunch to go running in addition to following a rigid low-calorie diet. She experienced considerable difficulties in getting to sleep. Her father was an impulsive, aggressive, excessive drinker. One brother stole money from the bank where he worked. On examination she was very thin reticent and distrustful. No physical signs of neurological disorder were noted. Diagnosis: Intensive alcohol and cocaine use, anorexia nervosa, personality disorder: antisocial.

Patient 17: Syncopal Epilepsy, Tension Headache, Intensified Use of
             Cocaine and Alcohol

A 31 year old female secretary was seen for a one year history of numbness on the occiput and neck. She used tobacco and marijuana at 19, and cocaine at 20. For two years her cocaine inhalation was very intense. She
changed completely, becoming talkative, restless, incoherent, irritable, and
sleepless. After daily use, she decided to stop and, with her mother's help
and a religious intervention, she was able to do it. Altogether, she had
used cocaine for nine years. During that time, she had repeated episodes of
syncopal epilepsy when she was intoxicated by alcohol, cocaine or both. It
was intended to do a complete neurological investigation, but the patient
did not return. There was a suspicion of a temporal lobe lesion.

Patient 18: Coca Paste and Alcohol Abuse, Death by his own Father

A 20 year old male high school student was brought to the emergency
room with stab wounds. He smoked cannabis at 14 and coca paste at 16. He
failed many courses at school and had to resort to a night school. During
week-ends, he drank excessive amounts of alcohol and smoked coca paste.
Usually he got the money from his relatives, selling his possessions, or
stealing things from neighbours. One night he got into his parent's house
by breaking in through the window. He went into his monther's bedroom and
demanded a large sum of money. He was excited and violent. The mother
pleaded with him to calm down and be satisfied with the amount she gave to
him, although it was less than he had requested, but he persisted. Hearing
shouts and weeping, his father came into the room and requested his son to
stop his demands. The young man then took a brick and a kitchen knife, threw
the brick at his father and then attacked him with the knife. There was a
struggle during which his father grabbed the knife and stabbed his son several times. When the patient was brought to the hospital, he was pronounced
dead on arrival.

DISCUSSION

In this series of 389 patients who abused coca paste or cocaine, 64.9%
had mental disturbances prior to the addiction. Although it is not always
possible to distinguish clearly between primary disorders and drug intoxication, the signs of deterioration are clearly differentiated in most subjects. Deterioration can be physical or sociological. Among the pre-drug
associations, cerebral syndromes were represented in these patients by mental retardation, brain injuries, encephalic malformations, epilepsy, and
minor brain dysfunction. In some patients major psychological disorders
such as schizophrenia, cyclothymia, and depression, preceded drug abuse.
More frequent were the character disorders. In this group, 38% had personality disturbances prior to the chemical dependence.

Another antecedent was the use of other drugs before ingesting cocaine.
Use of alcohol, tobacco, and cannabis usually preceded coca paste smoking
or cocaine inhalation.

When coca derivatives were used intensively, the main syndromes of
acute intoxication could develop in one single binge, the user passing rapidly from euphoria to dysphoria, hallucinosis, and psychosis. Psychosis usually was short-lived, but it may persist for several weeks. In other publications I have described these syndromes extensively. (4-6) When the individual uses coca paste or cocaine compulsively he may show three characteristic disturbances: dependence, amotivation, and psychopathetization. Dependence is manifested by tolerance, craving, and abstinence. In my experience, cocaine abusers can tolerate considerable doses of cocaine. Many patients can smoke more than 40 paste cigarettes in one evening and some pa-

tients can inhale several grammes of cocaine per day. (9) Experience shows
that the craving or irresistible intense desire to smoke paste, or inhale,
or inject cocaine daily happens to most patients, and is one of the main
difficulties in rehabilitation.

The authority and experience of the investigators who wrote that cocaine
does not produce an abstinence syndrome were so powerful that for many years
it was considered to be true. (11,12) However, after working with patients
who used cocaine in considerable amounts for several years, if the drug is
stopped, a peculiar syndrome is observed. This is characterized by fatigue,
depression, sleep and appetite disturbances, irritability, and anhedonia.
All of these symptoms are associated with intense craving. (2,12,13)

The amotivational syndrome is seen in intensive users of cocaine who may
not yet be entirely dependent. Here, the essential interests in personal
care, family, and social relations, efficiency at work or study, or the pleasure of the open air and physical and erotic activities are lost. The person
does not bathe or care for his or her hair, disregards food, becomes isolated
and is only interested in getting money to buy drugs. The rest of the time
is spent idly or working slowly and inefficiently. The psychopathetization
syndrome observed in coca paste or cocaine abusers is signaled by intellectual, social, and physical deterioration. As relationships with family and
friends are severed, the addict is alienated from the productive members of
the community. Associations are then established with drug dealers, addicts,
peddlers, and vagabonds. When money is needed to buy drugs they resort to
swindling, robbing, and assault. Men may become drug traffickers and set
aside some cocaine for themselves while women may practice prostitution to
obtain the drug. In Peru, these persons frequently become involved with
well-organized gangs, and die when they do not comply with the criminals'
laws. They may also become victims or accomplices of corrupt police officers
or others in authority who still believe in the fabulous riches attributed
to cocaine dealers by the mass media.

Somatic complications are frequent in coca paste and cocaine hydrochloride abusers. It is impossible to cover the whole gamut in a few words.
Malnutrition and immunodeficiencies were frequent among these patients.
When cocaine is injected intravenously, the risk of acquiring viral infections as well as developing sudden cardiorespiratory accidents is very high.
With excessive doses, serious crises may develop with the oral, nasal, intravenous or respiratory routes of administration.

Malnutrition and immunodeficiency are conditioning factors for infections. In Peru, there was a predominance of tuberculosis, septicemias,
typhoid fever, brucellosis, salmonellosis, pyodermitis, and intestinal para-
sitosis in these subjects. When cocaine began to be used by inhalation,
nose bleeds, sometimes uncontrollable, have been recorded, as well as have
chronic rhinitis and perforations of the nasal septum. Ocular lesions occur
when the user inhales cocaine in great quantities and with considerable intensity. When cocaine reaches the eye, it produces chemical conjunctivitis,
keratitis, and corneal ulcers, as seen in some of these patients.

The intravenous injection of cocaine, heavy smoking of coca paste, and
even cocaine snorting may cause death by cardiac or respiratory arrest,
which may or not be preceded by generalized convulsions.

Cerebral lesions associated with cocaine administration may be represented by tuberculous or cryptococcal meningitis, pyogenic or viral meningoencephalitis and cerebral abscesses. These infections may be either systemic or neurological at their onset, as for example happens with toxoplasmosis,
cytomegalic inclusion disease, and herpetic encephalitis, complicating the
acquired immunodeficiency syndrome.

As can be inferred from this series, some patients developed generalized convulsions before using cocaine. Others had seizures only when they had ingested considerable quantities of the alkaloid. The latter had no family or personal history of epilepsy and the EEG was entirely normal before and after the seizures. There are evidently cases of cocaine epilepsy. The exact biochemical mechanisms of the convulsions are unknown at present. (18)

Cardiovascular disturbances are frequent in cocaine abusers, especially in middle-aged men. This group verified sinus tachycardia and other arrhythmias, hyperpyrexia, and high blood pressure during acute intoxication. (1,3, 6) Hypertension can cause cerebral haemorrhages, not only in subjects who have aneurysms or vascular malformations, but also in normotensive and hypertensive individuals. (20) Vasoconstriction favours the appearance of anginal attacks or myocardial infarction in persons who have coronary artery disease and even in those who have patent coronaries. (19)

Respiratory disturbances are represented in cocaine abusers by ventilatory deficiency at the alveolar level, pneumomediastinum, acute and chronic bronchitis, asthmatiform reactions, viral and bacterial pneumonias, bronchopneumonias, and acute, potentially fatal, insufficiency. (20)

Haematological complications consist of diverse varieties of anemia, and leukopenias with lymphopenias and lymphoproliferative syndromes.

Women who abuse cocaine during pregnancy may have obstetrical disturbances as well as a high incidence of spontaneous abortions. Abruptio placenta has been documented immediately after the intravenous injection of cocaine. Babies of cocaine abusers show deficiencies in organization as well as behaviour disturbances. (21)

Traumatic lesions can occur in traffic accidents or by attacks from assailants, police officers, drug dealers, or relatives. Several of the patients have been seriously wounded or killed by their father, brother, or sister. Some subjects suffer severe burns while preparing coca paste or cocaine hydrochloride because they use highly inflammable fluids (gasoline, ether, acetone). Others are burned while preparing the freebase. Suicide acts, gestures, and attempts are quite frequent in the intensive or compulsive user when he is in a drug crash or in the depressive phase of abstinence. Some actually commit suicide either directly or indirectly through threatening actions. For example, one of my patients advanced menacingly towards his sister. She had a loaded gun and several times she threatened that she would shoot him if he did not stop. In Peru, many addicts die in street fights, during an assault, or at the hands of hired killers.

Follow-up of this series has been very difficult. Few subjects had telephones, many gave false addresses, and others simply disappeared. However, some information could be collected from 181 patients. Only 25.9% of them said that they had been abstinent for the last two years after being treated by diverse methods in Government, social security, and private institutions. Abstinence meant no use of coca paste or cocaine, but some allowance was made for the use of marijuana or alcohol. Absolute abstinence of all drugs would have been an unrealistic demand at this stage of evaluation , although such use should be stopped because of its conditioning effect. (2)

Regarding mortality, within two years, 44 persons had died (24.3% of the 181 persons), a very high figure considering that most were young men. The majority had died due to tuberculosis (lung, miliary, or meningeal), others from infections, acute toxic reactions, status epilepticus, heart arrest, respiratory insufficiency and accidents. Nine were victims of homicide and six committed suicide.

Table 2. Complications and associations observed in abusers of coca
paste and cocaine hydrochloride.

Disturbances previous to the
use of cocaine

Cerebral syndromes
Schizophrenic disorder
Major depression
Cyclothymic disorder
Bipolar disorder
Personality disorder

Syndromes of intensified use

Euphoria
Dysphoria
Hallucinosis
Psychosis
Atypical reactions

Organic disturbances associated
with compulsive use

Malnutrition
Immunodeficiency
Overdoses
Infections
Nasal lesions
Ocular lesions
Cerebral lesions
Cardiovascular disturbances
Respiratory disorders
Haematologic disorders
Obstetric disorders
Traumatic disorders
Suicides
Homicides

Disturbances of compulsive use

Cocaine dependence
Amotivational syndrome
Psychopathetization syndrome

In conclusion, the associations of coca paste and cocaine abuse are
manifold. It is important to establish four aspects in every case: The
pre-addictive disturbances, the syndromes of intensified use, the disturbances of compulsive use, and the somatic complications seen in dependent
users (See Table 2).

Many areas remain to be investigated, for example, the persistent biochemical changes produced by cocaine in the brain, and the pathophysiology
of craving and abstinence. There is need to develop an efficient preventative method to stop the cocaine epidemic, as well as reliable treatment procedures for the millions of dependents of coca and its derivatives.

REFERENCES

1.   D. Paly, P. Jatlow, C. Van Dyke, F. R. Jeri, R. Byck, Plasma cocaine
     concentrations during cocaine paste smoking, Life Science, 30:731-
     738 (1982).
2.   A. W. Washton, Cocaine abusers get outpatient help in special programme,
     Psychiat. News, 20(9):6-24 (1985).
3.   F. R. Jeri, C. C. Sanchez, T. Del Pozo, Consumo de drogas peligrosas
     por miembros de la fuerza armada y de la fuerza policial peruana,
     Rev. Sanid. Minist. Int.37:104-112 (1976).
4.   F. R. Jeri, C. C. Sanchez, T. Del Pozo, M. Fernandez, El sindrome de
     la pasta de coca: Observaciones en un grupo de 158 pacientes del
     area de Lima, Rev. Sanid. Minist. Int. 39:1-18 (1978).
5.   F. R. Jeri, C. Carbajal, C. C. Sanchez, T. Del Pozo, M. Fernandez, The
     syndrome of coca paste, J. Psychedel. Drugs, 10:361-370 (1978).
6.   F. R. Jeri, C. C. Sanchez, T. Del Pozo, M. Fernandez, C. Carbajal, Fur-
     ther experience with the syndromes produced by coca paste smoking,
     Bul. Narcotics, 30:1-11 (1978).
7.   N. Noya, Coca and Cocaine: A Perspective from Bolivia, in "The Interna-
     tional Challenge of Drug Abuse, NIDA Research Monograph No. 19,"

National Institute on Drug Abuse, Rockville, Md. (1978).

8. F. R. Jeri, Coca paste smoking in some Latin American countries, a severe and unabated form of addiction, Bul Narcotics, 36(2):15-31 (1984).

9. C. Carbajal, Psicosis producida por inhalacion de cocaina, in "Cocaina, 1980," F. R. Jeri, ed., Pacific Press, Lima, Peru (1980).

10. H. Isbell, W. White, Clinical characteristics of addictions, Am. J. Med. 14:558-565 (1953).

11. N. Eddy, H. Halbach, H. Isbell, M. Seevers: Drug dependence: its significance and characteristics, Bul World Health Organization, 32:721-733 (1965).

12. R. K. Siegel, Cocaine Smoking, J. Psychoactive Drugs 14:271-359 (1982).

13. E. J. Khantzian, Extreme case of cocaine dependence and marked improvement with methylphenidate treatment, Am. J. Psychiat., 140:784-785 (1983).

14. M. H. Seevers, Drug addiction problems, Am. Scientist, Sigma XI Quarterly, 27: 91-102 (1939).

15. M. Almeida, Contribucion al estudio de la historia natural de la dependencia a la pasta basica de cocaina, Rev. Neuropsiquiat.41:44-53 (1978).

16. M. Nizama, Sindrome de pasta basica de cocaina, Rev. Neuropsiquiat. 42:185-208 (1979).

17. F. R. Jeri, Los problemas medicos y sociales generados por el abuso de drogas en el Peru, Rev. Sanid. Fuerz. Policiales, 46:36-45 (1985)

18. J. A. Myers, M. P. Earnest, Generalized seizures and cocaine abuse, Neurol. 34:675-676 (1984).

19. J. S. Schachne, B. H. Roberts, P. D. Thompson, Coronary artery spasm and myocardial infarction associated with cocaine use, New Eng. J. Med., 310:1665-1666 (1984).

20. J. Itkonen, S. Schnoll, J. Glassroth, Pulmonary dysfunction in "freebase" cocaine users, Arch. Int. Med. 144:2195-2197 (1984).

21. I. J. Chasnoff, W. J. Burns, S. H. Scholl, K. A. Burns, Cocaine use in pregnancy, New Eng J Med, 313:666-669 (1985).

22. B. B. Resnick, E. Shuysten-Resnick, Clinical aspects of cocaine: assessment of cocaine behaviour in man, in "Cocaine: Chemical, Biological, Clinical, Social and Treatment Aspects," S. J. Mule, ed., CRC Press, Cleveland (1976).

23. D. Paly, P. Jatlow, C. Van Dyke, F. Cabieses, R. Byck, Niveles pasma ticos de cocaina en indigenas peruanos masticadores de coca, in "Cocaina, 1980," F. R. Jeri, ed., Pacific Press, Lima, Peru (1980).

COCA PASTE EFFECTS IN BOLIVIA

Nils D. Noya

Drug Rehabilitation Centre
Santa Cruz, Bolivia

INTRODUCTION

Cocaine has recently become a substance of serious concern not only because its abuse produces medical, psychiatric, and social disorders, but more especially because of its newly-recognized potentially epidemic nature. Mankind has a very selective memory concerning facts that have been confirmed by experience with dangerous drugs. Cocaine is dangerous, perhaps the most dangerous drug of abuse known to mankind.

When cocaine was discovered and identified in 1859, it was described as having anti-fatigue, strength-reinforcing, and anti-depressant effects, as well as being a sexual aphrodisiac. Cocaine was praised by 19th Century physicians as a miraculous drug that could cure several diseases and was a cerebral tonic and stimulant as well. The great boom in cocaine lasted about 40 years, from 1885 to 1925, when most of the pharmaceutical companies made cocaine hydrochloride easily available, not only for sniffing, but also for intravenous injections. One of the causes of the spread of cocaine use in the United States was Prohibition, an era when cocaine hydrochloride was used as an alternative to alcohol, or in many cases, as an additive mixed with alcoholic beverages.

However, in 1884, two of its principle promoters, Sigmund Freud and William Halstead, became addicted to cocaine, along with a number of their patients. Indeed, the numbers of patients describing their addiction to cocaine had risen so substantially, that by 1887 in Europe a public campaign against Freud and his promotion of cocaine was being waged under the label: "The Third Scourge of Mankind." In 1890, American physicians were recommending cocaine as an antidote for narcotic addiction and also to counteract the effects of alcohol. However, also in 1890, the first American statistics became available as cocaine abuse became more evident: 200 cases of cocaine intoxication were reported as well as deaths due to overdoses. By 1920, in the larger American cities, only 3%-8% of the cocaine hydrochloride produced was sold for medical purposes. A million and a half pounds of cocaine entered the United States in 1907 when the total number of known addicts was the same as in 1967, yet with a considerably smaller population in 1907. In 1906, the view was expressed in the popular press that it was better to let the (cocaine) fiend die, as treatment was hopeless, and cocaine was the most insidious of all known drugs. In 1914 cocaine was still considered as one of the most dangerous drugs, even though unfortunately misnamed in the

Harrison Act as a narcotic drug. As legislative controls grew in all countries, the abuse and use of cocaine declined around 1930.

With cocaine in third place, after morphine and heroin as the predominant drugs of abuse, rehabilitation experts and physicians forgot the seriousness of its effects. Books written from 1930 onwards make slight mention of cocaine, referring to its "psychological dependency" and minimizing the psychological and intellectual disorders as well as the addictive potential of the drug stemming from the self-gratification effect.

COCAINE USE IN BOLIVIA

Cocaine hydrochloride has been used in Bolivia for a long time in the major cities by certain social classes and wealthy persons. Used at parties, as an aphrodisiac, and to improve behaviour caused by the abuse of alcohol, the number of consumers was small and its dependents were not of general concern. Traditionally, Bolivians have chewed coca leaves. Coca leaf production was stable for this traditional chewing, with a surplus left for the illegal production of cocaine hydrochloride. Since 1970, coca leaf production has increased enormously to meet the demand for cocaine hydrochloride.

As a signatory of the Psychotropic Convention of 1961, Bolivia agreed to substitute this crop with other crops over a period of 20 years. To date, this agreement has not been fulfilled; indeed, coca leaf production has reached unimagined levels. Previously, the peasant producer of coca leaves chewed the coca leaf, but was not addicted. It was used on a cultural basis and use could be stopped at will. These peasants are now being paid for their crops of coca leaves with the drug and many have become consumers. These peasants now have a double incentive to continue to grow coca: the money and the possibility of having cocaine constantly available.

All of this changes the basic production and consumption patterns. It is especially true of productive areas like the Chapare in Cochabamba, Yungas in La Paz, and Yapacani in Santa Cruz. In addition, there is a powerful peasant union movement which influences the general policies about drug production and control. Coca plantations have spread to other areas of the country, and with better road infrastructure, new cocaine developments have been established. Thus cocaine production, crystallization, and marketing are carried out without any kind of effective legal controls.

COCA PASTE OR COCAINE SULFATE

During the last 14 years, as a consequence of the larger production of coca leaf and the increased production of cocaine hydrochloride, the use and abuse of coca paste or cocaine sulfate has risen alarmingly. The "pitillo", a cigarette made from tobacco and coca paste and then smoked, demonstrated the compulsive nature of coca paste, its growth potential for consumption, and the resultant cerebral disorders following addiction.

Coca paste or cocaine sulfate is a yellowish-brown powder with a distinct smell. It contains cocaïne sulfate, ecgonine, benzoic acid, alcohol, kerosene, sulfuric acid, tar, and other substances. It is used with tobacco mixed in a proportion of 50 percent. When mixed with marijuana, the cigarette is known as a "submarine,"and causes particular effects and consequences. The content of cocaine in the "pitillos" is about 50-60%. The coca paste is sold in little envelopes of 300 to 600 milligrammes or in matchboxes which contain 5 or 6 grammes of the substance. This amount allows the production of a number of cocaine sulfate cigarettes which are then smoked in a compulsive manner in a short time. People can smoke about 15 to 20 grammes of the substance in 8 to 10 hours continuously.

Comparing the use and abuse of cocaine hydrochloride with that of the sulfate or paste, it can be seen that the hydrochloride is absorbed very quickly by the nasal mucosa until anaesthesia and ischemia are produced. Therefore, cocaine hydrochloride absorption is limited by the vasoconstrictor effect and the numbness. Cardiac failures as well as deaths have been described when a large amount of hydrochloride is absorbed by sublingual paths in cases of oral administration. During the burning of a cigarette containing the paste or sulfate, all of the impurities are eliminated, and the cocaine is absorbed in a 90-95% purity. It enters the pulmonary alveoli in the same way as oxygen does, reaching the brain by circulating very rapidly through the blood stream. It has been proven that intravenous injection takes longer than coca paste smoking to obtain the same results.

SUBCULTURES

Following use, there is a psychological and physical need for the drug, accompanied by depression and paranoid ideation, what the user calls "the follower" (a sense of being followed, observed, threatened), that diminishes as soon as the paste is smoked again. Acute cases of paranoia were seen directly attributable to cocaine consumption. Therefore, it is very rare to see an addict smoking alone out in the open sky. Most of them have a tendency to hide themselves and feel protected only when they smoke in groups with other addicts.

As is common to cocaine producing and consuming countries, a subculture has been established by the coca paste users. Groups of addicts get together to share the drug where one's sex, social class, economic level, or other social characteristics are not important. In general, this form of utilization makes everyone share what they have. Sometimes, when one of them does not have enough money or drug, he can continue consumption in accordance with a special "word of honour" that implies that he will share his money and drug with others who lack them.

When there is no money or drugs, the modern clans also organize smaller gangs to steal things to exchange for drugs. They use their clothes, personal effects, housewares, and any other objects which can be easily exchanged or sold. Most of them have had previous treatment and have received detoxification and psychiatric aid. Most are forcibly taken by the family or the police for treatment without the consent of the patient.

PERSONALITY CHANGES

After using coca paste for more than three months, there is a deep change in the user's personality. Very educated persons present with a temporary loss of their intelligence and knowledge, as if they had never had any education or skills. They lose their good self-image and replace it with self-agression. They live by telling lies to those people around them and eventually to themselves. Most of these lies have a megalomaniacal quality and they live in a psychotic world, devoid of human relationships, affection, or expressions of love and tenderness. The person becomes apathetic with no sensibility for the human problems around him. Because of the cocaine gratification or rewarding sensation, everything else including intellectual faculties, affections, and spiritual interests are placed on a secondary level. The sense of responsibility and the instinct to fight for life disappear completely. According to police, it is very common for users to be found smoking coca paste practically naked or with a minimum of clothing to protect them from the cold. The only thing the user can think of is the drug, how to get it, and where and when to smoke it.

When users have been using drugs for more than three months, there is a severe loss of weight and the facial skin is hard and dry. After cocaine use, the appetite is lost; this lasts for more than eight hours. At the same time, their saliva contains a large amount of cocaine alkaloids, producing stomach numbness and a sensation of plenitude when they do smoke, thereby doubling the anorexic affect. During interviews of patients addicted to coca paste, they cannot follow an idea or conversation to the end, but instead jump from one subject to another. When the excitation is over, the thought process is slow, tired, and very poor. Among chronic users, this kind of thinking becomes chronic. Attention and the ability to maintain it for more than five minutes are practically lost. Mathematical skills are also dramatically impaired. Loss of memory is evident. Paranoid ideation is the main symptom of cocaine use. Users feel threatened and believe that police and members of their families are going to harm them. As a result, they prefer to use the drug in hidden places, or among other consumers where they feel protected. It is important to emphasize that the coca paste abuser is not aggressive; on the contrary, they are very quiet and calm.

In recent years, very important demographic changes have been observed in the addicted persons. Previously, users were dealt with on an out-patient basis. Youngsters, teenagers, and young adults of the upper-middle class backgrounds were hospitalized by parents or family, and underwent medical care forced on them. More recently, the extremes have been seen, from children between 8 and 10 years old who smoke "pitillos" and turn to drug dealing to maintain their dependency, to old people of about 70 years old who use the paste intensively. It is also alarming to see professionals become addicted to the drug, with its accompanying deterioration of interest and skill. Among the patients are physicians, lawyers, military men, policemen, professionals, and technicians, the human capital of an underdeveloped country which needs all its human resources, especially those who are well trained.

SYMPTOMATOLOGY

The last ten years have witnessed the development of new kinds of clinical features differing from the medical signs and symptoms of the hydrochloride abuse previously seen. When trying it for the first time, most coca paste users develop an immediate burning sensation in the eyes, tachycardia, headaches, and anxiety. With continuous use in high dosages, they can develop macropsia and micropsia, and the general sensations of their own body, of time and of hearing appear completely altered. Most consumers rapidly become addicts: that is, they quickly want to repeat the experience again to get the same excitement which is reinforced by another smoking session. These statements are confirmed by the work done in Peru and Bolivia in recent years. Most of the reported patients have been using this drug for ten years and have been treated in psychiatric hospitals, rehabilitation centres, and medical consultories.

One of the most important signs of cocaine dependence is the loss of weight. The police are very much aware of this fact and in their "round-ups" they arrest people by their looks. Coca paste produces the most intense weight loss effect. Patients coming for help who appear very thin are presumed to be cocaine abusers. This thinness is so serious that most of the fat tissue is completely gone. The bones are so marked that they resemble living skeletons. During treatment, when they are drug-free, they can regain 15 to 20 kilos in two weeks with regular nutrition. This is an important factor in regaining self-image.

The chronic abuser also has problems in muscle tone. One of the first symptoms of overdose, commonly called "jawing," is the falling of the lower jaw which cannot be closed because of the muscular relaxation; therefore, unintelligible speech results. Most abusers control this difficulty with diluted sugar and by drinking an alcoholic beverage.

Besides these symptoms, coca paste abuse leads to social and family problems because the addict becomes a disruptive influence on the family and society by stealing and being violent.

THEORETICAL APPROACH IN TREATMENT

The treatment of cocaine abuse/dependence presents a number of complex problems, many of which have not yet been solved by adequate scientific investigation. This complexity is due to a variety of factors which influence the presentation, course, and outcome of the disorder. These factors include the route of administration, the form of cocaine used, the frequency and intensity of the use, the simultaneous use of other drugs or impurities, the presence of diagnosable psychiatric illness, and the social class of the user. Many patients may be using cocaine to self-medicate various psychological states.

Treatment of the Acute Post-Cocaine Phase

Unlike narcotic or sedative dependence, there is no place for detoxification by gradual withdrawal. Attempts by cocaine users to gradually decrease their dosage have almost always ended in failure. Cocaine should be stopped abruptly.

There are three major psychiatric complications at this stage: dysphoric agitation, severe depresssion, and psychotic symptoms. Dysphoric agitation is best treated with oral or intravenous diazepam, and propanolol may be added for more persistent cases. Depressive symptoms and suicidal ideation often occur during the post-cocaine crash. Usually they are transient, requiring no acute treatment other than close observation, and resolve following sleep normalization. The effects of chronic cocaine use on neuroendocrine levels in the brain have led to the use of agents such as pyridoxine, tryptophan l-tyrosine, phenothiazines, and tricyclic antidepressants. However, controlled studies proving the efficacy of any of these agents are lacking and should be done. Psychotic symptoms are also usually transient (less than four days) and usually remit following sleep normalization. Neuroleptics such as chlorpromazine, haloperidol, and tioproperazine apparently have been used successfully to manage patients during these symptoms. Restoration of nutrition with oral or intravenous solutions has been tried and may be needed in individual cases, but has not yet been systematically studied.

In-Patient versus Out-Patient Treatment

In-patient stays of anywhere from a few days to one to three months or one year or more have been advocated by some professionals. There is general agreement that, if possible, out-patient treatment is preferable for the following reasons: lower costs and less disruption of the patient's life and that of the family. In addition, the tendency of patients to relapse, even after long hospital stays, suggests the importance of treating the patient in the environment in which he must live.

In-patient treatment is definitely recommended for severe depressions with suicidal ideation or psychotic symptoms if either lasts beyond three days of the post-cocaine crash, and for periods of homicidal ideation or multiple failures at out-patient treatment. Other reasons have been suggested, but have not been agreed upon as requiring hospitalization: heavy freebase use, coca paste, or IV use, concurrent dependence on alcohol and other drugs, serious psychiatric or medical problems, lack of motivation, or lack of family or social support.

Maintenance of Abstinence: Psychological Approaches

These approaches fall into three categories: behavioural, supportive, and psychodynamic. All three can be useful in the treatment of cocaine abusers and the importance of a flexible, individualized approach is emphasized. Severity of dependence, adequacy of psychological programmes, and presence of psychosocial support are but three factors among others that influence which approach should be used, and when.

Behavioural methods used include contingency contracting, desensitization, and relaxation training. Contingency contracting can be especially helpful early in therapy by focusing and magnifying the patient's perception of the harmful effects of continued cocaine use. Too severe contingencies may discourage patients from participating and also raise ethical questions about the therapist's role in bringing about the negative effect. More attention should be given both to the use of more graduated sanctions, and the possibility of using positive contingencies. Research done so far suggests good results early in therapy for patients willing to participate but a high relapse rate once the contract period has elapsed. Decentralization and relaxation training have not been systematically studied for cocaine abusers, but they have been used with other substance abusers with mixed results. Such training appears to be useful as one ingredient in a total package, but not when employed as the sole or main technique.

Urine monitoring on a frequent, random basis can be an important tool for treatment programmes. Knowledge gained from such monitoring may help to detect a patient who is wavering about use; it may also detect slips before they become full relapses, as well as pick up the use of other mood-altering chemicals. It must be remembered, however, that cocaine metabolites can be detected in the urine with certainty only within approximately 27 hours of the last use. Therefore, occasional use of cocaine may pass undetected for a period of time.

Supportive methods emphasize dissociating the abuser from cocaine-use situations and the cocaine sources. Helping the abuser manage impulsive behaviour in general, and in cocaine use in particular, are also important. The following have been tried with varying degrees of success: regular vigorous exercise, supportive psychotherapy sessions, education concerning cocaine-incompatible behaviours (including the necessity for the elimination of all paraphernalia and drug stashes) increasing contact with non-cocaine using friends, avoidance of social situations with cocaine users and dealers, and training the abuser to deal with high risk situations and craving. Patients should review in therapy potentially risky situations and how they might manage them, e.g., calling a supportive friend, vigorous exercise, postponing use for one hour at a time, etc.

The family should be involved in the treatment process. They also need education, both about the drug use as well as how to deal with the manipulations of the abuser. Limit setting, consistency of behaviour toward the drug user, and the avoidance of double-bind behaviour can all be profitably discussed. Some programmes refuse to treat a patient unless the family gets involved, because their support and positive pressure is so important. On a cautionary note, however, it is emphasized that in some cases the family is so disorganized or pathological that its involvement will do more harm than good. Again the importance of individualizing the treatment approach is stressed. Also, in different countries, geographic mobility may limit the possibility of family involvement. The specific methods to be used in treatment, e.g., single versus multi-family groups varies, depending upon cultural attitudes and therapist training.

Psychodynamic treatment approaches aim toward understanding the function that cocaine has played in the life of the abuser and to help him serve these functions without using drugs. Sexual problems, feeling of inadequacy, boredom, narcissistic needs, and a sense of inner emptiness are all areas that can be usefully explored for some cocaine abusers. Understanding these needs may provide an increased sense of control in the abuser which often limits his need to turn to cocaine euphoria for an illusory sense of power and control.

A combination of all three orientations, behavioural, supportive, and psychodynamic, is probably the most common form of treatment for both in-patient and out-patient settings. The optimal mixture is best determined by taking into account the needs of the abuser at the time of his seeking treatment, rather than by arbitrarily simple programme structure. Behavioural methods seem particularly useful for mild and moderate abusers and supportive measures for moderate and severe abusers. Psychodynamic treatment was most responded to by moderate abusers; however, severe users may benefit from this method at a later stage in their treatment. Some patients may need rehabilitation to return to previously satisfactory levels of functioning, while others may need habilitating, never having functioned adequately previously.

The Role of Abstinence

There is general agreement that the goal of treatment should be total abstinence from cocaine. It is highly unlikely that the compulsive abuser can ever return to occasional, controlled use, and the attempt to do so is often the reason for relapse. There is less agreement as to whether individuals need to abstain from all mood-altering drugs, e.g., alcohol, cannabis, etc., if they have never had a problem with such agents. While it is felt that such abstinence is desireable, both because relapse often occurs when under the influence of the other drugs, and because of the danger of developing a new addiction, it is also recognized that many patients resist total abstinence until they learn from their own experience the danger of partial abstinence.

Pharmacological Treatments

At present, all pharmacological treatments should be considered experimental because large, random assignment, double-blind studies have not been conducted. However, there has been an accumulation of clinical experience and open trials that suggest that certain agents may be useful, either as diagnostic specific agents, or for more general use. Neuroleptics and antidepressants are still our best weapons.

COCAINE AND BASUCO: AN OVERVIEW OF COLOMBIA, 1985

Augusto Perez Gomez

Advisor to the United Nations
Universidad de los Andes
Bogota, Colombia

ABSTRACT

In common with other countries of South America, Colombia is currently experiencing an impressive rise in the use of psychoactive substances. But in the last few years, one of these substances has taken first place on the market: "Basuco," a by-product of cocaine processing. This paper presents an overview of the situation, describing the state and private services offered to users, research results, clinical data about symptoms of intoxication and withdrawal, the consequences of chronic use, and a brief evaluation of two institutions working on different kinds of treatment. The need for more systematic and integrated research is stressed as otherwise some of the fundamental decisions in this field will be simply new versions of old mistakes.

INTRODUCTION

On March 11, 1947, a law was passed in Colombia which officially prohibited the cultivation, distribution, sale, and possession of coca leaves. However, in different parts of the country, the peasants maintained their habit of chewing these leaves ("mambeo"), and the authorities did not see the need to take action, as this use was very restricted and had little social importance.

In the 1970's, cocaine use started rising in privileged classes, but during the entire decade, its social importance was minimal. Until 1980, Colombia was merely a platform for trafficking, but due to overproduction and increased international controls, the traffickers began dealing inside the country. This change in the situation was further complicated by other developments such as:

1. The production of a local bush with low cocaine content, the Erythroxylum novogranatense tipica (0.6% cocaine), which grows at altitudes between 500 and 1500 metres.

2. The deterioration of the country's socio-economic conditions.

3. The increase in unemployment (20% for young people) and the lack of opportunities to improve living conditions.

4. The moral deterioration brought about by the "marijuana boom," which
    brought great amounts of currency into the country, and developed a
    parallel economy easily accepted by the authorities.

5. The prestige and power obtained by traffickers such as Carlos Lehder
    and Pablo Escobar, who appeared as "saviours" to the poor.

6. The connection between drug traffickers and some guerrilla groups,
    which today is undeniable.

All of these circumstances as well as others, paved the way for the
"basuco explosion." The word "basuco" comes from the word "base" - of
cocaine - and also refers to a weapon, the bazooka. Basuco was very inexpensive which made it accessible to the majority of the population. Indeed,
the first addicts were peasants on the payroll of the cocaine producers.

Basuco Production

There are two methods which are generally used to produce basuco. In
the first method, coca leaves are treated with gasoline, lime, alcohol,
ammonia, potassium permanganate and sulphuric acid. A very impure cocaine
sulphate is obtained in this method. The second method, rarely used because
of its great cost, involves mixing the residue from the process of refining
cocaine with a purer paste to which methaqualone or rivotril (an anti-con-
vulsant) is then added.

Basuco contains 40% to 80% cocaine sulphate, and about 10% to 15% pure
cocaine, although these amounts may vary a great deal.

Patterns of Use

In Columbia there seems to be no discrimination in the use of basuco by
social class. However, the more expensive cocaine hydrochloride powder is
generally used by artists, industrialists, and executives, who usually snort
the powder and alternate its use with alcohol. Injection or smoking cocaine
is rarely found in this group.

Basuco is smoked, usually after being mixed with tobacco, or in some
cases with marijuana (making a "diablito") or methaqualone. Its use is
also combined with alcohol. In most cases, the dealers will mix the coca
paste (basuco) with talcum powder, brick dust, and many other different
types of residues.

General characteristics of the pattern of basuco use include:

1. 80% of users smoke basuco only on the weekends.
2. The addicts smoke from three times a week to every day.
3. All users smoke several cigarettes each time from 5 to 200, perhaps
    even more.
4. The package of basuco costs between US$3 and $6 and contains from ½
    to 1 gramme of basuco; with this amount, 4 to 10 cigarettes may be made.
5. Most of the time, basuco, as well as cocaine, is used simultaneously
    or alternatively with alcohol. Because of exhaustion, insomnia, and
    severe anxiety, users frequently take other drugs, such as diazepines,
    and barbiturates.

There are a considerable number of articles that carefully describe
the entire process of absorption and elimination of cocaine and its by-products (Jeri, et al., 1978; Siegel, 1979; Paly, et al., 1982) and all of
these processes apply to basuco. The absorption rate of smoking basuco,
about 20 seconds, is quicker that that of snorting cocaine hydrochloride,

and when it appears, the resulting euphoria lasts 4 to 5 minutes. However, for the majority of users, there is only a dysphoric effect, which is appeased by smoking another cigarette. The effects of basuco are quicker and stronger than those of cocaine hydrochloride for three reasons:

1.   The content of alkaline substances accelerates absorption.

2.   In contrast to cocaine hydrochloride, coca paste does not decompose when subject to heat; when smoking basuco, therefore, the lung is subjected to and absorbs a greater amount of the alkaloid.

3.   Due to its alkalinity, coca paste is soluble in lipids; consequently, its action on the brain lipids is immediate.

Prevalence of Use

Some authors (Velasquez de Pabon, 1983) suggest that almost 25% of the population over 12 years of age has tried basuco at least once, and that between 3% and 5% are addicts. However, the data of prevalence and incidence in Colombia are extremely partial. Very often they are simply an extrapolation of consultations in hospitals or clinics, or the results of isolated surveys made in high schools. This makes compilation and interpretation of data very difficult. Research is needed to rectify this serious lack of accurate statistical and other data in order to realistically assess the current national situation. It will be necessary, in the near future, to develop systematic research. Unfortunately, in Colombia as well as in some other developing countries, authorities are more interested in action than in research. Obviously, decisions are often based on broad impressions not on facts, and the few existing resources are misused.

According to "Investigaciones Medicas", Journal of the Pan American Health Organization (OPS) (No. 21, 1985), the general aspect in Colombia would be the following:

--First experience with basuco: between 14 - 17 years of age for 60%.
--Prevalence in the sixth grade: 4-5%
--Prevalence in the twelfth grade: 20%
--Maximum prevalence in women: between 16 - 17 years of age.
--Maximum prevalence in men: between 18 - 24 years of age.
--Maximum use: in Government-sponsored night schools (ten times more than in private day schools).
--Total prevalence of occasional use: 4% in the school population. Of these users, 70% use more than one substance, in the following order: basuco, 90%; marijuana, 80%; alcohol, 50%; methaqualone, 10%; narcotics, 1%. Alarmingly, there has been a considerable increase in the number of recent consultations for heroin abuse (Prometeo, 1985).

Physical Effects of Basuco

The main physical effects identified as those directly associated to the use of basuco are summarized below:

1.   Hyperactivity of the Central Nervous System
2.   Anorexia
3.   Weight-loss
4.   Polydipsia
5.   Insomnia
6.   Hyperthermia
7.   Alteration of hepatic functions; hepatitis
8.   Polyneuritis
9.   High muscular tone

10. Respiratory pathology
11. Cycles of diarrhea and constipation
12. Hypertension
13. Saturnism (lead intoxication)
14. Tachycardia and tachypnea
15. Convulsions

The most relevant characteristics of the withdrawal syndrome include: restlessness, crying crisis, tachypnea, sweating, tachycardia (130-150p.m.), and tremors.

The psychological and psychiatric symptoms resulting from the persistent use of basuco are as follows:

1. Restlessness
2. Anxiety
3. Auditory hypersensitivity
4. Auditory hallucinations
5. Visual hallucinations
6. Anorgasmia
7. Loss of sexual drive
8. Insomnia
9. Social withdrawal
10. Depression
11. Persecution fears
12. Changes in time perception
13. Convulsions
14. Cocaine psychosis, paranoid type
15. Loss of every interest not related to the drug

In the case of chronic use, the probability of the appearance of a cocaine psychosis, with the risk of suicide or homicide, is extremely high.

Among the social consequences associated with the use of basuco are:

--Rise in traffic accidents
--Rise in delinquency (robbery, thievery, swindling)
--Rise in violent crimes
--Rise in school failures
--Rise in family problems
--Severe alterations in the economic field
--Corruption and demoralization

(Velasquez de Pabon, 1983; Jeri, 1984; Motta, 1984).

Treatment and Rehabilitation

In Colombia, the entity responsible for dealing with all aspects of the use of toxic substances is the National Narcotics Council (Consejo Nacional de Estupefacientes). This body was created in 1974 by Decree No. 1188, which included 87 articles. Of these, only one referred to enforcement; the other articles emphasize prevention, treatment, and rehabilitation. However, until very recently, the Council has dealt exclusively with enforcement.

Consequently, the problem of treatment fell initially onto the private institutions, and then later, onto the Mental Hospitals, but there was no coordination of efforts. During the past three years, many activities have been carried out which give a better view of the resources available for the treatment of cocaine and basuco users. It must be pointed out that psychiatric hospitals dedicate only a fraction of their services to basuco users.

Although there are a variety of therapy programmes, the main treatment is detoxification, followed by group and occupational activities. Family therapy or community education programmes were only recently begun. The critical point is that there are virtually no evaluations of these activities.

Comparisons of data relating to consultations at two well-known institutions in Colombia, one private and one public, show striking similarities. In 1981, the consultations relating to basuco at the two facilities were 25% and 31% respectively. These percentages had risen to 50% and 55% in 1982, and within the next two years, by 1984, the consultations relating to basuco had reached 80% and 99%. These figures exclude consultations for cocaine, marijuana, or alcohol abuse, although it is usual for a basuco user to use these other substances as well. However, these figures do exemplify the dramatic increase in basuco use from 1981 to 1984 as shown in the number of hospital consultations for treatment.

Description of Treatments

The most common treatments employed in Colombia may be grouped into several categories: individual, group, and family therapies, with different theoretical backgrounds. These therapies are time-limited, consisting of about ten to fifteen sessions. The detoxification programmes include: hospitalization of two to three weeks, sedatives, diuretics, vitamins, antidepressants, and rest. Motta (1984) underlines that in cases of acute intoxification and withdrawal syndrome, there is a remarkable resistance to phenothiazines, diazepines and barbiturates. But chlonidine (a hypotensor) gives good results with anxiety and tachycardia. Most of the institutions and services are currently working on the basis of community action, especially with youth self-help and self-management groups including the creative use of leisure time.

One of the most serious problems in all these treatments is the lack of systematic and methodologically acceptable evaluations. Quite recently, an evaluation of two well-known treatment centres for people using psychoactive substances was made (Perez, Cobos, and Echeverri, in press). The evaluations by the authors differed completely from those proposed by the people in charge of the two institutions, who were much more optimistic. In fact, 14 subjects considered "cured" according to the institutions were actually in prison, dead (due to drug abuse), or in a new treatment. The findings from this evaluation showed:

1. There is generally less use of drugs after treatment, with the exception of alcohol. However, the only drug with a statistically significant change was cocaine.
2. Unexpectedly, there were more subjects who were unemployed after treatment than before treatment.
3. Subjects in both institutions presented a significant decrease in criminal activities.
4. There is also a decrease in problems with the family and the neighbours although not statistically significant.
5. There appears to be a direct relationship between the completion of treatment and the successful recovery from drug use, i.e., those who completed the programme had significantly higher success rates than those who did not complete the programme.
6. Most of the subjects considered as recovered were between the ages of 23 and 27.

As a whole, these results require more research especially with regard to the definition of "success" in the treatments for drug addiction.

Finally, inspite of the efforts of the Government of Colombia to eliminate the cocaine business, the problem is still very extensive. Undoubtedly, only international cooperation, with local initiatives, will resolve the cocaine crisis existing in our country.

CONCLUSIONS

Within the general populations of other South American countries, there is an autonomous social control of a cultural origin of coca and its by-products. In Colombia, however, the situation facilitates the expansion of the use of the drug without such internal limitations. No cultural tradition of coca use exists in this country and the present consumption is the result of economic conditions and the lack of promising expectations for the future. Contributing to this is the fact that Colombia is a centre for cocaine trafficking and the economic benefits of this trade offer false improvements in living standards to certain strata of the population. Creating a sense of anomie, there is a total breakdown of values making the population more susceptible to the search for easy escape mechanisms.

Although the country has made a strong effort in the last two years to provide rehabilitation for those who need it, a unified, coordinated effort in treatments is lacking. If this is not soon developed, the indices of relapse will increase dramatically and we could lose the ground so arduously gained.

REFERENCES

Jeri, F. R., et al. (1978), The syndrome of coca paste, Journal of Psychedelic Drugs, 10:361-370.

Jeri, F. R., (1984), La practica de fumar coca en algunos paises de America Latina: una toxicomania grave y generalizada, Boletin de estupefacientes de las Naciones Unidas, XXXVI:17-34.

Ministerio de Salud de Colombia-Unfdoc, (1984), Seminario clinico internacional sobre adicciones a la hoja de coca y sus derivados, Bogota, Colombia.

Ministerio de Salud de Colombia-Unfdoc, IV Encuentro nacional de Servicios de Farmacodependencia y Alcoholismo, Bogota, Colombia.

Ministerio de Salud de Colombia, (1985) "Instituciones de Salud Mental y Servicios de Farmacodependencia," documento de circulacion restringida.

Motta, G., (1984), Una farmacodependencia epidemica: bazuco, Investigaciones Medicas, VI, #18.

O.P.S., (1985) "Investigaciones medicas," # 21.

Paly, D., et al., (1982), Plasma cocaine concentrations during cocaine paste smoking, Life Science, 30:731-738.

Perez Gomez, A., Cobos, L., and Echeverri de Pardo, G., "Modelo para evaluar la efectividad de los tratamientos de drogadiccion," in press.

Prometeo, (Nov. 1985), personal communication.

Siegal, R. K. (1979), Cocaine smoking, New England Journal of Medicine, 300-373.

Siegel, R. K., (1982), Cocaine smoking, Journal of Psychoactive Drugs, 14:271-359.

WORKING TOWARDS THE FUTURE: A CHALLENGE FOR CHANGE

COCAINE ADDICTION: A SOCIO-ETHICAL PERSPECTIVE

David F. Allen

National Drug Council
Nassau, Bahamas

Cocaine is the fastest growing drug of abuse in the Western world. The coca paste epidemic in South America, the freebase (crack) epidemic in The Bahamas, and the sudden upsurge in crack (rock) cocaine use in the U. S. have spurred a public outcry for bold initiatives to confront the problem on both national and international fronts.(1) Despite increased law enforcement and educational efforts, the cocaine crisis deepens daily, making the situation more acute and frustrating.

Faced with this devastating plague, what is our ethical responsibility? Ethics is the study of human behaviours and institutions as they relate to moral duty.(2) Ethical analysis of issues such as the cocaine crisis reveal underlying social problems which must be addressed. Cocaine's powerful addictive potential, high availability, its deleterious effects on individuals and families, the strain on the social, legal, and economic resources of a society create a situation involving the most basic socio-ethical issues of life. This chapter discusses the socio-ethical perspective of the cocaine problem by examining the facts, value beliefs, moral reasoning, and loyalties, as well as suggesting possible solutions to be implemented in the current cocaine crisis. (3)

THE FACTS

Good ethics begin with good data. Drug use in general and cocaine use in particular have been associated with strong denial of its harmful effects, misinformation concerning its dangers, misplaced idealism, simplistic reductionism, and projective defences such as blaming other factors. These factors have produced a state of confusion and inaction, making it difficult to unite community support or to mobilize a public outcry for the recognition of the inherent dangers of drug use and to deal with the problem. (4) To wage an effective war against cocaine, there must be an unflinching commitment by individuals and communities to obtain the facts, regardless of the political or social ramifications. Firth's Ideal Observer Theory provides an excellent model for gathering, categorizing, and analyzing facts. According to Firth, the Ideal Observer is omniscient (informed), omnipercipient (perceptive), disinterested and dispassionate (objective), consistent, and otherwise normal (recognized limitations). (5)

Being Informed

Being informed involves recognizing the scope of the problem. Cocaine abuse in the Western Hemisphere has reached epidemic proportions. With the present high availability and low price of the drug, a wide cross-section of society is vulnerable, from school children to professionals. Ignorance of cocaine's easy accessibility encourages gross denial and projective defences such as "It could never happen to me or my family," or "Only derelicts and losers fall prey to drugs anyway." Parental denial has been cited as the most detrimental factor in the child's initial drug use and in subsequent rehabilitation. (6,7)

Until recently, cocaine enjoyed the mistaken reputation as a non-addictive, "safe" recreational drug. With increased use by a wider population, its deleterious effects can no longer be swept under the carpet. Mounting hospital admissions bear testimony to the information now available from the research studies conducted by medical centres and laboratories across the country. Cocaine is the most reinforcing drug known to mankind. Promising euphoria, it leads to dysphoria and compulsive use. With ready to smoke freebase (the term used to denote crack as well as rock cocaine in this paper) the initial extremely intense high is created in such a way that the pleasure centres of the brain are damaged, perhaps permanently. Indeed, the neuroreceptors become refractory to dopamine, the substance responsible for the ability to feel pleasurable sensations and feelings. The reinforcing memory of that first intense yet now unattainable high, leads to subsequent hits, and ends by producing a state of anhedonia followed by depression.

Binges lasting up to three or four days of compulsive use occur, with little sleep, food, or hydration. Studies have shown that rats, when exposed to unlimited amounts of cocaine, exhibited the same compulsive using behaviour which continued until they died. However, when exposed to unlimited amounts of heroin, the rats used the heroin in a non-compulsive way, continuing to eat and sleep, although addicted. (8) It must be emphasized that freebase cocaine is a no-barrier drug (9) and once use is established, the only way to curtail its use is to reduce its availability.

This most addictive form of cocaine has become increasingly available and at a lower price. Delivering 80% pure cocaine to the brain in eight seconds the powerful reinforcing first high of the freebase hit carries a 70-80% probability of addiction on the first hit. The drug is now cheap, easy to come by, easy to use, and produces a powerful, almost instantaneous though short-lived high. Documenting the sudden rise of cocaine related hospital admissions in the Bahamas from 1983, Jekel et al., found the most significant variable was the switch from powder to freebase cocaine. Since 1983, its widespread use has wreaked havoc in all areas of Bahamian society, affecting some of the most talented and productive persons in the community.

Ironically, in 1983, at the time when mental health professionals in the Bahamas were observing the ravages of freebase cocaine, the popular literature was still proclaiming that cocaine was not addictive. Producing a state of ambiguity, this made it difficult to mobilize community action against the onslaught of cocaine.

The greatest obstacle to community action, however, is the tremendous amounts of money involved. The cocaine business is a multibillion dollar industry. Cash flow generated by cocaine affects the whole economy. Not only individual pushers, but storekeepers, car dealers, housing contractors and others are getting rich from "drug money" used to purchase luxuries with ready cash. Corruption of law enforcement officers and government officials is difficult to prevent when the bribes are so great. Individuals and the society as a whole seem to prosper from the cash flow generated by cocaine. Thus the monetary benefits seduce all segments of society into an unconscious collusion.

This blind commitment to financial gain must be weighed against the social and moral cost at stake. The erosion of individual and community life lies down the road, while the present prosperity provides easy rationalization for an entire society's compliance in the drug economy. Sadly, it is not until a member of our own family is affected or the sanctity of our home is destroyed by the criminal behaviour needed to obtain drug money, that we are jolted into concern. By then, it is often too late.

The treatment of cocaine addiction is arduous, expensive, and fraught with high recidivism rates. Inadequate knowledge about the drug, lack of proper treatment programmes and the high availability of the drug make lasting cures difficult to obtain.

For example, on returning to his community after treatment, a young man found that his "friends" had thrown a $200 rock of cocaine on his porch. He said, "I tried to throw it back at them, but it would not leave my hand, so I hit (smoked) it." The cycle of addiction began again. Even after a drug free period of six months or more, an addict will experience cravings for cocaine when faced with persons or things associated with previous drug use. While in a treatment programme, a young woman was visited by her pusher. Although she had not used cocaine for some time, she immediately left the programme to go with him and returned to her cocaine use.

The most frightening aspect of the cocaine plague is that children are at risk. In the Bahamas, South America, and the U.S., freebase cocaine is within the price range of school children with lunch money. Studies indicate that cocaine's low price, high availability, and powerful immediate addiction potential have already taken a toll on school-age children. (10) Johnson et al., studied 16,000 students in 132 American high schools. Seventeen percent of graduating seniors admitted thay had used cocaine. In Canada Reginald Smart of Ontario's Addiction Research Foundation studied 4,154 students from grades 7,9,11, and 12 in 132 Ontario schools. Findings showed an increase from 4.0% having used cocaine in 1983 to 4.5% in 1985. In Toronto, Canada's largest city, the increase over the same period was from 3.2% to 5.8%. (11) Nine and ten-year old addicts have come in for treatment in New York and Bolivia in many cases having been given cocaine by older siblings. Pushers often sell to 19 year olds, who then sell to 16 year olds, who sell to 12 year olds.... If the war against cocaine is to be won, a ring of protection must be put around our children.

Although vast sums of money have been spent on the interdiction process, there has been only marginal success in curbing the supply of cocaine. Stressing this point, Judge Irving R. Kaufman of the United States Court of Appeals for the Second Circuit and Chairman of the President's Commission on Organized Crime said, "Law enforcement has been tested to its utmost, but let's face it - it hasn't succeeded." (12) Dr. Sidney Cohen, a well-known expert in the field of drug abuse treatment, put it this way, "There have been massive seizures of cocaine and coca paste, laboratories were destroyed in Colombia and thousands of coca bushes uprooted and burned in Peru and other countries. But in spite of this the coca plantations have spread over the immense land areas of northwestern South America where police and military access is prevented by well-armed anti-government groups. In those areas where police control is possible, enormous sums of money are used to bribe officials to prevent interference with trafficking. And most sadly where corruption is not possible, murder may occur as in the case of the Colombian Minister of Justice in 1984. The end result of all this is that the price of cocaine has declined about 25% and the quality is improved."(13)

This depressing reality does not imply that enforcement activities should be curtailed, but that enforcement alone is not enough. There must simultaneously be a determined search for creative initiatives to impact the

problem.   Demand reduction strategies such as education and rehabilitation
should match our commitment to law enforcement.

Cocaine addiction has invaded the workplace, and along with the existing
high rates of alcoholism, leads to poor attitudes, decreased productivity,
high personnel turnover, increased absenteeism, higher medical benefit costs,
stealing and violence. The job is the last thing an addict wants to lose
because therein lies the support for his/her habit. Recognizing this, the
President's commission on Organized Crime recommended drug screening for
employees to deter and detect drug usage. Although the subject of much debate concerning the civil liberties of employees versus the protection of
society, urine screening has been implemented by such major companies as IBM,
Union Oil, and the Federal Aviation Administration. Mandatory random urine
testing has been recommended for employees whose alertness on the job may
affect public safety such as pilots, train captains, nuclear industry workers,
etc.

The invasion of cocaine into professional sports has had a dethroning
effect on this powerful alternative to drug abuse which is held in such high
esteem by young people. Seeing the drug scourge take hold of the sports
heroes, the ones who have made it against so many odds, is especially tragic
not only for the ones whose lives are destroyed, but for young people desperately looking for role models.

These and other issues provide a bird's eye view into the urgency and
complexity of the cocaine problem. Good ethics demand good data. There can
be no proper resolution without a commitment to examine all the issues involved.

Being Perceptive

Too often we as professionals and ordinary citizens seek the facts and
their analysis and then form our own particular point of view. The solution
to the drug problems depends not only on our being informed, but also on our
ability to empathize or identify with the pain of those who suffer. This
requires the ability to place ourselves in the position of the other, to
feel what they feel, and then move to treat them as we would like to be
treated. Such empathic projection requires patience, tolerance, and understanding. It means moving beyond the narcissistic calculus to be touched by
another. According to Alfred North Whitehead, this sensitivity transforms
a fact from being a mere fact to being invested with all its possibilities,
thus becoming "the architect of our purposes, and the poet of our dreams."
(14)

The fearsome spectacle of widespread drug abuse is more than a red line
on a statistical chart - it is destroying our families, prostituting our
daughters and robbing the manhood of our sons. Hear the mother bewail the
death of her child. They told her to do the best for her kid, and she tried.
They told her to send her child to school, and she did. They told her she
lived in a green and pleasant land, an idyllic paradise, and she believed
them. But the high went higher than high, the crash even deeper...the
bullet rang; the rope tightened, and her son, her child, was dead. As a
psychiatrist, I have mourned with five such grieving mothers in a period of
18 months: what a terrifying experience!

Is cocaine addiction the new slavery? Hear a pusher call a fellow Bahamian "my slave." Running out of money, the addict sells him or herself
to the pusher to obtain more cocaine. Is this not the classic character
defence of the repetition compulsion? Upon being freed, the slave chooses
to return to the bondage of a new slavery. In extricating ourselves from
an infantilizing colonization, have we chosen to be enslaved to a more

dehumanizing process of drug addiction.

Consider the effects on the addict's children. Listen to the poetry of
a ten year old girl whose mother is a cocaine addict and leaves home for a
week at a time:

> Life is nothing to me.
> Life don't mean a thing to me.
> With my life, it is terrible
> People pulling me apart
> The things I go through are horrible
> And some are breaking my heart.

Can we feel the violence destroying her life? Is her personalized pain
universalized in our heart?

Being Objective

Recognizing the complexity and emotional nature of the cocaine problem,
a sense of objectivity is necessary in working towards meaningful solutions.
Using the model of the legal jury system, this might best be achieved by
multidisciplinary formats to prevent particular interests or biases from
dominating the process. A microcosm of society itself, the cocaine crisis
requires input from all segments of the community. Noting the difficulties
in achieving this type of cooperation, Archibald says, "Researchers can't
communicate with treatment workers, physicians can't communicate with educa-
tors, and so on. At the same time, people in research, treatment, and prevention tend to look down on the police. And vice versa." (15)

Although there were growing pains, the multidisciplinary composition
of the National Drug Council of the Bahamas has provided a mutual educational
experience, broadly based analysis of the local cocaine situation and a
cooperative approach in working toward its solution. (16)

Being Consistent

In trying to deal with the enormous new threat posed by cocaine abuse,
other needy areas must not be forgotten. Being an illegal drug with dramatic
immediate repercussions on society, cocaine has overshadowed the serious
alcohol problem affecting all aspects of society. Consistency dictates that
even as we press to eliminate illegal drug use whether it be cocaine,
designer drugs, or marijuana, we simultaneously seek to affect healing related to the abuse of prescription drugs and alcohol. Abuse of any substance
is often related to abuse in other areas. Cocaine addicts admit increased
alcohol and marijuana use to control the "crash" after the cocaine high.
Early marijuana use in school age children is now seen as a precursor to
subsequent cocaine use by the mid-teen years, i.e., many high school students
undergoing treatment for cocaine addiction began smoking marijuana in pre-
or early adolescence. (17) If the cocaine epidemic provides the stimulus
for the development of a consistent and effective approach to all drug abuse
the community would benefit enormously.

Recognizing Limitations

It is impossible to understand all the issues related to the cocaine
crisis. It is magical thinking to believe that a short, simple, solution
exists. Healing must start with the development of an effective infrastructure which slowly brings about the desired changes. This process, frought
with ambivalence and frustration, confronts our limitations, taxes our patience, and often produces burn out. Confrontation with our limitations and
helplessness is painful. As professionals, we are so enmeshed in our "per-

ceived" role i.e., the great doctor or psychologist, that is is hard to relate to our actual role, that of a frail human being who is limited and vulnerable. To be effective in the field of cocaine addiction treatment requires that a meaningful balance be maintained between the perceived and
the actual role, manifested by a sense of competency with a realistic awareness of limitations. We must have the personal integrity to say, "I don't
know," the ability to tolerate frustration and failure and still find the
strength to persist, to give hope where there seems to be none.

Value Beliefs

Impacting the cocaine problem requires not only accurate knowledge and
technical expertise, but also an unflinching commitment to a humane value
system or moral centre. According to Arieti:

> Values always accompany and give special psychological signi-
> ficance to facts...When we deprive facts of their value, we
> fabricate artifacts which have not reality in human psychology.
> An individual may suspend his value judgment when he wants to
> examine a fact from a specific point of view, but then the
> ethical content has to be re-established if the fact is to have
> human significance. If we remove the ethical dimension, we
> reduce man to subhuman animal. (18)

Emphasizing the importance of a moral centre, Eisenberg argues that what
one believes about the nature of human beings exerts a subtle but controlling influence on the attitudes, behaviours, and treatment of individuals(19).

Western ethics are based on the Judeo-Christian tradition which, at
its core, views all human beings as having been made in the image of God
(Gen. 1:27). Providing the basis for personhood, dignity, and human rights,
this age-old concept is the operative force enhancing personal meaning, interpersonal relationships and human community. Elaborating on this concept,
Niebuhr says this "reverence" for all human beings is the quintessential
element for meaningful social reform. (20) Thus all individuals, regardless
of race, class, handicap, illness, sex or age, are persons with meaning and
dignity deserving the utmost respect and concern.

The principle of autonomy inherent in this respect for the uniqueness
of human personhood carries with it responsibilities and duties as well.
The person with a cocaine problem should not be seen as a "junkie", but as
a person who has the right to be respected, a right which may involve receiving proper treatment. Similarly, he/she has a responsibility as a person
in society to other persons in that society to work on his/her drug problem
by taking advantage of treatment opportunities offered. The basis of our
relationship in caring for or working with addicted persons is our mutual
personhood with mutual respective rights and responsibilities. By recognizing in each other the shared human qualities which transcend individual
differences or problems, we actualize the principle of reciprocity in a
practical way so that we treat others as we would want to be treated.

The moral responsibility inherent in this reciprocal, empathic, interpersonal relationship requires allegiance to other vital principles such as
trust, forgiveness, truth-telling, love, promise-keeping, justice, liberty,
and non-injury. Being absolutely germane to the human community, these
principles may be called constitutive imperatives, i.e., the underlying
principles upon which all laws governing society are made.

Johnson and Butler emphasize the importance of this concept:

> Respect for individuals requires that every individual be

treated in consideration of his uniqueness, equal to every
other, and that special justification is required for interference with their purposes, their privacy, or their behaviour. It implies sets of liberties, rights, duties, and
obligations especially of promise-keeping and truth-telling. (21)

The cocaine problem may be described as a crisis in values, as they relate to the community in general and the individual in particular. How
could the cocaine epidemic spread so rapidly? The sad truth is that persons
in the producer countries of South America, the transshipment areas of the
Caribbean and the massive consumer centres like the United States, are
willing to sacrifice basic human values for sordid gain through blood money.
The words of a crack addicted person ring true: "Money is more important
than people and principle does not count." (22)

Our value beliefs form a moral centre which garnishes faith and renews
the determination to fight for the healing of human society. One is struck
with the profound observation of a well known sports announcer, Howard
Cosell, who, when he was asked why drugs had invaded professional sports
replied, "There is no definable moral centre in America anymore, and that's
the problem of the culture." (23)

Moral Reasoning

The moral reasoning inherent in any process profoundly influences the
way persons are treated. Forms of moral reasoning range from the ethical
egoism of Kohlberg's Stage I to the more sophisticated formalism of Stage
VI. (24) Western ethics is mainly influenced by two ethical systems:
formalism and utilitarianism. (25)

Formalism is deontological and requires commitment to basic principles
such as justice, promise keeping, honesty, and non-injury. The major thrust
is being faithful to principles in spite of the consequences or outcome.
Utilitarianism, on the other hand, is teleological and has as its basic
tenet the facilitation of the best balance of pleasure over pain and the
greatest good for the greatest number.

Connected to utilitarianism is the prevailing value of instant gratification or success, as opposed to the formalistic approach which emphasizes
the need to struggle for long term and more enduring results. In the areas
of education, enforcement, or rehabilitation, the utilitarian perspective
would emphasize the immediate results. When and if such results are not
forthcoming, frustration, anger and burn-out follow. There are no magical
solutions to the cocaine problem and even if the short term results are not
what we wish, it is our duty to persist in working toward a society where
persons may chose a drug-free lifestyle and, if addicted, are able to receive
treatment with dignity and respect.

So often the emphasis is on the greatest number or the most effective
approach for the majority. That is well and good, but what about individual
responsibility? In the area of enforcement, it will be difficult to eliminate or effectively reduce the supply of cocaine in a short period of time.
This should not cause despair; through the principle of autonomy, persons
from childhood up could be taught effectively to take responsibility for
their lives. Is this not the hallmark of democracy, that people are free
to say "No" to destructive personal influences? (26)

Another major ethical issue is the utilitarian argument that drug addicts are the losers of society who choose to destroy themselves by their
personal choice of addiction. As a result, they should not receive attention and resources at the expense of the majority. Firstly, this is a mis-

understanding of the addictive process. The bane of addiction is that the user continues to use compulsively despite devastating adverse consequences. With cocaine addiction, a biological hunger drive is created which puts the brain on automatic pilot for cocaine which may be contrary to the desire of the addict. Secondly, this form of social utilitarianism ignores the pathos of the vicious world of the addicted person. Though appealing to the majority of the most powerful, this philosophy offers little for those who are in the minority and/or without power. When unchecked, this motivation has led to the atrocities inflicted on such disadvantaged groups as the mentally ill, the mentally retarded, and the racially despised - groups considered expendable for the greater good.

Connected to the dynamic of social utilitarianism is the common expression, "Drugs have brought economic advantages to the impoverished people in the developing world, and only a minority have been hurt by them." The argument of economic benefit in the light of human destruction and social disintegration illustrates a cold moral reasoning which emphasizes money and power over human compassion. Related to this is a type of social Darwinism expressed by pushers who claim that only weak persons get into trouble with cocaine.

Underlying the social utility concept is the assumption that only life of a certain quality has worth. There is a tendency to define personhood on the basis of relative social utility. Whenever one's utility/disutility ratio is affected, worth as a person is diminished. Being a cocaine addict for instance, reduces utility and basic worth in the society. Thus the utilitarian view of justice denies positive presumption and equality under the euphemism "for the public good." This strains the moral fibre of the society itself and undermines the meaning of such principles as promise-keeping, justice, and liberty for all its members.

Loyalties

Loyalties dictate the ultimate ends we serve. The war against drugs requires a clear understanding of loyalties in terms of ends and means. The ultimate goal is to create an environment in which individuals would freely choose a drug free lifestyle, or, if addicted, would receive treatment in programmes which respect the meaning and dignity of human personhood. Yet so often, whether in areas of enforcement, education, research, or rehabilitation, we become frozen in our own ideas and fused to our personal projects. Refusing to be flexible and open to the overall perspective we are subject to petty jealousies, destructive competition, and defensive communication. As a result, efficiency and creativity are compromised and the programme becomes an end in itself rather than the means to serve the best interests of those being treated. In some rehabilitation programmes, for example, the criteria for admission is so rigid that hard-core cocaine addicts become frustrated and re-enter the vicious cycle of further abuse. Despite the growing importance of outpatient therapy for cocaine addicts, few guidelines have been developed. However, Crowley and associates have developed a contingency contracting approach based on behavioural principles which uses negative reinforcement (e.g., sending information on an addicted physician's cocaine habit to the local Board of Registration if the patient returns to cocaine). (27) Rounsaville et al. claim that although the results of this programme are promising, its utility is limited because (a) many clients are initially unable to control drug use in treatment and so such a programme could deny treatment to an uncontrollably heavy user who is in dire need of help, and (b) the clinician could find him/herself in the ethical ambiguity of making the client's life even more stressful in addition to the pain of his/her cocaine addiction. (28)

Implementation

The burgeoning cocaine crisis, with its horror stories of threat to individuals, families, and countries, may evoke panic and a tendency to impulsive action. On the other hand, as we learn more about the severity and complexity of the situation, there is a parallel tendency to feel overwhelmed to despair, and to withdraw. Recognizing that both of these options are counterproductive, the clarification of goals and effective implementation are best served by thoroughly analyzing the data base (facts), value beliefs, moral reasoning, and loyalties inherent in one's plan of action. Even then, one should proceed cautiously with a sense of openess to further information and different approaches. In the light of the complexity of the situation, the war against cocaine requires a multiplicity of well-coordinated approaches. Archibald stresses that without coordination success will be at best isolated and temporary: "Drug traffickers are multinational corporations with highly developed systems including marketing specialists, promotion specialists, training for couriers. They have it in their power to change the political map of the world. In contrast, the addictions field is rife with territorialism and mutual disdain, specialty for specialty, group for group."(29)

Similarly, the Health Ministers' Conference in London (March, 1986) called for a coordinated international antidrug effort by the participating nations. Stressing that drug abuse affects not only the individual but the society as a whole, they called for a coordinated strategy involving:

(a) Coherent national policies for prevention, treatment, and rehabilitation.
(b) Improvement of data collection systems especially as they relate to epidemiology and trends of drug abuse.
(c) Development of community projects to reduce demands for drugs and provide appropriate treatment and rehabilitation services.
(d) Encouragement of international collaboration, training, and research around the use and abuse of psychoactive substances.(30)

Recognizing the importance of a well-thought out and coordinated plan, a strategy to deal with the cocaine crisis should address some of the following issues:

1.   Clarification of Commitment. As citizens in general and professionals in particular, we must break through the prevailing ambivalence and denial and face the problem head on. Requiring a total commitment, we should seek excellence not only in our own particular area of expertise, but be willing to work for effective change in all areas including enforcement, education, rehabilitation, research, and the creation of alternatives, etc.

2.   The Introduction of Strict Enforcement. The serious nature of cocaine addiction requires the implementation of tough measures which may restrict personal autonomy to protect the rights of society as a whole. No longer can we sit back and watch our families, neighbourhoods, and countries be destroyed. The price of cocaine trafficking or abuse should be made extremely high to deter involvement. Practical measures to accomplish this could include:
--mandatory random testing of urine for public safety personnel, e.g., doctors, teachers, pilots, control tower operators, etc,
--confiscation of goods and monies obtained from drug trafficking,
--mandatory long-term imprisonment for drug trafficking,
--stricter laws regarding driving under the influence of drugs or alcohol,
--the use of entrapment by police and security officers to ensnare drug dealers, and

--mandatory treatment of known cocaine addicts through mental health
committal.

3.   Establishment of Employee Assistance Programmes. Cocaine addiction in
the workplace results in low productivity, poor attitudes, chronic absenteeism, stealing, impaired judgment, etc. No country, especially a small
young country such as the Bahamas, can afford to lose talented employees
continuously. Once they lose their jobs, these onetime bright employees
are tempted into criminal activity, often involving the vicious cycle of
drug trafficking. An Employee Assistance Programme provides drug education
and referral opportunities for the treatment of addicted employees and their
families.

4.   Drug Education. There is much debate concerning the efficacy of drug
education, especially in the schools. Drug education should not focus solely
on drug information, but endeavour to promote emotional development by improving self-esteem, values clarification, decision-making strategies, and
of course, problem solving techniques. Such programmes should be instituted
in schools, Sunday Schools, and community organizations.

   More knowledge about drugs is not enough. What is needed is a change
in the fundamental way drugs are perceived. As the tobacco industry has
long been aware, the "image" associated with a product has a profound effect
on the desire the public will have for that product. Advertising techniques, the subtle messages conveyed in films, the image produced by Madison
Avenue is perhaps a far more persuasive education tool than textbook warnings in the schools. Perhaps Madison Avenue could devise a campaign to
affect attitudes toward drug use, showing young people that there is no
such thing as "recreational" drug use, and portraying it as a demeaning,
unattractive, out-of-it cop-out and not at all cool.

   Movies and other forms of media communication should convey the message
that using drugs never solves problems, it only complicates them. Even the
"high" itself is really a "down" because it tends to isolate persons and
destroy relationships in the long run, even though at first there seems to
be a jovial sense of camaraderie associated with its use. The momentary
good feelings do not compare with the satisfaction or enjoyment of a job
well done or other forms of pleasure which can be relived for years to come.

5.   Rehabilitation. Recognizing that individuals addicted to cocaine often
have developmental deficits manifested by poor ego strength, low frustration
tolerance, and lack of assertiveness, it will be difficult for them to cope
in society. Nevertheless, rehabilitation for cocaine addicted persons is
most effective for the majority of persons on an outpatient basis, where
coping skills can be integrated into everyday family and social life. Of
course, inpatient treatment is a must for addicted persons with severe psychopathology, inadequate support systems, or previous failure in outpatient
therapy. (31) Beyond inpatient or outpatient treatment, there is a need for
long term rehabilitation programmes where clients could remain for a year
or more. In such programmes, the focus would be on increasing ego strength
and preparing the patient to cope in the larger society.

6.   Organization of Parent Groups. Cocaine is a family issue and often a
neighbourhood problem. The formation of local parent groups provides support, education and increased "watchdog" awareness for the local neighbourhood. In Nassau, I have seen the emotional support parents have received
from such a group, leading to reduced family stress and increased family
unity. This is a powerful tool for prevention for families who get involved
in a programme before the need arises.

7.   Organization of the Religious Community. Research has repeatedly shown
that low religiosity is a powerful precursor of drug abuse. (32) The organ-

ization of meaningful programmes by the religious community could become a powerful force against drugs in the community. Examples of such activities include drop-in centres, counselling centres, drug education programmes, creative worship activities, halfway houses, or long term treatment projects. Treatment facilities organized by religious groups such as Teen Challenge and His Mansion in New Hampshire, have had a remarkable success in treating addicted persons, even from non-religious backgrounds.

8. **Family Life Education.** There is some credibility to the self-medication hypothesis in that often cocaine addicts suffer from depression, boredom, and fear of being alone. Considering the increasing disintegration of the family and the development of less bonded neighbourhoods, children themselves are less bonded, suffer from depression, low frustration tolerance, poor impulse control, and negative self-image. Family life education could teach child-rearing techniques to parents and enable them to become more integrated persons. Through the ripple effect, this would enhance development of their children, enabling them to be more creative, inner-directed, and hence, less vulnerable to drugs.

9. **Research.** Recognizing the inadequate knowledge base concerning cocaine, national and international research should be a key component of any major strategy to fight the cocaine problem. Particular issues requiring further understanding include:
--What are the main variables leading to cocaine use?
--What are key variables in the prevention of cocaine and other drug use, and how can these be effectively used to prevent further spread of the cocaine epidemic?
--Why do some children never take drugs, even though they live in high risk areas for drug abuse?
--Is cocaine psychosis an entity in itself, or is it a sequel to underlying emotional vulnerability or major psychiatric disorder (DSM-III axis I diagnosis)?
--What is the economic impact of cocaine addiction on the society, the country?
--Are inpatient programmes as effective as outpatient treatment programmes?
--What are the factors contributing to the high recidivism rate in cocaine addiction?

Beyond the answers to particular questions, a research attitude should permeate all aspects of an anti-cocaine strategy to review practices and to seek new approaches.

10. **Creation of a Community Groundswell.** After being involved in the struggle with perhaps the world's most devastating freebase cocaine epidemic in the Bahamas, I believe a community groundswell against drugs is essential to the success in the war against cocaine. The major variable creating the freebase (crack) cocaine epidemic is the easy availability of cheap, high quality cocaine. This frustrates attempts at enforcement, education, rehabilitation, and the creation of alternatives. For example, the police force is only as good as the cooperation it receives from the community. Thus the people of the community must see the drug problem as their own and realize that a quiet passive attitude is actively encouraging the destruction of the community by drugs. When the community rises to say "NO" to cocaine, all aspects of the antidrug strategies become increasingly effective. In the light of this, the total community must be mobilized with professional, non-professional, government leaders, and ordinary citizens working together. This requires meaningful dialogue which described by Paulo Freire is

> ...the way by which men (persons) achieve significance as men (persons). Dialogue is thus an existential necessity. And since dialogue is the encounter in which the united reflection

and action of the dialoguers are addressed to the world which is to be transformed and humanized, the dialogue cannot be reduced to the art of one person's "depositing" ideas in another...Because dialogue is an encounter among men (persons) who name the world, it must not be a situation where some men (persons) name on behalf of others. (33)

CONCLUSION

In conclusion, the cocaine phenomenon is a complex issue and requires technical knowledge buttressed with patience, courage, humility, and compassion. Perhaps none of us will fire the shot that wins the war against cocaine, but by working faithfully on all fronts, we can move the process one step closer to victory.

REFERENCES

1. New York Times (1986) Fighting Narcotics is Everyone's Issue Now, in This Week in Review, New York Times, Sunday, August 10, 1986.
2. M. Adams, (1984) Socioethical Issues in the Management of Developmental Disability in Perspectives and Progress in "Mental Retardation, Vol. I, Social Psychological and Educational Aspects," J. M. Berg, ed., International Association for the Scientific Study of Mental Deficiency, University of Hull, England.
3. R. Potter (1969) "War and Moral Discourse," John Knox Press, Richmond.
4. National Task Force Against Drugs, (1984), "Report of the National Task Force on Drug Abuse in the Bahamas, July, 1984," D. F. Allen, ed., Bahamas Government Publication, Nassau, Bahamas.
5. Firth, R., Ethical Absolutism and the Ideal Observer, Philosophy and Phenomenological Research, 12:317-345.
6. National Task Force Against Drugs, op. cit.
7. J. Jekel, D. F. Allen, H. Podlweski, S. Dean-Patterson, N. Clarke, P. Cartwright, C. Finlayson, (1986) Epidemic Freebase Cocaine Abuse, Case Study from the Bahamas, The Lancet March 1:459-462.
8. M. A. Bozarth, R. A. Wise, (1985), Toxicity associated with long term intravenous heroin and cocaine self-administration in the rat, JAMA 254:81-83.
9. J. Jekel, D. F. Allen, et al., op. cit.
10. Newsweek, March, 1986.
11. Addiction Research Foundation (1985), Teens doing fine, but for cocaine, The Journal, December 1, 1985, Vol. 14, No. 12.
12. New York Times, Thursday, 17 April, 1986.
13. S. Cohen, (1985) Drug Abuse and Alcoholism Newsletter, Vol. XIV, No. 2, April, 1985.
14. A. N. Whitehead, (1962) "The Aim of Education and Other Essays," Ernest Benn, London.
15. D. Archibald (1986) Coordinated anti-drug action is imperative, The Journal, 15(2):1, February 1, 1986.
16. National Drug Council, (1986), "The National Drug Council: A Summary of Activities, Findings, and Recommendations, 1985," D. F. Allen, ed., Bahamas Government Publication, in press.
17. M. B. Kandel, D. Murphy, D. Karus, (1985), Cocaine use in young adulthood: patterns of use and psychological correlates, in "Cocaine Use in America: Epidemiological and Clinical Perspectives, NIDA Research Monograph No. 61," N. Kozel and E. Adams, eds., National Institute on Drug Abuse, U. S. Government Printing Office, Washington, D. C.
18. S. Arietti, (1975), Psychiatric controversy: man's ethical dimension, Am. J. of Psychiatry, January 1975, 132.1
19. L. Eisenberg, (1972), The human nature of human nature, Science, Vol. 176, April 14, 1972.

20. R. Niebuhr, (1932) "Moral Man and Immoral Society," Scribners, New York, 1932, 1960.
21. A. R. Johnson, L. H. Butler, (1975), Public ethics and policy making, Hastings Centre Report, Vol. 5, August, 1975.
22. Remark made by cocaine addicted patient in Community Psychiatry Clinic, Nassau, Bahamas.
23. Miami Herald, Friday, September 20, 1985, p. 23A.
24. L. Kohlberg, (1973) The claim of moral adequacy of a highest state of moral judgment, J. of Philosophy, Vol. 70, Oct. 25, 1973, pp. 603-646.
25. M. Adams, op. cit.
26. S. Cohen, (1985), The story of cocaine is scary, The Journal, Nov. 1, 1985, p. 2.
27. A. L. Anker, T. J. Crowley, (1982), Use of contingency in specialty clinics for cocaine abuse, in "Problems of Drug Dependence, 1981, NIDA Research Monograph No. 41," National Institute on Drug Abuse, Rockville, Md.
28. B. J. Rounsaville, F. Gawin, H. Kleber, (1985) Interpersonal Psychotherapy adapted for ambulatory cocaine abusers, Am. J. Drug and Alcohol Abuse, 11(3 & 4):171-191.
29. D. Archibald, (1986) op. cit.
30. Addiction Research Foundation, (1986) Health chiefs want coordinated drug policies, The Journal, Vol. 15, No. 4, p.1. April 1, 1986.
31. For further discussion of treatment, see A. M. Washton "Cocaine: The Drug Epidemic of the 1980s" in this volume.
32. F. A. Tennant, Jr., R. Detels, V. Clarke, (1975), Some childhood antecedents of drug and alcohol abuse, Am. J. Epidemiology, 102:377-385.
33. P. Friere, (1973) "The Pedagogy of the Oppressed," The Seabury Press, New York.

EPILOGUE: A VISION OF HOPE

David F. Allen

National Drug Council
Nassau, Bahamas

"If thou art privy to thy countr's fate, which happily
foreknowing may avoid, O speak."

Horatio, Act I, scene i
Hamlet by William Shakespeare

The cocaine crisis facing the Western world bodes ill for all of us.
Broken lives, increased mortality, burgeoning crime rates and the dethroning
of cherished values confront the very heartbeat of society. The major concern among the public in general and professionals in particular is cocaine's
ability to disrupt basic responsibilities and priorities that provide structure and meaning to our lives. This stresses the need for a new understanding of physical and psychological addiction.

Wikler has developed a meaningful model for the psychological understanding of addictive illnesses on a behavioural approach using classical and
operant conditioning. (1) This model is helpful in understanding cocaine
abuse. Initially the addict uses cocaine due to any number of factors,
including curiosity, peer pressure, boredom, partying, and recreation. However, the physiological reward effects of the distinctive, overwhelming high
(especially in the case of crack or freebase cocaine) lead to repeated use
of the drug to obtain further pleasure. This is clearly demonstrated by the
tendency of experimental animals to self-administer cocaine to the point of
death or serious toxicity.(2) The euphoria is a positive reinforcement and
the crash (including dysphoria, depression, and craving) occurring on cessation of use is a negative reinforcement for compulsive cocaine use. Lacking
the severe physical abstinence effect of opiate or sedative/hypnotics, cocaine is characterized by severe craving. Cocaine addiction then according
to Dakis and Gold, is an interplay of positive and negative reinforcements
(3). Thus physical addiction may be defined in terms of "physiological reinforcement" rather than obvious withdrawal symptoms.

The physiological mechanisms of cocaine abuse override basic instinctual
survival drives such as hunger, thirst, and sex. In response, addicts turn
to compulsive use of cocaine to obtain pleasure and relief from craving.
Thus traditional psychotherapeutic methods to resolve emotional conflicts
may prove ineffective in treating cocaine addiction. The entrenched pleasure
reinforced compulsion requires education, peer therapy, and a major commitment to a drug-free lifestyle.

We live in a time of family disintegration, permissiveness and amorality circumscribed by a crass and empty materialism. Our prevailing state of mind might be best expressed in the lines of William Butler Yeats, written in 1921:

> Turning and turning in the widening gyre
> The falcon cannot hear the falconer;
> Things fall apart; the centre cannot hold.
> Mere anarchy is loosed upon the world,
> The blood-dimmed time is loosed, and everywhere
> The ceremony of innocence is drowned;
> The best lack all conviction, while the worst
> Are full of passionate intensity. (4)

This recognition has made us sceptical and produced a crisis of faith in our belief in human nature. Prime Minister Margaret Thatcher, speaking at the Commonwealth Heads of Government Conference held in October, 1985, in the Bahamas, said that when she was a university student, it was believed that if governments provided a good education, comprehensive health care, and a high standard of living, all the problems of society would be solved and people would be kind to one another. However, she added, life is not like that, because we are still left with the problem of human nature, the problem of drugs and the consuming desire to obtain power by any means. She said that our major task is to deal with the problems of human nature, the difficulty of living with one another. (5)

This dilemma is described by T. S. Eliot in this way:

> Has the Church failed mankind, or has
>   mankind failed the Church?
> When the Church is no longer regarded, not
>   even opposed,
>   and men have forgotten
> All gods except Usury, Lust and Power.
>
> Where is the Life we have lost in living?
> Where is the wisdom we have lost in knowledge?
> Where is the knowledge we have lost in information?
> The world turns and the world changes,
> But one thing does not change.
> In all my years, one thing does not change...
> The perpetual struggle of Good and Evil. (6)

Vaclav Havel, the playwright and essayist who spent five years in a Czech jail, puts it this way:

> It seems to me that man has what we call a human heart, yet
> he also has something of the baboon within him. But the mod-
> ern age treats the heart as a pump and denies the presence
> of the baboon within us. And so again and again, this offi-
> cially non-existent baboon goes on the rampage, either as
> the personal bodyguard of a politician or as a member of
> the most scientific police force in the world. (7)

Caught between the gods of Usury, Lust, and Power and those humans with baboon hearts, we easily fall prey to the drug pushers who promise euphoric panaceas with one hand while dealing death with the other. The modern situation is truly one of caveat emptor....

What we need more than anything else at this time of crisis is a vision of hope, emanating from a meaningful faith in life and leading to renewal within ourselves and a reorientation of our values. According to Whitehead, "To experience this faith is to know that in being ourselves, we are more than ourselves: To know that our experience, dim and fragmentary as it is, yet sounds the utmost depths of reality." (8)

It seems to me, as paradoxical as it might appear, this faith is the healing factor in the broken lives of many freebase, crack, or coca paste addicts. Recognizing their powerlessness over the drug and the bankruptcy of their lives, they cry out in despair and yet in hope beyond themselves, to each other, to their therapist, and to that ultimate meaning of life, that Higher Power of the Alcoholics or Narcotics Anonymous, or the Redeeming Christ of the believing Christians. This faith is then experienced in a change of lifestyle, a re-creation of values, and a deeper sense of human community. Could it be that this was what Immanual Kant meant when he wrote that the critical analysis of human experience stresses the need for belief in God, freedom, and immortality? (9)

In the midst of a paralyzing freebase cocaine epidemic, I have seen this faith in God or the Ultimate Meaning of life provide a vision of hope and a new courage to live. I know pushers who no longer sell drugs...I have seen young women having destroyed their lives through cocaine and prostitution, turn around and live meaningful lives. I have seen young men caught in the vice of crime and cocaine determine in their heart to go straight and experience the reality of starting life afresh. The cocaine crisis challenges all of us, both addict and non-addict, to re-evaluate our life and come to the commitment of a deeper faith in the Ultimate Meaning of life.

James Joyce at the very end of "A Portrait of the Artist as a Young Man" has his young hero write in his diary, "Welcome, O Life! I go to encounter for the millionth time the reality of experience and to forge in the smithy of my soul the uncreated conscience of my race." (10)

Let us welcome life and be open to its challenges, regardless of the difficulties! It is only through pain and frustration that we can truly empathize with others, grow as individuals, become as one with ourselves, and coalesce into a caring, meaningful society.

Let us go over and over again to encounter the reality of our experience. Standing together, we can face our drug problems and stop denying them through pointless projections and innuendo. For it is only by accepting our responsibility both as individuals and as a society that we can move in new directions towards meaningful solutions. Let us forge in the smithy of our souls the uncreated conscience of our race and our people. Looking beyond the problems of reality and the disappointments of our dreams, we can move courageously to create a new synthesis in culture, society, and our lives as a whole, and say with Tennyson's Ulysses:

> Come my friends,
> 'Tis not too late to seek a newer world...
>
> 'Though much is taken, much abides; and though
> We are not now that strength which in old days
> Moved earth and heaven; that which we are, we are -
> One equal temper of heroic hearts
> Made weak by time and fate, but strong in will
> To strive, to seek, to find, and not to yield.

(11)

REFERENCES

1. A. Wikler, R. Pescor (1967), Classical conditioning of a morphine abstinence phenomenon reinforcement of opiod drinking behaviour and "relapse" in morphine-addicted rats, Psychopharmacologia 10:255-284.
2. G. A. Denean, T. Yawagita, M. H. Seevers (1969) Self-administration of
   psychoactive substances by the monkey, Psychopharmacologia, 16:30-48.
3. C. A. Dakis, M. Gold (1986) New concepts in cocaine addiction: the dopamine depletion hypothesis, The Clinical Psychiatry Quarterly Newsletter of the American Academy of Clinical Psychiatrists, Vol. 9,No.3.
4. W. B. Yeats, The Second Coming in "Immortal Poems" O. Williams, ed.,
   Washington Square Press, p. 489. (1958)
5. Nassau Guardian, Tuesday, 22 October, 1985.
6. T. S. Eliot, Chorus from the Rock, in "The Great Thoughts," G. Seldes,
   ed., Ballantine/Random House, New York (1985) p. 121.
7. Harpers, (oct. 1985) Evil in a Rational Age, adapted from "Thriller"
   by Vaclan Havel in the June/July issue of The Idler.
8. A. N. Whitehead, "Science and the Modern World," p. 18.
9. W. D. Geoghegan, an article taken from a Phi Beta Kappa invitation speech
   delivered in May, 1984, at Bowdoin College and reported in Key Reporter, Summer 1984, Vol. XLIX, No. 4.
10. J. Joyce, "A Portrait of the Artist as a Young Man," as quoted by R. May
    in "The Courage to Create," W. W. Norton, New York.
11. A. L. Tennyson (1958) Ulysses, in "Immortal Poems," O. Williams, ed.,
    Washington Square Press, New York, p. 377.

ABOUT THE EDITOR

Dr. David F. Allen trained in medicine at St. Andrew's University, Scotland, in psychiatry at Harvard Medical School and in public health at Harvard School of Public Health. He was Joseph P. Kennedy, Jr., Foundation Fellow in Medical Ethics at Harvard, 1973-1974. He was Clinical Research Fellow at Massachusetts General Hospital, 1974-1976, and Assistant Clinical Professor at Yale Medical School, 1978-1980. Presently he is the Chairman of the National Drug Council, Commonwealth of the Bahamas. He also lectures in Public Health at Yale School of Public Health.

CONTRIBUTORS

BEAUBRUN, Prof. Michael
 Professor of Psychiatry, Head of the Department of Medicine,
 University of the West Indies, Port of Spain, Trinidad and Tobago.

CARROLL, Ms. Kathleen M.
 Researcher, Clinical Studies Unit, Department of Psychiatry, Yale
 Medical School, New Haven, Connecticut, U. S. A.

CARTWRIGHT, Paul
 Chief of Psychological Services, Community Psychiatry Clinic,
 Princess Margaret Hospital, Nassau, Bahamas.

CLARKE, Dr. Nelson
 Consultant Psychiatrist, Director of Drug Abuse, Ministry of
 Health, Bahamas.

COHEN, Prof. Sidney
 Professor of Psychiatry, Neuropsychiatric Institute, UCLA,
 Los Angeles, California, U. S. A.

DEAN-PATTERSON, Dr. Sandra
 Chief of Social Work, Ministry of Health and Coordinator,
 National Drug Council, Bahamas.

FENTON, Ms. Lisa R.
 Coordinator of Clinical Services, Department of Psychiatry,
 Yale Medical School, New Haven, Connecticut, U. S. A.

GAWIN, Dr. Frank
 Assistant Professor of Psychiatry, Substance Abuse Treatment Centre,
 Yale Medical School, New Haven, Connecticut, U. S. A.

GAY, Hon. Dr. Norman R.
 Minister of Health, Commonwealth of the Bahamas.

HUMBLESTONE, Dr. Brian G.
 Chief Psychiatrist, Ministry of Health, Bahamas.

JEKEL, Prof. James F.
 Winslow Professor of Public Health, Yale Medical School,
 New Haven, Connecticut, U. S. A.

JERI, Prof. F. Raul
 Professor of Clinical Neurology, Universidad Nacional Mayor
 de San Marcos, temporary consultant to W. H. O. (P.A.H.O.),
 Lima, Peru.

JOHNSON, Ms. Elaine
    Acting Deputy Director, National Institute on Drug Abuse,
    Rockville, Maryland, U. S. A.

KELLER, Daniel S.
    Researcher, Clinical Studies Unit, Department of Psychiatry, Yale
    Medical School, New Haven, Connecticut, U. S. A.

KHANTZIAN, Dr. Edward J.
    Associate Professor of Psychiatry, Harvard Medical School, at
    Cambridge Hospital, Cambridge, Massachusetts, U. S. A.

McCARTNEY, Dr. Timothy
    Chief of Psychology, Ministry of Health, Bahamas

MANSCHREK, Dr. Theo
    Associate Professor of Psychiatry, Harvard Medical School, at
    Massachusetts General Hospital, Boston, Massachusetts, U. S. A.

NEVILLE, Dr. Michael
    Consultant Psychiatrist, Ministry of Health, Bahamas

NOYA, Dr. Nils D.
    Chief, Medical Department of Rehabilitation, Treatment, and
    Prevention of Drug Abuse, Santa Cruz, Bolivia.

PEREZ-GOMEZ, Dr. Augusto
    Professor, Universidad de los Andes, and advisor to the
    United Nations, Bogota, Colombia.

PODLEWSKI, Dr. Henry
    Consultant Psychiatrist, Ministry of Health, Bahamas.

TURNER, Dr. Carlton E.
    Director, Drug Abuse Policy, Deputy Assistant to the President,
    The White House, Washington, D. C., U. S. A.

WASHTON, Dr. Arnold M.
    Director of Research and Treatment, Regent Hospital, New York City;
    Founder and Research Director of the National Hotline, 800-COCAINE,
    Fair Oaks Hospital, Summit, New Jersey, U. S. A.

BIBLIOGRAPHY

Adams, E. II., Kozel, N. J. (1985) Cocaine use in America, in "Cocaine Use
   in America: Epidemiological and Clinical Perspectives; NIDA Research
   Monograph No. 61," N. J. Kozel and E. H. Adams, eds., Government
   Printing Office, Washington, D. C.

Adams, M. (1984) Socioethical issues in the management of developmental disability in perspectives and progress, in "Mental Retardation, Vol. I.,
   Social Psychological and Educational Aspects," J. M. Berg, ed., International Association for the Scientific Study of Mental Deficiency,
   University of Hull, Hull, England.

Addiction Research Foundation (1986) Health chiefs want coordinated drug
   policies, The Journal, Vol. 15, No. 4, p.1, April 1, 1986.

Addiction Research Foundation (1985) Teens doing fine, but for cocaine, The
   Journal, Vol. 14, No. 12, December 1, 1985.

Androvny, A., Magnusson, P. (1985) Pneumopericardium from cocaine inhalation,
   New England Journal of Medicine 313:48-49.

Aigner, T. G., Balster, R. L. (1978) Choice behaviour in rhesus monkeys:
   cocaine vs food, Science 201:534-535.

Allen, D. F., et. al., (1984) "National Task Force Report on Drugs," Bahamas
   Government Publication, Nassau, Bahamas.

Almeida, M. (1978) Contribucion al estudio de la historia natural de la dependencia a la pasta basica de cocaina, Rev Neuropsiquiat 41:44-53.

American Psychiatric Association (1980) "Diagnostic and Statistical Manual
   of Mental Disorders, Third Edition," American Psychiatric Association,
   Washington, D. C.

Anker, A. L., Crowley, T. J. (1982) Use of contingency contracts in special-
   ity clinics for cocaine abuse, in "Problems of Drug Dependence, 1981,
   NIDA Research Monograph Series No. 41," L. S. Harris, ed., National
   Institute on Drug Abuse, Rockville, Md.

Archibald, D. (1986) Coordinated anti-drug action is imperative, The Journal
   15(2):1 February 1, 1986.

Arietti, S. (1975) Psychiatric controversy: man's ethical dimension, Amer
   Journal of Psychiatry January 1975, 132.1.

Ashley, R. (1975) "Cocaine: Its History and Effects," St. Martin's Press,
   New York.

Austin, G. (1978) Cocaine U.S.A. in "Perspectives on the History of Psychoactive Substance Use, NIDA Research Monograph No. 24," G. A. Austin,
   ed., National Institute on Drug Abuse, Rockville, Md.

Bacon, S. (1962) Alcohol and complex society, in "Society, Culture, and
   Drinking Patterns," D. J. Pittman and C. R. Snyder, eds., John Wiley
   and Sons, New York.

Baldinger, R., Goldsmith, B. M., Capel, W. C., and Stewart, G. T. (1972)
   Potsmokers, junkies, and squares: A comparative study of female values,
   International Journal of Addictions, 7:153-166.

Bandura, A. (1977) Self-efficacy: toward a unifying theory of behavioural change, Psychological Review, 84:191-215.

Barton, A. H. (1963) Social organization under stress, in "Disaster Research Study No. 17," National Academy of Sciences, National Research Council Disaster Research Corp., Washington, D. C.

Bechman, L. J. (1976) Alcoholism problems and women: An overview, in "Alcohol Problems in Women and Children," M. Greenblatt and M. A. Schukit, eds., Grune and Stratton, New York.

Bechman, L. J. (1975) Women alcoholics: A review of social and psychological studies, Journal of Studies on Alcoholism 36:797-824.

Beck, A. T., Emery, G. (1977) "Cognitive Therapy of Substance Abuse," unpublished manuscript.

Beck, A. T., Rush, A. J., Shaw, B. F., Emery, G. (1979) "Cognitive Theory of Depression," The Guildford Press, New York.

Bejerot, N. (1970) A comparison of the effects of cocaine and sympathetic central nervous system stimulants, British Journal of Addictions 65:35-37.

Berger, P. L., Luckman, T. (1967) "Social Construction of Reality," Anchor Books, New York.

Beschner, G. M., Reed, B. C., Mondanaro, J. (eds.) (1981) "Treatment Services for Drug Dependent Women, Vol. 1, NIDA Treatment Research Monograph Series," National Institute on Drug Abuse, Government Printing Office, Washington, D. C.

Beschner, G., Thompson, P. (1981) "Women and Drug Abuse Treatment Needs and Services: NIDA Services Research Monographical Series," National Institute on Drug Abuse, Government Printing Office, Washington, D. C.

Blatt, S. (1974) Levels of object representation in anaclitic and introjective depression, Psychoanalytic Study of the Child, 29:107-157.

Blatt, S. J. , Berman, W., Bloom-Feshback, S., et al. (1984) Psychological assessment of psychopathology in opiate addicts, J Nerv Ment Dis , 156-165.

Blatt, S. and Shichman, S. (1983) Two primary configurations psychopathology, Psychoanalysis and Contemporary Thought 6:187-254.

Blum, R. H., et. al. (1970) "Students and Drugs," Jossey-Bar, San Francisco.

Boggs, J. (1970) unpublished manuscript quoted in "Women's Liberation," S. Stambler, Ace Books, New York.

Boothroyd, W. E. (1963) The female drug addict: a profile, Addictions 10:36-40.

Boudin, H. M. (1972) Contingency contracting as a therapeutic tool in the deceleration of amphetamine use, Behaviour Therapy 3:604-608.

Bozarth, M. A., Wise, R. A. (1985) Toxicity associated with long-term intravenous heroin and cocaine self-administration in the rat, JAMA 254(1):81-83.

Branch, M. N., Dearing, M. E. (1982) Effects of acute and daily cocaine administration on performance under a delayed-matching-to-sample procedure, Pharmacol Biochem Behav 16(5):713-718.

Brandon, N. (1969) "Psychology of Self-Esteem," Bantam Books, New York.

Brownell, K. D., Marlatt, G. A., Lichtenstein, E., Wilson, G. T. (1986) Understanding and preventing relapse, American Psychologist 41:765-782.

Bry, R. (1983) Empirical foundations of family-based approaches to adolescent substance abuse, in "Preventing Adolescent Drug Abuse: Intervention Strategies, NIDA Research Monograph No. 47," T. J. Glynn, C. G. Leukefeld, J. P. Ludford, eds., National Institute on Drug Abuse, Rockville, Md.

Burt, M. R., Glynn, T. J., Sowder, B. (1979) "Psychosocial Characteristics of Drug Abusing Women," National Institute on Drug Abuse, Rockville, Md.

Butterworth, A. T. (1971) Depression associated with alcohol withdrawal: imipramine therapy compared with placebo, Q J Stud Alcohol 32:343-348.

Byck, R. (1975) "Cocaine Papers: Sigmund Freud," Stonehill Publishing Co. New York.

Byck, R. (1979) Testimony before the Select Committee on Narcotics Abuse and Control, House of Representatives, 96th Congress, July 24,26, Oct.10,1979.

Carbajal, C. (1980) Psicosis producida por inhalacion de cocaina, in "Cocaina 1980," F. R. Jeri, ed., Pacific Press, Lima, Peru.

Carroll, K. (1985) "Manual for Relapse Prevention in the Treatment of Cocaine Abuse," unpublished manuscript.

Celetano, D. D., McQueen, D. V., Chee, E. (1980) Substance abuse by women: a review of the epidemiological literature, Journal of Chronic Diseases 33(6):383-394.

Chambers, C. D., Griffey, M. S. (1975) Use of legal substances within the general population, the sex and age variables, Addictive Diseases: An International Journal 2(1):7-19.

Chasnoff, I. J., Burns, W. J., Scholl, S. H., Burns, K. A. (1985) Cocaine use in pregnancy, New England Journal of Medicine 313:666-669.

Chasnoff, I. J., Burns, W. J., Schnoll, S. H., Burns, K. A. (1986) Effects of cocaine on pregnancy outcome, in "Problems of Drug Dependence, 1985, NIDA Research Monograph Series No. 67," L. S. Harris, ed., National Institute on Drug Abuse, Rockville, Md.

Chein, I., Gerard, D. L., Lee, R. S., et al. (1964) "The Road to Narcotics, Delinquency, and Social Policy," Basic Books, New York.

Chitwood, D. D. (1985) Patterns and consequences of cocaine use, in Cocaine Use in America: Epidemiological and Clinical Perspectives, NIDA Research Monograph Series No. 61," E. H. Adams and N. J. Kozel, eds., National Institute on Drug Abuse, Rockville, Md.

Chodorow, N. (1978) "The Reproduction of Mothering: Psychoanalysis and the Socializing of Gender," University of California Press, Burbank.

Clayton, R. R., Ritter, C. J. (1983) Cigarette, alcohol, and drug use among youth: selected consequences. Paper read at a meeting of the National Council on Alcoholism and Research Society on Alcoholism, Houston, Tx.

Clayton, R. R., Voss, H. L. (1982) Marijuana and cocaine: the causal nexus. Paper read at the meeting of the National Association of Drug Abuse Problems, New York.

Clinard, M. B. (1973) "Crime in the Developing World," Wiley and Sons, New York.

Cohen, S. (1984) Cocaine: acute medical and psychiatric complications, Psychiatric Annals 14:747-749.

Cohen, S. (1984) "Cocaine Deaths," lecture delivered at the Seminario Clinico Internacional sobre Cocaina, Bogota, Colombia, December 11-14, 1984.

Cohen, S. (1985) "Cocaine, The Bottom Line," The American Council on Drug Education, New York.

Cohen, S. (1981) "Cocaine Today," American Council on Drug Education, New York.

Cohen. S. (1985) Drug Abuse and Alcoholism Newsletter, Vol. XIV, No. 2, April, 1985.

Cohen, S. (1985) Drug abuse: predisposition and vulnerability, in "Psychiatry, the State of the Art, Vol. 6," P. Pichot, P. Berner, R. Wolf, K. Than, eds., Plenum Publishing Press, New York.

Cohen, S. (1982) Health hazards of cocaine, reprinted from "Cocaine Today," Drug Enforcement, Fall, 1982.

Cohen, S. (1985) Reinforcement and rapid delivery systems: understanding adverse consequences of cocaine, in "Cocaine Use in America, NIDA Research Monograph Series No. 61," E. H. Adams and N. J. Kozel, eds., National Institute on Drug Abuse, Rockville, Md.

Cohen, S. (1985) The story of cocaine is scary, The Journal, November 1,1985.

Colten, M. E. (1979) Description and comparative analysis of self-perceptions and attitudes of heroin-addicted women, in "Addicted Women: Family Dynamics, Self-Perceptions, and Support Systems," National Institute on Drug Abuse, Rockville, Md.

Crowley, T. J. (1984) Contingency contracting treatment of drug abusing physicians, nurses, and dentists, in "Behavioural Intervention Techniques in Drug Abuse Treatment, NIDA Research Monograph No. 46," J. Grabowski, M. L. Stitzer, and J. E. Henningfield, eds., National Institute on Drug Abuse, U. S. Government Printing Office, Washington, D. C.

Crowley, T. J. (1982) Quoted in "Reinforcing Drug-Free Lifestyles," <u>ADAMHA News</u>, p. 3, August 27, 1982.

Cummings, C., Gordon, J. R., and Marlatt, G. A. (1980) Relapse: prevention and prediction, <u>in</u> "The Addictive Behaviours," W. R. Miller, ed., Pergamon Press, New York.

Cuskey, W., Berger, L., Densen-Gerber, J. (1977) Issues in Treatment of female addiction: a review and critique of the literature, <u>Contemporary Drug Problems</u> 6(3):307-371.

Dackis, C. A., Gold, M. S. (1985) Bromocryptine as a treatment for cocaine abuse, <u>The Lancet</u> 2:1151-1152.

Dackis, C. A., Gold, M. S. (1986) New concepts in cocaine addiction: The dopamine depletion hypothesis, <u>The Clinical Psychiatry Quarterly Newsletter of the American Academy of Clinical Psychiatrists</u>, Vol. 9, No. 3.

DeLeon, G., Bascher, G. M. (eds.) (1977) "The Therapeutic Community: Proceedings of Therapeutic Communities of America Planning Conference," National Institute on Drug Abuse, Rockville, Md.

Denean, G. A., Yawagita, T., Seevers, M. H. (1969) Self-administration of psychoactive substances by the monkey, <u>Psychopharmacologia</u>, 16:30-48.

DeWit, H. Stewart, J. (1981) Reinstatement of cocaine-reinforced responding in the rat, <u>Psychopharmacology</u> 75:134-143.

Dorus, W., Senay, E. C. (1980) Depression, demographic dimension, and drug abuse, <u>American Journal of Psychiatry</u> 137:699-704.

Dougherty, J., Pickens, R. W. (1973) Fixed-interval schedules of intravenous cocaine presentation in rats, <u>Journal of the Experimental Analysis of Behaviour</u>, 20:111-118.

Dumont, M. (1968) "The Absurd Healer," Science House, New York.

Dupont, R. L. (1977) Foreward <u>in</u> "Cocaine, 1977, NIDA Monograph No. 13," R. C. Petersen and R. C. Stillman, eds., U. S. Government Printing Office, Washington, D. C.

Eckholm, E. (1986) Cocaine Treatment, Experts Dialogue, <u>New York Times</u>, Monday, September 8, 1986.

Eddy, N., Halback, H. Isbell, H., Seevers, M. (1965) Drug dependence: its significance and characteristics, <u>Bulletin of the World Health Organization,</u> 32:721-733.

Eisenberg, L. (1972) The human nature of human nature, <u>Science</u> Vol. 176, April 14, 1972.

Eldred, C. A., Washington, M. N. (1976) Interpersonal relationships in heroin use by men and women and their role in treatment outcome, <u>International Journal of Addictions</u> 11(1):117-130.

Eliot, T. S. (1985) Chorus from "The Rock," in "The Great Thoughts," G. Seldes, ed., Ballantine/Random HOuse, New York.

Elkind, D. (1979) "The Child and Society," Oxford University Press, New York.

Ellinwood, E. H. (1971) Assault and homicide associated with amphetamine abuse, <u>American Journal of Psychiatry</u> 127:1170-1175.

Epstein, P. N., Altshuler, H. L. (1978) Changes in the effects of cocaine during chronic treatment, Res Com Chem Pathol Pharmacol 22(1):93 105

Firth, R. (1952) Ethical Absolutism and the Ideal Observer, Philosophy and <u>Phenomenological Research</u> 12:317-345.

Fischman, M. W., Schuster, C. R. (1981) Acute tolerance to cocaine in humans, <u>in</u> "Problems of Drug Dependence, 1980, NIDA Research Monograph Series No. 34," L. S. Harris, ed., National Institute on Drug Abuse, Rockville.

Fischman, M. W., Schuster, C. R. (1980) Cocaine effects in sleep-deprived humans, <u>Psychopharmacology</u> 72(1):1-8.

Fischman, M. W., Schuster, C. R. (1982) Cocaine self-administration in humans, <u>Fed. Proc.</u> 41:241-246.

Fischman, M. W., Schuster, C. R. (1980) Experimental investigations of the actions of cocaine in humans, <u>in</u> "Cocaine, 1980: Proceedings, Inter-American Seminar on Medical and Sociological Aspects of Coca and Cocaine" F. R. Jeri, ed., Pacific Press, Lima, Peru.

Fischman, M. W., Schuster, C. R., Hatana, Y. (1983) A comparison of the subjective and cardiovascular effects of cocaine and lidocaine in humans,

Pharmacol Biochem Behav 18(1):123-127.

Fischman, M. W., Schuster, C. R., Resnekov, L., Shick, J. F. E., Krasnegor, N. A., Fennel, W., Freedman, D. X. (1976) Cardiovascular and subjective effects of intravenous cocaine administration in humans, Arch. Gen. Psychiatry 33:983-989.

Foltin, R. W., Preston, K. L., Wagner, G. C., Schuster, C. R. (1981) The aversive stimulus properties of repeated infusions of cocaine, Pharmacol Biochem Behav 15(1):71-74.

Forno, J. J., Young, R. T., Levitt, C. (1981) Cocaine abuse - the evolution from coca leaves to freebase, J. Drug Education 11(4):311-315.

Frankl, V. (1963) "Man's Search for Meaning," Washington Square Press, New York.

Freud, S. (1912) "The Dynamics of Transference, Standard Edition, Vol. 12" Hogarth Press, London.

Friere, P. (1973) "The Pedagogy of the Opressed," The Seabury Press, New York.

Gale, K. (1984) Catecholamine-independent behaviour and neurochemical effects of cocaine in rats, in "Mechanisms of Tolerance and Dependence, NIDA Research Monograph Series No. 5," C. W. Sharp, ed., National Institute on Drug Abuse, Rockville, Md.

Gawin, F. H., Kleber, H. D. (1986) Abstinence symptomatology and psychiatric diagnosis in chronic cocaine abusers, Arch Gen Psychiat 43:107-113.

Gawin, F. H., Kleber, H. D. (1984) Cocaine abuse treatment: an open trial with lithium and desiprimine, Archives of General Psychiatry 41:903-910.

Gawin, F. H., Kleber, H. D. (1985) Cocaine use in a treatment population: patterns and diagnostic distinctions, in "Cocaine Use in America: Epidemiologic and Clinical Perspectives, NIDA Research Monograph Series No. 61," N. J. Kozel and E. H. Adams, eds., National Institute on Drug Abuse, Washington, D. C.

Gaylin, W. (1979) "Caring," Avon, New York.

Geller, J. D., Schwartz, M. D. (1972) "An Introduction to Psychotherapy: A Manual for the Training of Clinicians," unpublished manuscript.

Geoghegan, W. D. (1984) an article taken from a Phi Beta Kappa invitation speech delivered in May, 1984, at Bowdoin College and reported in the Key Reporter, Summer, 1984, Vol. XLIX, No. 4.

Gerard, D. L., Kornetsky, C. (1954) Adolescent opiate addiction: a case study, Psychiatr. Q. 28:367-380.

Gerard, D. L., Kornetsky, C. (1985) Adolescent opiate addiction: a study of control and addict subjects, Psychiatr. Q. 29:457-486.

Gerber, G. J., Stretch, R. (1975) Drug-induced reinstatement of extinguished self-administration behaviour in monkeys, Pharmacology, Biochemistry, and Behaviour 3:1055-1061.

Goina, C., Byrne, R. (1975) Distinctive problems of the female drug addicts: experiences at IDAP, in "Developments in the Field of Drug Abuse," E. Senay, ed., Schenkman, Cambridge, Mass.

Gold, M. S. (1984) "800-COCAINE," Bantam, New York.

Gold, M. S., Washton, A. M., Dackis, C. A. (1985) Cocaine abuse: neurochemistry, phenomenology, and treatment, in "Cocaine Use in America: Epidemiologic and Clinical Perspectives, NIDA Research Monograph Series No. 61," N. J. Kozel and E. H. Adams, eds., National Institute on Drug Abuse, U. S. Government Printing Office, Washington, D. C.

Goldberg, S. R. (1973) Comparable behaviour maintained under fixed-ratio and second-order schedules of food presentation, cocaine injection, or d-amphetamine injection in the squirrel monkey, Journal of Pharmacology and Experimental Therapeutics 186:18-30.

Gomberg, E. S. (1976) Alcoholism in women, in "The Biology of Alcoholism, Vol. 4," B. Kissin and H. Begleiter, eds., Plenum Press, New York.

Gomberg, E. S., Franks, W. (eds.) (1979) "Gender and Disorderly Behaviour: Sex Differences in Psychopathology," Brunner Megel, New York.

Gordon, A. (1908) Insanities caused by acute and chronic intoxications with opium and cocaine, JAMA 51:97-101.

Green, L. (1986) Speech to the Ministry of Education (Bahamas) as quoted in the _Nassau Guardian_, April, 1986.

Greenblatt, M., Schuckit, M. A. (eds.) (1976) "Alcoholism Problems in Women and Children," Grune and Stratton, New York.

Greenson, R. R. (1967) "The Technique and Practice of Psychoanalysis', Vol. I," International Universities Press, New York.

Griffiths, R. R., Findley, J. D., Brady, J. V., Dolan-Gutcher, K., Robinson, W. W. (1975) Comparison of progressive-ratio performance maintained by cocaine, methylphenidate, and secobarbital, _Psychopharmacologia (Berlin)_ 43:81-83.

Grinspoon, L., Bakalar, J. (1976) "Cocaine: A Drug and its Social Evolution," Basic Books, New York.

Gutierrez-Noriega, C. (1947) Alteraciones mentales producides por la coca, _Rev Neuropsiquiat_ 10:145-176.

Gutierrez-Noriega, C., Zapata, V. (1950) La inteligencia y la personalidad en los habituades a la coca, _Rev Neuropsiquiat_ 13:22-60.

Havel, V. (1985) Evil in a rational age, _in_ _Harpers_, October, 1985, adapted from "Thriller," in the June/July issues of _The Idler_.

Helfrich, A., Crowley, T. J., Atkinson, C. A., Post, R. D. (1983) A clinical profile of 136 cocaine abusers, _in_ "Problems of Drug Dependence, 1983, NIDA Research Monograph Series No. 41," L. S. Harris, ed., National Institute on Drug Abuse, Rockville, Md.

Hershan, H. (1972) The Disease concept of alcoholism: a reappraisal, _in_ "Proceedings, 30th International Congress on Alcoholism and Drug Dependence," Lausanne International Council on Alcoholism and Addictions, Amsterdam, 1972.

Howard, R. E., Hueter, D. C., David, G. J. (1985) Acute myocardial infarction following cocaine abuse in a young woman with normal coronary arteries, _JAMA_ 254(1):95-96.

Hunt, W. A., Matarazzo, J. D. (1970) Habit mechanisms in smoking, _in_ "Learning Mechanisms in Smoking," W. A. Hunt, ed., Aldine, Chicago, Ill.

Isbell, H., White, W. (1953) Clinical characteristics of addictions, _Amer. Journal of Medicine_ 14:558-565.

Itkonen, J., Schnoll, S., Glassroth, J. (1984) Pulmonary dysfunction in freebase cocaine users, _Archives of Internal Medicine_ 144:2195-2197.

Jekel, J., Allen, D. F., Podlewski, H., Dean-Patterson, S., Clarke, N., Cartwright, P., Finlayson, C. (1986) Epidemic Freebase Cocaine Abuse: Case Study from the Bahamas, _The Lancet_, March 1:459-462.

Jellinek, E. M. (1960) "The Disease Concept of Alcoholism," College and University Press, New Have, Conn.

Jeri, F. R. (1984) Coca paste smoking in some Latin American countries, a severe and unabated form of addiction, _Bulletin Narcotics_ 36(2):15-31.

Jeri, F. R. (1984) La practica de fumar coca en algunos paises de America Latina: una toxicomania grave y generalizada, _Boletin de estupefacientes de las Naciones Unidas_, XXXVI:17-34.

Jeri, F. R. (1985) Los problemas medicos y sociales generados por el abuso de drogas en el Peru, _Rev Sanid Fuerz Policiales_ 46:36-45.

Jeri, F. R. Carbajal, C., Sanchez, C. C., Del Pozo, T., Fernandez, M. (1978) The syndrome of coca paste, _J. Psychedel Drugs_ 10:361-370.

Jeri, F. R., Sanchez, C. C., Del Pozo, T. (1976) Consumo de drogas peligrosas por miembros de la fuerza armada y de la fuerza policial peruana, _Rev Sanid Minist Int_ 37:104-112.

Jeri, F. R., Sanchez, C. C., Del Pozo, T. Fernandez (1978) El sindrome de la pasta de coca: observaciones en un grupo de 158 pacientes del area de Lima, _Rev Sanid Minist Int_ 39:1-18.

Jeri, F. R., Sanchez, C. C., Del Pozo, T. Fernandez, Carbajal, C. (1978) Further experience with the syndromes produced by coca paste smoking, _Bulletin Narcotics_ 30:1-11.

Johanson, C. E. (1984) Assessment of the dependence potential of cocaine in animals, _in_ "Cocaine, Pharmacology, Effects, and Treatment of Abuse, NIDA Research Monograph No. 50," J. Grabowski, ed., U. S. Government

Printing Office, Washington, D. C.

Johnson, A. R., Butler, L. H. (1975) Public ethics and policy making, Hastings Centre Report Vol 5, August, 1975.

Johnson, D. (1973) "Marijuana Users and Drug Subcultures," Wiley Press, New York.

Johnston, L. D., O'Malley, P. T., Bachman, J. G. (1986) "Drug Use Among American High School Students, College Students, and Other Young Adults, National Trends Through 1985," U. S. Printing Office, Washington, D. C.

Jones, R. (1986) Cocaine increases natural killer cell activity, Journal of Clinical Investigation 77:1387-1390.

Jones, R. T. (1984) The pharmacology of cocaine, in "Cocaine: Pharmacology, Effects, and Treatment of Abuse, NIDA Research Monograph Series No. 50," J. Grabowski, ed., U. S. Government Printing Office, Washington, D. C.

Joyce, J. "A Portrait of the Artist as a Young Man," as quoted by R. May in "The Courage to Create," W. W. Norton, New York. (1975)

Kandel, D. B. (1978) Convergencies in prospective longitudinal surveys of drug use in normal populations, in "Longitudinal Research on Drug Use: Empirical Findings and Methodological Issues," Hemisphere, New York.

Kandel, D. B. (1975) Stages in adolescent involvement in drug use, Science 190:912-914.

Kandel, D. B., Murphy, D., Karun, D. (1985) Cocaine use in young adulthood: patterns of use and psychosocial correlates, in "Cocaine Use in America: Epidemiologic and Clinical Perspectives, NIDA Research Monograph Series No. 61," N. Kozel and E. Adams, eds., U. S. Government Printing Office, Washington, D. C.

Kandel, D. B., Treiman, D., Faust, R., Single, E. (1976) Adolescent involvement in legal and illegal drug use: a multiple classification analysis, Social Forces 55:438-458.

Kaplan, H. B. (1977) Antecedents of deviant responses: predicting from a general theory of deviant behaviour, J. Youth Adolescence 6:89-101.

Kernberg, O. (1975) "Borderline Conditions and Pathological Narcissism," Jason Aaronson, New York.

Kernberg, O. (1976) "Object Relations Theory and Clinical Psychoanalysis," Jason Aaronson, New York.

Khantzian, E. J. (1980) An ego-self theory of substance dependence: a contemporary psychoanalytic perspective, in "Theories on Drug Abuse, NIDA Research Monograph Series No. 30," D. J. Lettieri, M. Sayers, H. W. Pearson, eds., National Institute on Drug Abuse, Rockville, Md.

Khantzian, E. J. (1983) An extreme case of cocaine dependence and marked improvement with methylphenidate treatment, American Journal of Psychiatry 140:784-785.

Khantzian, E. J. (1979) Impulse problems in addiction: cause and effect relationships, in "Working with the Impulsive Person," H. Wishnie, ed., Plenum Press, New York.

Khantzian, E. J. (1974) Opiate addiction: a critique of theory and some implications for treatment, Am. J. Psychother. 28:59-70.

Khantzian, E. J. (1972) A preliminary dynamic formulation of the psychopharmacologic action of methadone, in "Proceedings of the Fourth National Methadone Conference, San Francisco, 1972," National Association for the Prevention of Addiction to Narcotics.

Khantzian, E. J. (1982) Psychological (structural) vulnerabilities and the specific appeal of narcotics, Ann N. Y. Acad Science 398:24-32.

Khantzian, E. J. (1985) The self-medication hypothesis of addictive disorders: focus on heroin and cocaine dependence, Amer Journal of Psychiatry 142:11 November, 1985.

Khantzian, E. J. (1975) Self-selection and progression in drug dependence, Psychiatry Digest 10:19-22.

Khantzian, E. J., Gawin, F., Kleber, H. D., et al. (1984) Methylphenidate treatment of cocaine dependence - a preliminary report, J. Substance Abuse Treatment 1:107-112.

Khantzian, E. J., Khantzian, N. J. (1984) Cocaine addiction: is there a psychological predisposition? <u>Psychiatric Annals</u> 14(10):753-759.

Khantzian, E. J., Mack, J. E., Schatzberg, A. F. (1974) Heroin use as an attempt to cope: clinical observations, <u>Am. J. Psychiatry</u> 131:160-164.

Khantzian, E. J., Treece, C., DSM-III psychiatric diagnosis of narcotic addicts:recent findings, <u>Arch. Gen. Psychiatry</u> (in press.)

Khantzian, E. J., Treece, C. (1977) Psychodynamics of drug dependence: an overview, <u>in</u> Psychodynamics of Drug Dependence: NIDA Research Monograph No. 12," J. D. Blaine, D. A. Julius, eds., National Institute on Drug Abuse, Rockville, Md.

Kleber, H. D., Gawin, F. H. (1984) Cocaine abuse: a review of current and experimental treatments, <u>in</u> "Cocaine: Pharmacology, Effects, and Treatment of Abuse, NIDA Research Monograph Series No. 50," J. Grabowski, ed., National Institute on Drug Abuse, Rockville, Md.

Kleber, H. D., Gawin, F. H. (1985) Cocaine use in a treatment population: patterns and diagnostic distinctions, <u>in</u> "Cocaine Use in America: Epidemiologic and Clinical Perspectives, NIDA Research Monograph Series No. 61," N. Kozel and E. Adams, eds., U. S. Government Printing Office, Washington, D. C.

Kleber, H. D., Gawin, F. H. (1984) The spectrum of cocaine abuse and its treatment, <u>Journal of Clinical Psychiatry</u> 45(12, Section 2):18-23.

Kohlberg (1973) The claim of moral adequacy of a highest state of moral judgment, <u>J. of Philosophy</u> 70:603-646.

Kohut, H. (1971) "The Analysis of the Self," International Universities Press, New York.

Kohut, H. (1977) "The Restoration of the Self," International Universities Press, New York.

Kosten, T. R., Hogan, I., Jalali, B., Steidl, J., Kleber, H. D., The effect of multiple family therapy on addict family functioning: a pilot study, <u>Adv Alcohol Subs Abuse</u> 5:51-62.

Kozel, N. J. (1985) "Reports of coca paste smoking field investigations in South Florida and New York," internal report, National Institute on Drug Abuse, Rockville, Md.

Krystal, H., Raskin, H. A. (1970) "Drug Dependence: Aspects of Ego Functions" Wayne State University, Detroit.

Ladner, J. (1971) "Tomorrow's Tomorrow," Anchor Doubleday, New York.

<u>Lancet, The,</u> (1983) Images of cocaine (editorial) <u>The Lancet</u> II:1231-1232.

Lauer, R. H. (1973) "Perspectives in Social Change," Allwyn and Bacon, Boston.

Levy, S. J., Doyle, K. M. (1974) Attitudes towards women in a drug abuse treatment programme, <u>Journal of Drug Issues</u> 4:428-435.

Lieber, J. (1986) Coping with Cocaine, <u>The Atlantic Monthly</u>, January, 1986.

Lichtenfeld, P.J., Rubin, D., Feldman, R. S. (1984) Subarachnoid Haemorrhage precipitated by cocaine smoking, <u>Arch Neurol</u> 41:223-224.

Luborsky, L. (1984) "Principles of Psychoanalytic Therapy: A Manual for Supportive-Expressive Treatment," Basic Books, New York.

Maddox, George (1962) Teenage drinking in the United States, <u>in</u> "Society, Culture, and Drinking Patterns," D. J. Pittman and C. R. Snyder, eds., John Wiley and Sons, New York.

Madsen, W. (1973) "The American Alcoholic: Nature-Nurture Controversy in Alcohol Research and Treatments," Charles C. Thomas, Springfield, Ill.

Maier, H. W. (1926) "Der Kocainismus," G. Thieme, Leipzig.

Mandonaro, J. (1977) Women: pregnancy and children and addiction, <u>Journal of Psychedelic Drugs</u> 9(1).

Manschreck, T. C., Laughery-Flesche, J. A., Weisstein, C. C., Allen, D. F., Mitra, N., Characteristics of cocaine psychosis (in preparation).

Marlatt, G. A., Gordon, J. R. (1980) Determinants of relapse: implications for the maintenance of behaviour change, <u>in</u> "Behavioural Medicine: Changing Health Lifestyles," P. O. Davidson and S. M. Davidson, eds., Brunner Mazel, New York.

Marlatt, G. A., Gordon, J. R. (1985) "Relapse Prevention: Maintenance Strategies in Addictive Behaviour Change," Guildford Press, New York.

Mausner, J., Kramer, S. (1985) "Epidemiology: An Introductory Text, Second Edition," W. B. Saunders, Philadelphia.

McCartney, T. (1971) "Neurosis in the Sun," Executive Printers, Nassau, Bahamas.

McLellan, A. T., Lubosky, L., Woody, G. E., O'Brien, C. P. (1980) An improved diagnostic instrument for substance abuse patients: the addiction severity index, Journal of Nervous Mental Diseases 168:26-33.

McLellan, A. T., Luborsky, L. Woody, G. E., O'Brien, C. P., Druley, K. A. (1983) Predicting response to alcohol and drug abuse treatment, Archives of General Psychiatry 40:620-625.

McLellan, A. T., Woody, G. E., O'Brien, C. P. (1979) Development of psychiatric illness in drug abusers, N Eng J Med 201:1310-1314.

Milkman, H., Frosch, W. A. (1973) On the preferential abuse of heroin and amphetamine, Journal of Nervous Mental Diseases 156:242-248.

Miller, J. D., Cisin, I. H., Gardner-Keaton, H., Harrell, A. V., Wirtz, P. W., Abelson, H. I., Fischburne, P. M. (1983) "National Survey on Drug Abuse: Main Findings, 1982," U. S. Government Printing Office, Washington, D. C.

Miller, W. R., Hester, R. K. (1986) Inpatient alcoholism treatment: Who benefits? American Psychologist 41:794-805.

Minkoff, K., Bergman, E., Beck, A. (1973) Hopelessness, depression, and attempted suicide, American Journal of Psychiatry 130:455-460.

Ministerio de Salud de Colombia (1985) "Instituciones de Salud Mental y Servicios de Famacodependencia," documento de circulacion restringida.

Ministerio de Salud de Colombia-Unfdoc, IV Encuentro Nacional de Servicios de Farmacodependencia y Alcoholismo, Bogota, Colombia.

Ministerio de Salud de Colombia-Unfdoc (1984) Seminario Clinico Internacional Sobre Adicciones a la Hoja de Coca y sus Derivados, Bogota, Colombia.

Minuchin, S., Fishman, H. C. (1981) "Family Therapy Techniques," Harvard University Press, Cambridge, Mass.

Modell, A. (1976) The "holding environment" and the therapeutic action of psychoanalysis, Journal of the American Psychoanalytic Association, 24:285-308.

Moise, R., Reed, B. C., Connell, C. (1981) Women in drug abuse treatment programmes: factors that influence retention at very early and later stages in two treatment modalities, International Journal of Addictions 16(6):1295-1300.

Moise, R., Reed, B. G., Ryan, V. S. (1982) Issues in the treatment of heroin addicted women: a comparison of men and women entering two types of drug abuse programmes, International Journal of Addictions 17(1):109-139.

Mortimer, W. G. (1901) "Peru History of Coca, The Divine Plant of the Incas, with an Introductory Account of the Incas and of the Andean Indians of Today," J. H. Vail and Co., New York.

Motta, G. (1984) Una farmacodependencia epidemica: bazuco, Investigaciones Medicas, VI, No. 18.

Mule, S. J. (1984) The pharmacodynamics of cocaine abuse, Psychiatric Annals 14:724-727.

Musto, D. F. (1973) "The American Disease: Origins of Narcotic Control," Yale University Press, New Haven.

Musto, D. F. (1986) Lessons of the first cocaine epidemic, Wall St. Journal June 11, 1986.

Myers, J. A., Earnest, M. P. (1984) Generalized seizures and cocaine abuse, Neurol 34:675-676.

Nassau Guardian (1986) Cocaine freebase houses sweeping South Florida, January 6, 1986.

National Commission on Marijuana and Drug Abuse (1973) "Drug Use in America: Problem in Perspective, Second Report of the National Commission on Marijuana and Drug Abuse," National Institute on Drug Abuse, U. S. Government Printing Office, Washington, D. C.

National Drug Council (1986) "The National Drug Council: A Summary of

Activities, Findings, and Recommendations, 1985," D. F. Allen, ed., Bahamas Government Publication, in press.

National Task Force Against Drugs (1984) "Report of the National Task Force on Drug Abuse in the Bahamas, July, 1984," D. F. Allen, ed., Bahamas Government Publication, Nassau, Bahamas.

National Institute on Drug Abuse (1984) "Drug Abuse and Drug Abuse Research, The First Triennial Report to Congress," U. S. Government Printing Office Washington, D. C.

National Institute on Drug Abuse (1986) "National Drug Abuse Survey, 1985 Preliminary Report," internal report.

Negrete, J. C., Murphy, H. B. (1967) Psychological deficit in chewers of coca leaf, Bulletin Narcotics 4:11-17.

Neville, M. (1985) Drug abuse in the Bahamas, a report to the Commission of Inquiry, Nassau, Bahamas.

Neville, M. (1985)"Treatment issues, medical and criminal perspectives in the realities of the provision of treatment to cocaine addicts in the Bahamas" presented at the First International Drug Symposium, Nassau, Bahamas, 20-22 November, 1985.

New York Times (1986) Fighting narcotics is everyone's issue now, in This Week in Review, New York Times, Sunday, 10 August, 1986.

Newsweek Magazine (1986) Crack and crime, June 16, 1986.

Nicholson, B., Treece, C. (1981) Object relations and differential treatment response to methadone maintenance, Journal of Nervous Mental Diseases 169:424-429.

Niebuhr (1932) "Moral Man and Immoral Society," Scribners, New York.

Nizama, M. (1979) Sindrome de pasta basica de cocaina, Rev Neuropsiquiat 42:185-208.

Noya, N. (1978) Coca and cocaine: a perspective from Bolivia, in "The International Challenge of Drug Abuse, NIDA Research Monograph No. 19," R. C. Petersen, ed., National Institute on Drug Abuse, Rockville, Md.

O. P. S. (1985) "Investigaciones Medicas," No. 21.

O'Malley, P. M., Bachman, J. G., Johnston, L. D. (1984) Period, age, and cohort effects on substance abuse among American youth, American Journal of Public Health 74:682-688.

Overall, J. E., Brown, D., Williams, J. D., et al. (1973) Drug treatment of anxiety and depression in detoxified alcoholic patients, Arch Gen Psychiatry 29:218-221 (1973).

Paly, D., Jatlow, P., Van Dyke, C., Cabieses, F., Byck, R. (1980) Niveles pasmaticos de cocaina en indigenas peruanos masticadores de coca, in "Cocaina, 1980," F. R. Jeri, ed., Pacific Press, Lima, Peru.

Paly, D., Jatlow, P., Van Dyke, C., Jeri, F. R., Byck, R. (1982) Plasma cocaine concentrations during cocaine paste smoking, Life Science 30:731-738.

Parsons, T. (1957) "The Social System," Free Press, Glencoe, Ill.

Patterson, S. D. (1976) "Alcohol Use and Abuse in the Bahamas: A Socio-Cultural Study," unpublished substantive paper, Brandeis University, Waltham, Mass.

Patterson, S. D. (1978) "A Longitudinal Study of Changes in Bahamian Drinking Habits, 1969-1977," Dissertation, Brandeis University, Waltham,Mass.

Patterson, S. D. (1986) "The Mental Health of the Bahamian Woman," in press.

Patterson, S. D. (1985) "Violence and the Bahamian Woman," paper presented to the Caribbean Federation Mental Health Conference, Nassau, Bahamas.

Paton, S., Kessler, R., Kandel, D. (1977) Depressive mood and adolescent illicit drug abuse: a longitudinal analysis, J Genet Psychol 131:267-289.

Pechacek, T. F., Danaher, B. G. (1979) How and why people quit smoking: a cognitive-behavioural analysis, in "Cognitive-Behavioural Interventions: Theory, Research, and Procedures," P. C. Kendall and S. D. Hollon, eds., Academic Press, New York.

Perez-Gomez, A., Cobos, L., Echeverri de Pardo, G., "Modelo para evaluar la efectividad de los tratamientos de drogadiccion," in press.

Perry, C. L., Jessor, R. (1983) Doing the cube: preventing drug abuse through adolescent health promotion, in "Preventing Adolescent Drug Abuse: Intervention Strategies, NIDA Research Monograph Series No. 47," T. J. Glynn, C. G. Leukefeld, J. P. Ludford., eds., U. S. Government Printing Office, Washington, D. C.

Pickens, R. W., Harris, W. C. (1968) Self-administration of d-amphetamine by rats, Psychomarmacologia (Berlin) 12:158-163.

Pickens, R. W., Thompson, T. (1971) Characteristics of stimulant drug reinforcement, in "Stimulus Properties of Drugs," T. Thompson and R. W. Pickens, eds., Appleton-Century-Crofts, New York.

Pickens, R. W., Thompson, T. (1968) Cocaine-reinforced behaviour in rats: effects of reinforcement magnitude and fixed ratio size, Journal of Pharmacology and Experimental Therapeutics 161:122-129.

Pittman, D. J., Snyder, C. R. (eds.) (1962) "Society, Culture, and Drinking Patterns," John Wiley and sons, New York.

Pollin, W. (1985) The danger of cocaine (editorial) JAMA 254(1):98.

Post, R. M. (1975) Cocaine psychosis: a continuum model, Am. J. Psychiatry 132:225-231.

Post, R. M., Kopanda, R. T. (1976) Cocaine, kindling, and psychosis, Amer Journal of Psychiatry 133(6):627-633.

Post, R. M., Kopanda, R. T., Black, K. E. (1976) Progressive effects of cocaine on behaviour and central amine metabolism in rhesus monkeys: relationship to kindling and psychoses, Biol Psychiatry 11:403-419.

Potter, R. (1969) "War and Moral Discourse," John Knox Press, Richmond.

Quitkin, F. M., Rifkin, A., Kaplan, J., et al. (1972) Phobic anxiety syndrome complicated by drug dependence and addiction, Arch Gen Psychiatry 27:159-162.

Radford, P., Wiseberg, S., Yorke, C. (1972) A study of "main line" heroin addiction, Psychoanal Study Child 27:156-180.

Rawson, R. A., Obert, J. L., McCann, M. J., Mann, A. J. (1986) Cocaine treatment outcome: cocaine use following inpatient, outpatient, and no treatment, in "Problems in Drug Dependence: 1985, NIDA Research Monograph Series No. 67," L. S. Harris, ed., National Institute on Drug Abuse, Rockville, Md.

Reed, B. G., Kovach, J., Bellows, N., Moise, R. (1981) The many faces of addicted women: implications for treatment and future research, in "Drug Dependence and Alcoholism, Vol. I: Biomedical Issues," A. J. Schecter, ed., Plenum Press, New York.

Resnick, R. B., Kestenbaum, R. S., Schwartz, L. K. (1977) Acute systemic effects of cocaine in man: a controlled study by intranasal and intravenous routes, Science 195:696-698.

Resnick, R. B., Resnick, E. B. (1984) Cocaine abuse and its treatment, Psychiatric Clinics of North America, Vol 7,pp 713-728.

Resnick, R. B., Shuysten-Resnick, E. (1976) Clinical aspects of cocaine: assessment of cocaine behaviour in man, in "Cocaine: Chemical, Biological, Clinical, Social and Treatment Aspects," S. J. Mule, ed., CRC Press Cleveland.

Robins, L. N. (1984) The natural history of adolescent drug use (editorial) American Journal of Public Health 74(7):656-657.

Robins, L. N., Smith, E. M. (1980) Longitudinal studies of alcohol and drug problems and sex differences, in "Alcohol and Drug Problems in Women, Vol. V, Research Advances in Alcohol and Drug Problems," O. J. Kalant, Plenum Press, New York.

Rosenstock, I. M. (1960) What research in motivation suggests for public health, American Journal of Public Health 50(2):295-302.

Rounsaville, B. J., Gawin, F., Kleber, H. (1985) Interpersonal Psychotherapy adapted for ambulatory cocaine abusers, Amer J Drug and Alcohol Abuse 11(3 & 4):171-191.

Rounsaville, B. J., Glazer, W., Wilbur, E. H., Weissman, M. M., Kleber, H. (1983) Short-term interpersonal psychotherapy in methadone maintained opiate addicts, Archives of General Psychiatry 40:629-636.

Rounsaville, B. J., Spitzer, R. L., Williams, J. B. W. (1986) Proposed changes in DSM-III substance use disorders: description and rationale, American Journal of Psychiatry 143:463-468.

Rounsaville, B. J., Tierney, T., Crits-Christoph, K., Weissman, M. M., KLeber, H. D. (1982) Prediction of outcome in treatment of opiate addicts: evidence for the multidimensional nature of addicts' problems, Comprehensive Psychiatry 23:462-478.

Rounsaville, B. J., Weissman, M. M., Crits-Christoph, K., et al. (1982) Diagnosis and symptoms of depression in opiate addicts: course and relationship to treatment outcome, Arch Gen Psychiatry 39:151-156.

Rounsaville, B. J., Weissman, M. M., Kleber, H., et al. (1982) Heterogeneity of pscyhiatric diagnosis in treated opiate addicts, Arch Gen Psychiatry 39:161-166.

Saenz, L. N. (1941) El coqueo: factor de hiponutricion, Rev Sanid Policia 1:129-147.

San Quentin Prison Report, 1981.

Schachne, J. S., Roberts, B. H., Thompson, P. D. (1984) Coronary artery spasm and myocardial infarction associated with cocaine use, New England Journal of Medicine 310:1665-1666.

Schiorring, E. (1981) Psychopathology induced by "speed drugs," Pharmacology, Biochemistry, and Behaviour 14(suppl 1):109-122.

Schnoll, S. H., Karrigan, J., Kitchen, S. B., Daghestani, A., Hansen, T. (1985) Characteristics of cocaine abusers presenting for treatment, in "Cocaine Use in America: Epidemiology and Clinical Perspectives, NIDA Research Monograph Series No. 61," E. H. Adams and N. J. Kozel, eds., U. S. Government Printing Office, Washington, D. C.

Schuster, C. R., Fischman, M. W. (1985) Characteristics of humans volunteering for a cocaine research project, in "Cocaine Use in America: Epidemiologic and Clinical Perspectives, NIDA Research Monograph Series No. 61," E. H. Adams and N. J. Kozel, eds., U. S. Government Printing Office, Washington, D. C.

Schuster, C. R., Johanson, C. E. (1980) The evaluation of cocaine using an animal model of drug abuse, in "Cocaine, 1980: Proceedings of the Inter-American Seminar on Medical and Sociological Aspects of Coca and Cocaine" F. R. Jeri, ed., Pacific Press, Lima, Peru.

Seevers, M. H. (1939) Drug addiction problems, Amer Scientist, Sigma XI Quarterly 27:91-102.

Shalloo, J. P. (1941) Some cultural factors in aetiology of alcoholism, Quarterly Journal of Studies in Alcoholism 2:464-478.

Siegel, R. K. (1979) Cocaine smoking, New England Journal of Medicine pp. 300-373.

Siegel, R. K. (1982) Cocaine smoking, Journal of Psychoactive Drugs 14(4):271-359.

Silverman, H. S., Smith, A. L. (1985) Staphylococcal sepsis precipitated by cocaine sniffing, New England Journal of Medicine 312:1706.

Smith, D. E., Buxton, M., Daminom, G. (1979) Amphetamine abuse and sexual dysfunction, in "Amphetamine Use, Misuse, and Abuse," D. E. Smith and D. R. Wesson and M. Buxton, eds., G. K. Hall, Boston.

Smith, G. M., Fogg, C. P. (1979) Psychological Antecedents of Teenage Drug Abuse, in "Research in Community and Mental Health, Vol. 1," R. G. Simmons, ed., JAI Press, Greenwich, Conn.

Spencer, D. J. (1971) Cannabis induced psychosis, International Journal of the Addictions (6) pp.322-326.

Spencer, J. (1970) Alcoholism: How serious is the problem? Bahamas Mental Health Association Magazine, November, 1970.

Spotts, J. V., Shontz, F. C. (1984) Drug-induced ego states: I. Cocaine: phenomenology and implications, Int J Addictions 19(2):119-151.

Spotts, J. V., Shontz, F. C. (1977) "The Lifestyles of Nine American Cocaine Users," National Institute on Drug Abuse, Washington, D. C.

Stanton, M. D., Todd, T. C. (1982) "Family therapy of drug abuse and addiction," Guildford Press, New York.

Stewart, J. (1983) Conditioned and unconditioned drug effects in relapse to opiate and stimulant drug self-administration, Progress in Neuropsychobiology 7:591-597.

Stone, N. S., Fromme, M., Kagan, D. (1984) "Cocaine: Seduction and Solution," Clarkson N. Potter, Inc., New York.

Strategy Council on Drug Abuse (1973) "Federal Strategy for Drug Abuse and Drug Traffic Prevention, 1973," U. S. Government Printing Office, Washington, D. C.

Stripling, J. S., Hendricks, C. (1981) Effect of cocaine and lidocaine on the expression of kindled seizures in the rat, Pharmacol Biochem Behav 14(3):397-403.

Stropp, H. (1981) Toward the refinement of time-limited dynamic psychotherapy, in "Forms of Brief Therapy," S. H. Budman, ed., Guilford Press, New York.

Swisher, J. D., Hu, Teh-wei (1983) Alternatives to drug abuse: some are and some are not, in "Preventing Adolescent Drug Abuse: Intervention Strategies, NIDA Research Monograph Series No. 47," T. J. Glynn, C. G. Leukefeld, J. P. Ludford, eds., National Institute on Drug Abuse, U.S. Government Printing Office, Washington, D. C.

Szaz, T. (1961) "The Myth of Mental Illness," Delta, New York.

Tennant, F. A., Jr., Detels, R., Clarke, V. (1975) Some childhood antecedents of drug and alcohol abuse, Amer Journal of Epidemiology 102:377-385.

Tennant, F. S., Rawson, R. A. (1982) Cocaine and amphetamine dependence treated with desipramine, in "Problems of Drug Dependence, 1982, NIDA Research Monograph Series No. 43," L. S. Harris, ed., National Institute on Drug Abuse, Rockville, Md.

Tennyson, A. L. (1958) Ulysses in "Immortal Poems," O. Williams, ed., Washington Square Press, New York.

Toohey, J. V. (1985) Activities for the Clarification of Values in Drug and Substance Abuse Education: A Manual for the Instructor," Arizona State University, Tempe.

Treece, C., Nicholson, B. (1980) DSM-III personality type and dose levels in methadone maintenance patients, Journal of Nervous Mental Diseases 168:621-628.

Vaillant, G. E. (1973) A twenty-year follow-up of New York narcotic addicts, Archives of General Psychiatry 29:237-241.

Van Dyke, C., Byck, R. (1982) Cocaine, Scientific American, 246: 128-146, March, 1982.

Van Dyke, C., Byck, R. (1977) Cocaine: 1884-1974, in "Cocaine and Other Stimulants," E. Ellinwood and M. Kilbey, eds., Plenum Press, New York.

Van Dyke, C., Ungerer, J., Jatlow, P., Barash, P., Byck, R. (1982) Intranasal cocaine: dose relationships of psychological effects and plasma levels, Int J Psychiatry Med 12(1):1-13.

Vereby, K. (ed.) (1982) Opiods in Mental Illness: Theories, Clinical Observations, and Treatment Possibilities, Ann N. Y. Academy of Sciences 398:1-512.

Vereby, K., Martin, D., Gold, M. S. (1986) Drug Abuse: interpretation of laboratory tests, in "Diagnostic and Laboratory Testing in Psychiatry," M. S. Gold, A. C. Pottash, eds., Plenum Press, New York.

Walsh, M. (ed.) "Interdisciplinary approaches to problems of drug abuse in the workplace," National Institute on Drug Abuse, U. S. Government Printing Office, Washington, D. C., in press.

Washton, A. M. (1985) The cocaine abuse problem in the U. S., Testimony presented before the Select Committee on Narcotics Abuse and Control, U. S. House of Representatives, Washington, D. C., July 16, 1985.

Washton, A. M. (1985) Cocaine abuse treatment, Psychiatry Letter 3:51-56.

Washton, A. M. (1985) Cocaine abusers get outpatient help in special programme, Psychiat News 20(9):6-24.

Washton, A. M. Crack, the latest threat in addiction, Medical Aspects of Human Sexuality, in press.

Washton, A. M., Gold, M. S, Changing patterns of cocaine use in America: a view from the National Hotline, 800-COCAINE, Advances in Alcohol and

Substance Abuse, in press.

Washton, A. M., Gold, M. S., (1984) Chronic cocaine abuse: evidence for adverse effects on health and functioning, _Psychiatric Annals_ 14:733-743.

Washton, A. M., Gold, M. S., (1986) "Cocaine Treatment Today," American Council on Drug Education, Bethesda.

Washton, A. M., Gold, M. S., Pottash, A. C. (1983) Intranasal cocaine addiction, _The Lancet_ 11:1378.

Washton, A. M., Gold, M. S., Pottash, A. C., Semlitz, L. (1984) Adolescent cocaine abusers, _The Lancet_ 11:letter.

Washton, A. M., Stone, N. S. (1984) The human cost of chronic cocaine use, _Medical Aspects of Human Sexuality_ 18:122-130.

Washton, A. M., Tatarsky, A. (1984) Adverse effects of cocaine abuse, _in_ "Problems of drug dependence, 1983, NIDA Research Monograph Series No. 44," L. S. Harris, ed., U. S. Government Printing Office, Washington, D. C.

Weill, A. (1972) "The Natural Mind," Houghton Mifflin, Boston.

Weiss, R. D., Mirin, S. M., Michael, J. L., Sollogub, A. (1983) "Psychopathology in Chronic Cocaine Abusers," paper presented at the 136th Annual Meeting of the American Psychiatric Association, New York.

Weissman, M. M., Slobetz, F., Prusoff, B., et al. (1976) Clinical depression among narcotic addicts maintained on methadone in the community, _Amer Journal of Psychiatry_ 133:1434-1438.

Wesson, D. R., Smith, D. E. (1985) Cocaine: treatment perspectives, _in_ "Cocaine Use in America: Epidemiologic and Clinical Perspectives, NIDA Research Monograph Series No. 61," E. H. Adams and N. J. Kozel, eds., U. S. Government Printing Office, Washington, D. C.

Wetli, C. V., Fishbain, D. A., Cocaine-induced psychosis and sudden death in recreational cocaine users, _Journal of Forensic Sciences,_ JFSCA, Vol. 30, No. 3, July, 1985.

Wetli, C. V., Wright, R. K. (1979) Death caused by recreational cocaine use, _Journal of the American Medical Association_ 241:2519-2522.

W. H. O. (1980) "Drug Problems in the Socio-Cultural Context: A Basis for Policies and Programme Planning," G. Edwards and A. Aris, eds., W. H. O. Paris, France.

W. H. O. (1952) Technical Report Series, World Health Organization No. 48, Expert Committee on Mental Health, A/C Sub. Committee, 2nd Report.

Whitehead, A. N. (1962) "The Aim of Education and Other Essays," Ernest Benn, London.

Whitehead, A. N. (1925) "Science and the Modern World, MacMillan, New York.

Wieder, H., Kaplan, E. H. (1969) Drug use in adolescents: psychodynamic meaning and pharmacogenic effect, _Psychoanalytic Study of the Child_ 24:399-431.

Wikler, A., Pescor, R. (1967) Classical conditioning of a morphine abstinence phenomenon reinforcement of opiod drinking behaviour and "relapse" in morphine-addicted rats, _Psychopharmacologia_ 10:255-284.

Wilbur, R. (1986) A drug to fight cocaine, _Science_ 86:42-46.

Wilsnack, S. C. (1973) Sex role, identity, in female alcoholism, _Journal of Abnormal Psychology_ 82:253-261.

Wilson, E. (1983) "What Is To Be Done About Violence Against Women," Penguin, London.

Wise, R. (1984) Neural mechanisms of the reinforcing action of cocaine, _in_ "Cocaine: Pharmacology, Effects, and Treatment of Abuse, NIDA Research Monograph Series 50," J. Grabowski, ed., U. S. Government Printing Office, Washington, D. C.

Woods, J. (1977) Behavioural effects of cocaine in animals, _in_ "Cocaine: 1977, NIDA Research Monograph Series No. 13," R. C. Petersen and R. C. Stillman, eds., National Institute on Drug Abuse, Rockville, Md.

Wood, J. (1982) Boys will be boys, _New Socialist,_ No. 5, May/June, 1982.

Woody, G. E., Luborsky, L., McLellan, A. T., O'Brien, C. P., Beck, A. T., Blaine, J., Herman, I., Hale, A. (1983) Psychotherapy for opiate addicts: Does it help? _Arch Gen Psychiatry_ 40:639-645.

Woody, G. E., O'Brien, C. P., Rickels, K. (1975) Depression and anxiety in heroin addicts: a placebo-controlled study of doxepin in combination with methadone, Amer Journal of Psychiatry 132:447-450.

Wurmser, L. (1978) "The Hidden Dimension," Jason Aaronson, New York.

Wurmser, L. (1972) Methadone and the craving for narcotics: observations of patients on methadone maintenance in psychotherapy, in "Proceedings of the Fourth National Methadone Conference, San Francisco, 1972," National Association for the Prevention of Addiction to Narcotics, New York.

Wurmser, L. (1974) Psychoanalytic considerations of the aetiology of compulsive drug use, J Amer Psychoanal Assoc 22:820-843.

Yeats, W. B., (1958) The Second Coming, in "Immortal Poems," O. Williams, ed., Washington Square Press.

Yalom, I. D. (1975) "The Theory and Practice of Group Psychotherapy, Second Edition," Basic Books, New York.

Zahler, R., Wachtel, P., Jatlow, P., Byck, R. (1982) Kinetics of drug effect by distributed lags analysis: an application to cocaine, Clin Pharm and Therap 31(6):775-782.

Zuckerman, M., Neary, R. S., Brustman, B. A. (1970) Sensation-seeking scale correlates in experiences (smoking, drugs, hallucinations, and sex) and preference for complexity, in "Proceedings of the 78th Annual Convention of the American Psychological Association, Vol. 5," American Psychological Association, Washington, D. C.

INDEX

Abortion, 18, 188
Absorption efficiency, 16, 17, 27,
    36, 112, 167, 193, 200, 201
Absenteeism, 24, 52, 56, 210, 216
Abstinence, 56, 57, 58, 59, 60, 72,
    76, 77, 78, 79-80, 81, 83ff,
    96, 97, 98, 101, 102, 122,
    161, 162, 167, 177, 179, 186,
    187, 188, 189, 195, 197, 221
Abstinence Violation Effect, 84, 86
Abuse potential, 16, 17, 41
Academic motivation, 22, 136, 185
Acupuncture, 4, 161, 164
Addiction, xii, 3, 4, 12, 15, 21,
    66, 67, 69, 71, 76, 122, 127,
    131, 135, 163, 188, 223
Addiction potential, xi, 16, 17, 21,
    36, 37, 50, 54, 55, 56, 61,
    65, 108, 111, 112, 120, 121,
    122, 125, 126, 140, 164, 167,
    168, 169, 192, 194, 207, 209
Addiction Research Foundation, 40,
    209
Addiction Severity Index, 78
Administration routes, 15-17, 27,
    47, 77, 112, 195 (see also
    Modes of Use)
Adolescents, 24, 51, 52-53, 112,
    115, 194, 211
Adulterants, 27, 28, 47, 141, 195,
    200
Adverse consequences, 33, 45, 50,
    56, 76, 82, 214
Affect states, 66, 67, 68, 69, 70,
    73, 87, 89, 92, 101, 121
Affective disorders, 67ff, 78
Aftercare, 57, 95
Aggression, 18, 24, 68, 70, 71,
    120, 133, 179, 193, 194
AIDS, 16, 107, 112, 113, 172
Alcohol, 35, 37, 39, 49, 52, 54, 55,
    56, 57, 58, 70, 71, 77, 119,
    126, 130, 134, 135, 137, 141,
    146, 151, 161, 163, 180,
    182ff, 191, 197, 201, 203,
    210

alcohol addiction, 9, 69, 75, 76, 82,
    119, 122, 127, 128, 145, 146,
    147, 150, 155, 156, 162, 167,
    181, 184, 210
alcohol consumption, 121, 133, 148-159,
    185ff, 188, 191, 194, 200
Alcoholics Anonymous, 40, 60, 163, 173,
    223
Alternatives to drug use, 80, 98, 210,
    215, 217
Ambivalence, 80, 92, 139, 163, 211, 215
Ammonia, 27, 200
Amphetamines, 36, 38, 46, 58, 67, 70,
    141, 167, 180
Anaesthetic, 9, 33, 46, 112, 167, 193
Angel Dust, 120
Anhedonia, 17, 19, 55, 56, 81, 112,
    187, 208
Anorexia, 18, 130, 177, 178, 179, 185,
    194, 201
Antidepressants, 40, 68, 111, 122, 197,
    203
Antisocial personality, 69, 78, 79, 93,
    95, 179, 181, 182, 185
Anxiety, 15, 19, 20, 47, 49, 60, 68,
    92, 96, 121, 170, 171, 177, 178,
    179, 200, 202, 203
Aphrodisiac, 9, 46, 120, 191, 192
Argentina, 177
Armed Forces, U. S., 41
Assessment, 25, 42, 46, 77, 156, 179
Associations, 24, 49, 53
Associative conditioning, 58, 86 (see
    also conditioning)
Attention patterns, 22, 47, 49, 53, 60,
    72, 139, 143, 170, 179, 184,
    185, 194
Attitudes, 24, 29, 62, 87, 115, 154,
    162, 172, 210, 212, 216
Availability, 23, 34, 45, 46, 61, 62,
    65, 77, 107, 108, 110, 113, 115,
    121, 125, 131, 140, 167, 168,
    178, 207, 208, 209, 217
Avitaminosis, 18, 24, 121, 129, 172
Awareness, 29, 42, 48, 99, 167

245

B cells, 39
Bahamas, 3, 4, 11, 16, 17, 18, 19, 25, 54, 62, 107, 108, 109, 110 119, 126, 209, 216, 217, 222
Baking soda, 16, 25, 28, 54
Barbiturates, 35, 58, 77, 180, 200, 203
Base house, 17, 110, 121, 129, 135
Basuco, 4, 5, 15, 168, 199-204
Behaviour, 70, 72, 76, 85, 96, 169, 170, 179, 188, 192, 196, 197, 212
Behaviour modification, 24, 40, 82ff, 172, 196, 214
Bellak and Hurvich Interview Rating Scale for Ego Functioning, 67
Bellak Ego Functioning Interview, 69
Benzodiazepines, 58, 180, 185
Bias, Len, 11
Binge, 17, 18, 19, 20, 24, 28, 38, 49, 55, 56, 76, 79, 96, 137, 177, 180, 182, 183, 186, 208
Bio-psycho-social, 170
Blacks, 9, 10, 155
Blood levels, 37, 55, 112, 167, 168
Bolivia, 3, 5, 10, 15, 167, 168, 169, 191ff, 209
Borderline personality, 78, 93, 94
Brazil, 177
Bromocryptine, 60, 172
Bush, George, 12

Caffeine, 9
Camoke, 3, 16, 20, 120, 129
Canada, 108, 133, 209
Cannabis, 120, 130, 180, 181, 182, 184, 185, 197
Cardiovascular effects, 18, 29, 37, 38, 55, 171, 177, 188, 189
Caribbean Institute on Alcohol and Other Drug Problems, 173
Central de Policia, 178
Central nervous system, 18, 38, 112, 141, 201
Chewing, 7, 8, 15, 192, 199 (see also Modes of Use, chewing)
Child abuse, 21, 139, 151, 153
Children, 113, 194, 208, 209, 211, 217
Chlorpromazine, 195
Civil liberties, 42, 210, 215
Clinical observation, 126, 142, 201
Clinical research, 126
Clonidine, 203
Coca-Cola, 9
Coca dollars, 5, 192, 200, 208
Coca leaf, 7, 8, 12, 15, 27, 46, 192, 199, 200
Coca paste, 4, 5, 11, 12, 15, 27, 36, 168, 177ff, 192-195, 207 209

Coca plant, 7, 46
cultivation, 5, 10, 11, 177, 192
Cocaine, 7, 70, 112, 161, 209
alkaloid, 7, 8, 12, 15, 27, 167
"bug," 24, 129, 171, 183
camp, 17, 19, 20, 21, 121, 135
cartel, 11, 12, 62
cost, 10, 12, 16, 24, 27, 28, 34, 45, 46, 49, 51, 54, 61, 62, 107, 108, 110, 113, 121, 125, 131, 133, 134, 137, 142, 152, 167, 268, 183, 200, 208, 209
freebase, vii, xi, 4, 11, 12, 16, 20, 24, 25, 27, 28, 54, 107, 112, 113, 129, 131, 133, 137, 139, 140, 141, 142, 152, 168, 208, 221
history, vii, 4, 7ff, 12, 33ff, 107ff, 119, 120, 191ff
hydrochloride, 3, 4, 11, 12, 15, 16, 25, 27, 36, 46, 54, 83, 107, 108, 110, 113, 120, 121, 125, 129, 131, 133, 137, 140, 167, 168, 177, 191, 192, 200
laws, 10, 12, 30, 114, 122, 192, 199
as narcotic, 10, 192
and narcotic addiction, 9, 33, 191
psychosis, 3, 4, 19, 55, 60, 130, 139, 202, 217
sulphate, 15, 27, 168, 192, 200
Cocaine Anonymous, 40, 58, 59, 60, 101
Cognitive behaviour, 4, 40, 81, 84
Cognitive effects, 141
Cohen, Sidney, 4, 5, 16, 27, 161, 162, 209
Cohort effect, 34
Colombia, 3, 10, 15, 108, 168, 199-204
Community commitment, xii, 157, 207, 208, 210, 215, 217
Community Psychiatry Clinic, 108, 109, 126, 127, 128, 134
Compensation, 22
Compulsive treatment, 78, 122, 215
Compulsive use, 15, 19, 28, 29, 40, 42, 49, 55, 56, 65, 67, 73, 96, 126, 170, 177, 178, 179, 180, 186, 189, 192, 197, 208, 214, 221
Conditioning, 77, 82, 83, 89, 96, 99, 173, 188
Confidentiality, 100, 114
"Conning," 142
Consumer countries, 5 (see User Countries; individual countries)
Consumption patterns, 27, 192, 193 (see also Drug Use Patterns)
Contaminants, 15, 27, 168, 192, 200 (see also Adulterants)
Contingency contracting, 40, 80, 82, 95, 196, 214
Contingency management, 101
Contracting, 57, 58

Convulsions, 24, 25, 28, 38, 171, 179, 184, 187, 188, 202
Coping, 29, 58, 76, 81, 83, 84, 85, 86, 101, 137, 138, 147, 154, 164, 216
Corruption, 5, 29, 202, 208, 209
Counselling, 56, 58, 156
Court referrals, 78, 109, 129
Crack, xi, xii, 3, 4, 11, 16, 19, 20, 25, 27, 28, 36, 46, 54-56, 62, 66, 77, 107, 110, 120, 129, 139, 140, 207, 208
Crack house, 17 (see also Base House)
Crash, 46, 49, 55, 56, 77, 79, 81, 122, 137, 179, 188, 210, 211
Craving, 17, 19, 28, 36, 39, 46, 50, 57, 58, 59, 60, 65, 76, 77, 80, 81, 83, 85, 86, 95, 96, 97, 101, 114, 115, 121, 122, 135, 136, 137, 151, 164, 170, 172, 178, 179, 186, 187, 189, 193, 195, 196, 209, 221
Creativity, 10
Crime, 10, 139, 154, 168, 185, 223
Crime rates, 3, 21, 28, 202
Criminal behaviour, 28, 29, 154, 187, 193, 195, 203, 209, 216
Crisis, 129-130
Crops, destruction of, 11, 192, 209
Cues, 77, 81, 82, 84, 85, 86, 89, 94, 96, 98, 99, 101, 155, 179, 209

Dealing, 21, 28, 50, 52, 53, 54, 101, 136, 194
Deaths, drug-related, 9, 11, 16, 18, 33, 34, 37, 38, 47, 61, 111, 120, 126, 130, 131, 139, 169, 171, 177, 186, 187, 188, 193, 210, 221
Decompensation, 19, 22
Delivery rate, 16, 17, 27, 39, 46, 55, 129, 140, 208
Delusion, 20, 22, 55, 139, 141
Denial, 56, 58, 59, 68, 78, 95, 98, 130, 154, 155, 162, 207, 208, 215, 223
Dependence
  cocaine, 56, 128, 133, 137, 164, 167, 169, 189
  definition, 36, 186
  drug, 127, 128, 141, 155, 161, 169
Depressants, 52
Depression, 15, 20, 21, 22, 60, 66, 68, 69, 71, 72, 78, 85, 88, 89, 97, 101, 111, 112, 115, 122, 128, 130, 135, 137, 139, 154, 162, 177, 179ff, 186, 189, 193, 195, 202, 217
  cocaine, 17, 47, 49, 51, 55, 56, 57, 77, 170, 187, 188, 195,

Depression
  cocaine (continued) 208, 221
Desipramine, 114, 122
Detoxification, 60, 76, 162, 167, 172, 193, 195, 202, 203
Diagnosis, 25, 38, 40, 41, 55, 60, 68, 71, 78, 89, 93, 129, 178, 179
Diazepam, 195, 200, 203
Dilutants, 27 (see also Adulterants, Contaminants)
Disease theory of drug abuse, 163, 170, 178
Disco Biscuit, 120
Dopamine, 17, 20, 60, 112, 122, 172, 208
Dos de Mayo, 178
Dose, 36, 37, 46, 50, 55, 61, 69, 140, 141, 168, 171, 179, 186, 187, 195
Doxepin, 68
Drinking patterns, 149-159
Driving behaviour, 18, 35, 50, 52, 53, 180, 188, 202, 215
Drug abuse, 127, 128, 129, 130, 136
Drug Abuse Warning Report (DAWN), 39, 126
Drug of choice, 65-73, 107
Drug dealers, 17, 27, 52, 54, 59, 79, 86, 137, 151, 187, 188 (see also Pushers)
Drug education, 42, 57, 58, 62, 81, 114, 196, 202, 207, 210, 213, 216, 217 (see also education)
Drug Enforcement Agency, 11, 82
Drug-free lifestyle, 57, 59, 62, 151, 157, 162, 167, 196, 214, 222
Drug-free treatment, 57, 194
Drug paraphernalia, 25, 27, 59, 79, 80, 84, 196
Drug progression, 35, 185ff, 211 (see also Gateway drugs, Stepping-stone hypothesis)
Drug use detection, 42, 194
Drug use patterns, 29, 40, 48, 49, 50, 50, 51, 52, 53, 55, 58, 73, 126, 129, 131, 134, 135, 147, 150, 151, 178, 180, 192, 195, 200
DSM-III, 69, 76, 77, 169, 178, 179, 217
DSM-IIIr, 76
Dysphoria, 15, 17, 19, 46, 49, 55, 60, 66, 69, 70, 72, 77, 112, 121, 122, 141, 170, 171, 177, 179, 180, 186, 189, 195, 201, 208, 221

Economic assistance, 11
Economics, xii, 29, 41, 145, 147, 202, 208, 217
Edison, Thomas, 8
Education, 11, 12, 22, 29, 111, 170, 214, 215, 221 (see also Drug education)
Effects, 33, 37, 40, 46, 53, 55, 80,

Effects (continued) 112, 140, 170ff,
        179, 186, 187, 189, 192
  chewing, 8, 15
  freebasing, 23-25, 36, 39, 46,
        55, 65, 172, 187
  intravenous, 16, 29, 38, 46, 72,
        81n, 187
  smoking, 15, 177ff, 193-195, 201-
        202
  snorting, 16, 23, 24, 38, 184,
        187
Ego, 22, 66, 67, 68, 93
Empathy, 210
Employee drug use, 24, 25, 40, 41,
        52, 53, 56, 78, 96, 109, 129,
        135, 136, 150, 154, 203, 210,
        215, 216
Enablers, 58, 99, 100
Endorphins, 69
Enforcement, viii, xii, 12, 17, 25,
        30, 62, 108, 110, 202, 207,
        209, 210, 213, 214, 215, 217
Epidemic, vi, vii, xi, 3, 4, 12, 45,
        48, 50, 61, 108, 125, 130,
        156, 161, 167, 168, 177,
        189, 191, 211
  coca paste, 11, 207
  definition, 110
  freebase, xii, 3, 11, 65, 120,
        125, 131, 140, 207, 217, 223
Epidemiology, 4, 34, 107, 109, 126,
        145, 215
Equador, 7, 168
Erlenmeyer, Albrecht, 9
Erythroxylon coca, 7, 46, 167
Erythroxylon novagranatense, 7, 199
        Truxillense, 7
Esterases, 18
Ether, 27, 36, 54, 168, 188
Ethical fragmentation, 19, 24, 141,
        142, 154, 187, 193
Ethics, 42, 196, 207-219
Euphoria, 10, 15, 17, 28, 46, 47,
        54, 55, 56, 60, 66, 72, 80,
        112, 113, 122, 129, 139, 162,
        170, 186, 189, 197, 201, 208,
        221
Europe, 8, 9, 33, 108, 167
Evaluation, 58, 70, 75, 79, 115
Experience seeking, 22
Experiments, animal, xi, 33, 37, 38,
        81, 82, 83, 126, 170, 208,
        221

Fair Oaks Hospital, 172
Familization Therapy of Co-depen-
        dence Programme, 161
Family centred treatment, 25, 57,
        58, 59, 98, 156, 162, 195,
        202, 203
Family drug use, 23, 99, 145, 149

Family, effects on, 21, 41, 98, 109,
        136, 145, 146, 153, 202, 203,
        210, 211, 215, 216, 217
Family influence, 71, 98, 129, 130,
        136, 162
Family relationships, 21, 77, 98-100,
        121, 136, 137, 195, 222
Family violence, 21, 70, 71, 195
Father influence, 71, 147
Female treatment needs, 145-159
First International Drug Symposium,
        vii, xi, xiii, 4, 65
Follow-up, 59, 61, 137, 172, 188
Food and Drug Act, 10
Frank, Jerome, 173
Frankl, Victor, 15
Freebase, see Cocaine, freebase
Freebasing, 3, 4, 17, 19, 54, 55, 61,
        77, 79, 120, 126, 136, 137,
        151, 154, 168, 195
Freud, Sigmund, 8, 9, 191

Gaedecke, Friedrich, 8
Gangs, 193
Gastrointestinal system, 18
Gateway drugs, 35, 211 (see also Drug
        progression, Stepping-stone
        hypothesis)
General Hospital, Trinidad, 168
Geographic factors, 10, 12, 50, 51,
        108, 119
        in trafficking, 11, 12, 108, 149
Glandular system, 18
Grandiose self, 133
Group therapy, 57, 59, 70, 95-102, 115,
        122, 154, 162, 164, 202, 203

Hallucinations, 15, 19, 22, 55, 109,
        129, 130, 137, 141, 171, 177,
        179, 183, 184, 186, 189, 202
Hallucinogens, 10
Haloperidol, 195
Halstead, William, 9, 191
Hammond, W. A., 9
Harrison Act, 10, 192
Health belief model, 111-112, 114
Heroin, 18, 29, 35, 39, 46, 47, 57,
        65ff, 107, 113, 121, 126, 161,
        168, 170, 180, 181, 182, 192,
        201, 208
High, viii, 3, 16, 17, 21, 22, 23, 28,
        36, 42, 46, 58, 96, 113, 121,
        171, 208, 210, 216, 221
High-base craziness, 19
Hit, 19, 22, 121, 129, 152, 208
Holding environment, 94
Homicide, 18, 20, 56, 130, 171, 177,
        188, 189, 195, 202
Hospitalization, 40, 56, 57, 58, 76,
        77, 95, 120, 136, 148, 151, 194,
        195, 203

Hotline, 114
  800-COCAINE, 4, 11, 45-62, 107
Hypersensitivity, 18

Ibsen, Henrik, 8
Ideal Observer, 207ff
Imipramine, 114
Impotence, 9, 47, 49, 55, 120, 135,
  170, 202
Incas, 7, 167
Indians, 8, 15, 33, 167
Inderal, 162
Indicators of abuse, 15, 121 (see
  also Signs, Symptoms)
Infections, 18, 19, 39, 177, 179,
  187, 188, 189
Illicit drugs, viii, 34, 66, 107,
  115, 125
Inpatient treatment, 40, 57, 76, 77,
  79, 101, 122, 133, 137, 149,
  152, 164, 195, 196, 216, 217
Insomnia, 28, 56, 57, 129, 135,
  170, 171, 178, 179, 200,
  201, 202
Instinctual fragmentation, 19, 47,
  49, 142, 170, 187, 193, 221
Interpersonal relationships, 86, 87,
  97, 99, 122
Intervention, 41, 79, 88ff, 100,
  101, 108, 161
Intoxication, 9, 19, 52, 76, 78,
  120, 140, 141, 143, 179, 180,
  184, 185, 186, 191, 203
Intravenous use, 12, 77, 113, 140,
  154, 168, 172, 178, 180,
  187, 188, 191, 195
Isolation, 24, 47, 55, 67, 69, 81,
  97, 101, 139, 141, 153, 187,
  202, 216
Itching, 24, 129 (see also Skin,
  Cocaine "bug")

Jamaica, 3

Kandel, Denise, 35
Killer cells, 39
Kindling, 139

LSD, 119, 133, 180, 184
Labeling, 86
Larco Herrera, 178
Lawn, John, 11
Legree, Simon, 167
Lidocaine, 46
Lifestyle, 60, 71, 79, 85, 95, 121,
  142, 162, 172
Lincoln Hospital Acupuncture Pro-
  gramme, 161
Lithium, 40, 122, 141
Locus of control, 79, 80, 83
Loevinger Sentence Completion, 69

Longitudinal study, 149, 155
Low, 23, 121
Low-base craziness, 19, 20
Low-low depression, 20 (see also
  Depression, cocaine)

Malnutrition, 25, 172, 177, 179, 187,
  189
Mantegazzo, Pablo, 8
Manufacture, freebase, 28, 36, 54, 55,
  107, 108, 129, 188, 200
Mariani, Angelo, 8
Marijuana, 10, 12, 15, 22, 34, 35, 36,
  39, 41, 54, 55, 56, 57, 58, 107,
  119, 126, 130, 133, 134, 135,
  136, 137, 141, 148, 151, 152,
  161, 184, 188, 192, 200, 201,
  203, 211
Marketing, 16, 27, 28, 46, 54, 108,
  110, 113, 129, 131, 140, 168,
  192, 199, 215
Media, viii, 34, 36, 107, 111, 120,
  191, 216
Medical treatment, 66, 162, 191, 194
Mental illness, 22, 69, 134, 195
Mental patients, 22, 142
Meperidine, 180
Mescaline, 180
Methadone, 66, 69
Methamphetamine, vi, 47
Methaqualone, 120, 133, 200, 201
Methylphenidate, 40, 72
Modes of use, 4, 12, 15, 37, 39, 46,
  50, 107, 126, 129, 130, 131, 171
  chewing, 15, 126, 167
  freebase, 38, 49, 50, 95, 107, 126,
  129
  intravenous, 16, 49, 50, 95, 107, 126
  smoking, 177, 193ff
  snorting, 16, 49, 50, 95, 107, 126,
  167
Moral fragmentation, see Ethical fragmentation
Morbidity, 5, 33, 178
Morphine, 9, 192
Mortality, 5, 188
Motivation, 56, 61, 98, 102, 114, 135
  for treatment, 56, 57, 59, 60, 77, 78,
  89, 114, 161, 162, 163, 195
  for use, 42, 65ff, 155, 169, 170, 179,
  180, 197, 204, 221
Multidrug use, 23, 35, 37, 39, 47, 49,
  52, 54, 58, 70, 72, 77, 109, 120
  127, 128, 130, 134, 138, 141,
  143, 151, 162, 170, 177, 180ff,
  192, 195, 200, 203
Multi-family therapy, 98-100, 101, 102,
  196

Narcissism, 93, 197, 210
Narcissistic personality, 78, 93-94

Narcotic Drug Import and Export
    Act, 10
Narcotics, 10, 66ff, 201
Narcotics Anonymous, 163, 173,
    223
National Drug Council, 11, 211
National Institute on Drug Abuse,
    4, 34, 39, 40, 42, 126
National Institute of Mental
    Health, 141
National Narcotics Council, 202
National Task Force, 3, 11
Negative reinforcement, 17, 49,
    214, 221
Neurosis, 88, 170, 173
Neurotransmitters, 55, 65, 112,
    122, 139, 172
New York Anti-Cocaine Bill, 10
Nicotine, 126
Niemann, Albert, 8, 167
No-barrier drug, 20, 21, 34, 47,
    142, 208
Noradrenaline, 112, 172
North America, 11, 108, 110, 121,
    167
Nutrition, 121, 162, 194, 195

Occasional use, see Social use
On Coca, 9
Opiates, 35, 38, 49, 54, 58, 66ff,
    75, 76, 77, 221
Opium, 9, 10, 162
Outpatient treatment, 22, 38, 40,
    56, 57, 76, 77, 78, 87, 115,
    122, 125, 126, 194, 195,
    196, 214, 216, 217
Overdose, 18, 55, 61, 139, 189,
    191, 194

Palm Beach Institute, 161
Panic behaviour, 20, 70
Paranoia, 49, 53, 56, 115, 129,
    136, 139, 179, 182, 193, 202
Paranoid
  behaviour, 24, 142
  delusions, 15, 119, 171
  ideation, 16, 18, 19, 29, 47,
    130, 135, 179, 182, 184,
    193, 194
Parental drug use, 23
Pathology, 22, 93, 163, 170, 178
Peer pressure, 23, 66, 112, 149,
    152, 221
Pemberton, John Styth, 9
Pentazocine, 180, 181, 182
Personality
  addiction-prone, 93, 171, 179
  change, 24, 193
  disorder, 69, 72, 78, 88ff, 93,
    101, 115, 122, 177, 179, 189
Peru, 3, 5, 7, 10, 15, 167, 168,

Peru (continued) 169, 178ff
Pharmacology, 37, 39, 40, 46, 78, 101,
    122, 197
Pharmacotherapy, 162
Phencyclidine, 35, 39, 58, 107, 120,
    161
Phenothiazine, 195, 203
Philip II of Spain, 8
Physiological effects, 21, 38, 40, 46,
    121, 122, 146
Pitillo, 5, 15, 192, 194
Pizarro, Francisco, 8
Pleasure centres, 17, 112, 122, 169,
    170, 172, 208
Pleasure threshold, 17
Poma de Ayala, Felipe Guaman, 8
Pope Leo XIII, 8
Potency, 65, 140 (see also Purity)
Precursers to drug abuse, 4, 15, 21,
    211, 216
Predisposition, 65, 66, 68, 72
Pregnancy, 18, 81n, 129, 188, 189
Prevalence, 34, 35, 41, 45, 48, 54,
    61, 101, 121, 126, 127, 128,
    130, 133, 191, 201, 203, 208
Prevention, 5, 29, 41, 42, 112, 113-
    115, 170, 189, 202, 211, 215,
    217
Princess Margaret Hospital, 109
Prison, 70, 120, 145, 162, 163, 203,
    215
Prisoners, 70, 162, 178
Procaine, 47
Producer countries, vii, xi, 3, 132,
    193
Production, 5, 61, 108, 177, 192, 199
Progression, drug use, 178, 179ff, 186
Prohibition, 9, 191
Promiscuity, 24, 51, 153
Propanolol, 162, 195
Prostitution, 17, 19, 20, 137, 142,
    150, 152, 153, 154, 210
Pryor, Richard, 107
Psychiatric disorders, 66, 129, 140,
    141, 186, 191, 195
Psychoactive drugs, 80, 113, 203, 215
Psychoanalysis, 66, 94
Psychological addiction, 29, 34, 121,
    126, 164, 169, 172, 192, 221
Psychological effects, 20, 37, 171,
    178, 179
Psychopathetization, 186, 187, 189
Psychopathology, 19, 22, 40, 65, 68,
    69, 77, 78, 88, 122, 139, 141,
    143, 184, 216
Psychopharmacology, 40, 67, 68, 69
Psychosis, 69, 89, 140, 142
  drug-induced, 19, 47, 77, 119, 120,
    133, 135, 141, 142, 143, 177,
    179, 186, 189, 193, 195, 202
Psycho-social process, 50, 55, 56, 77,

Psycho-social process (continued)
        78, 95, 101, 145, 149, 156
Psychotherapy, 4, 69, 75-106, 115,
        122, 196, 197
Psychotropic Convention, 192
Psychotropic drugs, 22, 66, 69, 122
Public health, 4, 5, 42, 61
    models, 108, 110-115
Pulmonary effects, 29, 55, 81n,
        172, 182
Purity, 37, 108, 113, 121, 131,
        140, 168, 170, 200, 209, 217
Pushers, 21, 22, 24, 108, 110, 113,
        125, 129, 131, 151, 209,
        210, 214
Pyridoxine, 195

Quaaludes, 120, 133

Ready to smoke freebase, xii, 11,
        16, 28, 46, 54, 77, 120,
        129, 168, 208
Reagan, Ronald, 11
Recovery, 57, 58, 60, 96, 139
Recreational use, see Social use
Rehabilitation, 114, 137, 139, 161,
        162, 164, 172, 187, 197,
        204, 210, 213, 214, 214, 216
        (see also individual subject)
Reinforcing effect of cocaine, 17,
        29, 34, 36, 41, 45, 80, 81-
        82, 113, 170, 191, 192, 193,
        194, 208, 221
Relapse, 57, 59, 60, 77, 80, 81,
        82, 85, 86, 97, 122, 137,
        139, 141, 151, 153, 154,
        161, 164, 167, 195, 196,
        197, 204, 209, 217
Relapse Prevention, 82-87
    strategies, 59, 76, 79, 82-87,
        94, 96, 99, 101, 137, 196
Relationships, 56, 59, 68, 70,
        88, 90, 92, 94, 96, 136,
        147, 150, 152, 153, 155,
        156, 164, 187, 193, 202, 212
Religiosity, 23, 115, 173, 186, 216
Religious uses, 7, 12, 15, 167
Research, 38, 39, 40, 41, 48, 76,
        78, 86, 121, 126, 145, 150,
        153, 162, 169, 197, 201,
        211, 214, 215, 217
Residential setting, 69, 79, 137
Resistence, 89, 90, 100
Resnick, 178
Respiratory effects, 17, 38, 55,
        177, 179, 188, 189, 202
Reward centres, 17, 20, 38, 169
        (see also Pleasure centres)
Risk factors, 23, 34, 35, 36, 45,
        112, 156, 170
Risk situations, 83, 84, 85, 96, 196

Rivotril, 200
Rock, 16, 18, 25, 54, 108, 110, 113,
        208
    house, 17, 28 (see also Base house)
Rogers, Donald, 11
Role models, 95, 210
Rorschach, 69
Royal Commission of Inquiry, 11
Run, 29 (see also Binge)
Rush, see Hit, Euphoria

Sadism, 70, 71
St. Ann's, Trinidad, 168
St. Augustine's Monastery, 164
San Antonio, 178
San Isidro, 178
San Juan de Dios, 178
San Quentin, 162, 163
Sandilands Rehabilitation Centre, 109,
        119, 126, 127, 128, 130, 133,
        134, 142, 143
Schizophrenia, 89, 128, 133, 140, 142,
        143, 148, 177, 179, 186, 189
Schuyton-Resnick, 178
Sedatives, 52, 54, 55, 60, 68, 70,
        161, 203, 221
Selection, drug of choice, 65, 66, 70,
        107, 112
Self-administration, 36, 46, 54
Self-esteem, xii, 22, 29, 59, 66, 71,
        72, 87, 93, 97, 135, 152, 153,
        154, 155, 156, 193, 194, 216
Self-help, 40, 59, 66
    groups, 40, 57, 58, 59, 100, 173, 203
    self-medication, 4, 22, 55, 58, 65-73,
        78, 87, 89, 111, 115, 195, 217
Sexual practices, 137, 184
Sexuality, 97, 120, 142, 154, 171, 185,
        196
Shanghai Conference, 10
Signs, 24, 42, 55, 108, 121, 194
Single parents, 153, 155
Skin, 18, 24, 25, 135, 193
Sleeping pills, 49
Slip, 58, 80, 81, 82, 83ff, 101
Snorting, 3, 12, 16, 19, 36, 37, 46, 54,
        56, 61, 72, 112, 120, 126, 129,
        133, 136, 137, 152, 168, 170,
        172, 177, 178, 183, 185, 191,
        200
Social costs of drug use, 8-12, 21, 34,
        45, 53, 109, 129, 131, 139, 145,
        202, 207, 211
Social skills, 40, 97
Social user, 29, 30, 45, 47, 57, 58,
        80, 111, 133, 135, 143, 178,
        180, 208, 216
Socio-cultural factors, 15, 39, 62,
        146-159, 192, 196, 199, 203
Socio-economic factors, 20, 39, 62,
        199

Socio-ethical factors, 5, 207ff
Sodium bicarbonate, 16, 36, 54, 129
Sodium hydroxide, 27
South America, 4, 10, 11, 46, 108,
    121, 131, 155, 167, 199,
    203, 209
South Florida Task Force, 11, 12
Spanish conquistadores, 8, 167
Speed, vi, 107
Speedball, 18, 46
Splitting, 68
Sports, 42, 107, 210
Stealing, 20, 21, 24, 28, 41, 49,
    53, 54, 109, 114, 129, 136,
    142, 152, 154, 182, 186,
    193, 195, 210, 216,
Stepping stone hypothesis, 35 (see
    also Gateway drugs, Drug
    progression)
Stereotyped behaviour, 20, 139, 153,
    154, 155, 156
Stevenson, Robert Louis, 8
Stimulants, 46, 68, 70, 71, 154,
    167
Stress, 70, 78, 84, 101, 142, 147
Students, 11, 24, 34, 35, 136, 201,
    209, 211
Subcultures, 193
Subpopulations, 40, 79, 81n, 153
Suicide, 16, 19, 20, 55, 130, 171,
    177, 188, 189, 202
suicide ideation, 49, 53, 56, 77,
    130, 135, 139, 152, 162,
    182, 188, 195
Superego, 19, 24, 120
Support, family, 57, 77, 78, 162,
    195
Support systems, 57, 94, 101, 122,
    195, 216
Supportive-expressive therapy, 87-
    102
Survey, 34, 35, 39, 45, 48, 53,
    201, 209
Symptoms, 15-21, 42, 47, 68, 78,
    80, 101, 108, 109, 129, 140,
    141, 170, 171, 178, 194-195
  chewing, 15
  freebase
    physical, 17-19, 55, 81n,
    135ff, 139
    psychological, 19-20, 55,
    135ff, 139
    social, 20, 21
  intravenous, 16, 81n
  smoking, 15, 16, 199-202
  snorting, 16, 170, 171

Tetracaine, 47, 86
Therapeutic Learning Process, 163,
    164
Termination, 92, 102, 154

Therapist, 22, 77, 80, 82-105, 154,
    157, 162, 163, 196, 223
Therapy, 114
Third Scourge of Humanity, 9, 191
Thorazine, 162
Tioproperazine, 195
Tobacco, 15, 27, 36, 55, 180, 184, 192,
    200
Tolerance, 37, 47, 76, 169, 171, 178,
    186
Toxicity, 36, 47, 55, 140, 161, 167,
    168, 177, 188, 221
Traffickers, 16, 108, 110, 115, 163,
    187, 200, 215
Trafficking, 11, 36, 130, 133, 149,
    156, 168, 191, 199, 209, 216
Tranquilizers, 35, 49
Transference, 88ff, 94, 100
Trans-shipment countries, vii, xi, 3,
    4, 11, 108, 121, 132, 149
Treatment, 25, 39, 40, 41, 42, 47, 56,
    58-60, 61, 66, 67, 70, 73, 91,
    114, 121, 141, 143, 145, 156,
    161, 162, 163, 170, 172, 189,
    191, 193, 195ff, 202, 203, 204,
    209, 212, 214, 215, 217
  applicant, 45
  clinic, 20, 21, 46, 54, 66, 75, 108,
    109, 121
  difficulties, 25, 75, 115, 209
  drug substitution, 40, 60, 72, 100,
    122, 164
  failure, 59, 77, 79, 80, 161, 162,
    163, 164, 195
  issues, 56, 75, 79, 89, 122
  mandatory, xii (see also Compulsive
    treatment)
  outcome, 25, 69, 73, 78, 195, 203
  personnel, 46, 173, 192, 211
  programmes, 47, 75, 81-105, 163,
    196, 197
  selection, 76, 94, 98
  success, 60, 61, 203
Trends of use, 41, 48, 62, 120, 126,
    127, 130, 131, 215
Tricyclic anti-depressants, 40, 111,
    114, 115, 195
Trinidad, 4, 167-173
Tropane, 167
Tryptophan 1-tyrosine, 195

Uniform State Narcotic Act, 10
Unit time, 16
United Kingdom, 125, 132, 148
United States, 3, 4, 10, 11, 16, 108,
    125, 132, 142, 148, 150, 155,
    168, 209
Urinalysis, 25, 41, 42, 55, 57, 58,
    80, 82, 122, 164, 168, 196,
    210, 215
User countries, vii, xii, 3, 193

User profiles, 34, 35, 36, 39, 42,
        45, 47, 48, 49, 50, 51, 61,
        88, 133, 134, 135, 150, 151,
        152, 178, 179, 193, 194, 200,
        201
Users, vii, viii, xi, 17, 40, 163,
        171

Valium, 162
Values, 114, 115, 167, 172, 204,
        207, 212, 213, 215, 216, 223
Vasoconstrictor effect, 18, 33, 46,
        112, 167, 188, 193
Venezuela, 177
Vespucci, Amerigo, 8
Vin Mariani, 8
Violence, 5, 22, 24, 69, 70, 71,
        141, 188, 195
  criminal, 70, 188, 202
  family, 70, 152, 186, 188
  social, 70
  workplace, 210
Violent behaviour, 16, 21, 53, 55,
        120, 129, 130, 136, 139,
        142, 171
Vitamin deficiencies, 18, 172
von Anrep, Vassili, 8
Vulnerability, 21, 22, 29, 36, 42,
        45, 53, 66, 111, 142, 143,
        146, 208, 209, 217

Water pipe, 3, 16, 25, 54
Weight loss, 19, 24, 53, 135, 143, 185,
        193, 194, 201 (see also anorexia)
Wise, Roy, 170
Withdrawal, 47, 55, 67, 69, 76, 129,
        139, 141, 169, 178, 195, 202
Withdrawal symptoms, 17, 19, 20, 36,
        37, 57, 60, 72, 76, 86, 161, 170,
        179, 187, 202, 203, 221
Women, 4, 134, 135, 137, 145-159
Workplace, 53-54, 56
Wright, Harrison, 10

Yale Cocaine Abuse Treatment Programme,
        75
YAVIS, 79